History of Computing

The *History of Computing* series publishes high-quality books which address the history of computing, with an emphasis on the 'externalist' view of this history, more accessible to a wider audience. The series examines content and history from four main quadrants: the history of relevant technologies, the history of the core science, the history of relevant business and economic developments, and the history of computing as it pertains to social history and societal developments.

Titles can span a variety of product types, including but not exclusively, themed volumes, biographies, 'profile' books (with brief biographies of a number of key people), expansions of workshop proceedings, general readers, scholarly expositions, titles used as ancillary textbooks, revivals and new editions of previous worthy titles.

These books will appeal, varyingly, to academics and students in computer science, history, mathematics, business and technology studies. Some titles will also directly appeal to professionals and practitioners of different backgrounds.

Author guidelines: springer.com > Authors > Author Guidelines

For further volumes:
http://www.springer.com/series/8442

Jack Williams

Robert Recorde

Tudor Polymath, Expositor
and Practitioner of Computation

 Springer

Jack Williams, B.Sc., D.Sc. (Wales)
jwarchives@btinternet.com

ISSN 2190-6831   e-ISSN 2190-684X
ISBN 978-0-85729-861-4   e-ISBN 978-0-85729-862-1
DOI 10.1007/978-0-85729-862-1
Springer London Dordrecht Heidelberg New York

British Library Cataloguing in Publication Data
A catalogue record for this book is available from the British Library

Library of Congress Control Number: 2011940969

Printed on acid-free paper

Springer is part of Springer Science+Business Media (www.springer.com)

Foreword: Robert Recorde and the History of Computing

John V. Tucker

Robert Recorde was born *c.*1510 in the port of Tenby, in west Wales, and he died in in Southwark, just south of the Thames, in the summer of 1558. Recorde lived through the tumultuous reigns of Henry VIII, Edward VI and Mary; Elizabeth I was crowned 6 months after Recorde's death. Jack Williams' monograph examines the historical evidence and scholarship we have on Recorde and gives us an excellent account of what is known of his life and times. Such a scholarly study has been long awaited. At last, we have *terra firma* on which to base all sorts of historical studies involving Recorde and his achievements.

One such study is the history of computing. I am a computer scientist interested in the history of computing, which I view as a new field within the discipline of history of science and technology. My interest in Recorde, and his European contemporaries writing on practical mathematics, arose from my curiosity about the history of data. In my lectures on the history of computation I wanted to explore the development of

(i)   quantification and measurement;
(ii)   *data and computation;* and
(iii) *technical education.*

I found these topics played a significant role in the history of computation and, indeed, in the history of science and technology generally. However, I also found they seem to be neglected in the history of computation and marginal in the history of science and technology. Recorde is a particularly important representative of European writers on mathematical sciences in the Early Modern period. I hope we will see increased scholarly interest in Recorde and his contemporaries, including further volumes on the period in this Springer series on the History of Computation.

J.V. Tucker
Swansea University, Swansea, UK

Jack Williams' book is biographical and would be at home among works on the history of mathematics, education, currency, or the Tudor world in general. In this foreword I will explain why it is natural to welcome this invaluable work into our series on the History of Computing. We are grateful to Jack Williams for enabling us to broaden our coverage by introducing biography and a new period.

Recorde's Books

Recorde was a scholar and polymath, active in several fields and professions. He is remembered because of his mathematical works, which form a programme of instruction in English designed to teach some advanced ideas to a broad audience. Here is a summary of the four extant books[1]:

The Grounde of Artes

A first edition appeared circa 1543. It is a commercial arithmetic and covered arithmetic with pen, counters and fingers. Computations involve only natural numbers. It was long thought to be the earliest surviving mathematical work in English to introduce the Hindu-Arabic numbers.[2] New editions followed in 1549 and 1550 and an edition of 1552 added fractions and alligation. Over 40 editions followed over 150 years, the last in 1699 when pages about the abacus were omitted.

The Pathway to Knowledge

The first edition appeared in 1551. It is an introduction to geometry, the first in English. It simplifies material from the first four books of Euclid, which are about plane geometry, with a view to applications.[3]

---

[1] Fortunately, all four books are readily available in facsimile as follows:

Recorde R (2009) The Grounde of Artes 1543. Renascent Books, Derby
Recorde R (2009) Pathway to knowledge 1551. Renascent Books, Derby
Recorde R (2009) The castle of knowledge 1556. Renascent Books, Derby
Recorde R (2010) The Whetstone of Witte 1557. Renascent Books, Derby

[2] An earlier anonymous text surfaced at auction in 2005; it is published in facsimilie, *An Introduction 1539*, Renascent Books, Derby, 2009. Of course, Hindu-Arabic numbers were to be seen earlier. In St Mary's Church, Tenby, the date 1496 is in stone relief, close to a memorial to Recorde.

[3] The first complete English translation of Euclid was Henry Billingsley's Euclid of 1570.

The Castle of Knowledge

The first edition appeared in 1556. It is an introduction to astronomy, the first in English. It covers Ptolemy's sphere and mentions Copernicus's new (1543) theory.

The Whetstone of Witte

The first edition appeared in 1557, intended as a companion to the *The Grounde of Artes*. It is an introduction to algebra, the first in English. The algebra is in the German cossick tradition. It covers roots, equations and surds. It introduces the equals sign.

Recorde is commonly mentioned but rarely celebrated in histories of mathematics, which are dominated by the technical development of pure mathematics. He founded modern mathematics in the British Isles with a grand exposition of mathematical thought in which numbers are abstract and arguments are important. But his works are expository; they are directed toward practical activities that are ignored in pure mathematics; and they are based on advanced scholarship rather than original discoveries. Their detachment from the academic mathematical tradition began in Recorde's lifetime. To appreciate Recorde one needs to be interested in more than the technicalities of pure mathematics.[4] One needs to be interested in the *reflexive relationship between mathematics and the world's work*.

The view of Recorde from the perspective of computing is different from that of mathematics. Computing is fundamentally a mathematical science that is *intimate* with the world's work because computing is largely about collecting, analysing and creating data. Thus, if we are to seek the origins of computing, we must think about *quantification, measurement* and *data*. To be concrete, one need only ask a simple question such as:

> When, how and why did the Hindu-Arabic numbers, with their purely symbolic data representations and algorithms, develop and become standard in a country?

Data connects computing directly with knowledge, expertise and professional practice. In the history of computing, if we follow the data, we are led to Recorde and his many European contemporaries. The growing awareness of quantification and data in the Tudor period led to new conceptions of knowledge and science, and to the education of people in the ways of mathematics. Quantification, data and computation played a truly significant role in the undisputed cultural transformations of the sixteenth and seventeenth centuries.

---

[4] For a portrait of the mathematics of the period see Stedall JA (2002) A discourse concerning algebra: English algebra to 1685. Oxford University Press, Oxford

Commerce and the Rise of Computation

The sixteenth century saw trade in new international markets and products; increasing dissemination of information through printing; an increase in travel and the development of international networks; and a growing reliance on technical and expert knowledge in practical activities. European society and economy was becoming dependent on measurement and calculation in its organization and activities. The collection and use of data was reshaping the way in which knowledge and money were employed and distributed in Europe.

The origin of these innovations was Italy, from where they spread to Germany, the Low Countries, France and the British Isles.[5] From the fourteenth century, Italy had developed international banks; financial and accounting services; schools for calculation and universities; and printing presses. Italian business was trusting of the data and computation and so the conduct of business was becoming increasingly abstract.

The mathematical tradition at the beginning of the sixteenth century was surveyed by Luca Pacioli (1445–1517) in his great work *Summa de arithmetica, geometria, proportioni et proportionalita* (Venice, 1494, second edition 1523). Written in Italian, not Latin, it covered arithmetic; elements of algebra; tables of monies and weights; double-entry bookkeeping; and Euclidean geometry.

Thus, computation for commercial purposes was essential in Italian commerce. It was taught in schools to 8–10 year old children by *maestri d'abbaco* and led to a vernacular manuscript and book tradition called *Libri d'abbaco* – the so called *abbacus texts*.[6] Now it is important to note that the abbacus texts have nothing to do with the abacus; the word "abbacus" has been taken from the Italian name of the tradition. The abbacus texts have been studied in depth by Warren van Egmond who has revealed the nature of this important mathematical tradition.[7] Abbacus texts use Hindu-Arabic number systems only and modern methods of calculation; there are large collections of sample problems and wide varieties of problems; practical situations are used as exemplars; there are meticulous step-by-step explanations; and algebraic methods are introduced.

---

[5] For example, in the case of Germany, one thinks of works by Johannes Widmann (*c.* 1462–1498), Gregor Reisch (*c.* 1467–1525), Jacob Köbel (1470–1533), Michael Stifel (1487–1567), Adam Riese (1492–1559) and Christoff Rudolff (1499–1545).

[6] The tradition descends from Leonardo of Pisa's *Liber abbaci*, 1202, with its origins in the medieval Arab world; see Sigler LW (2002) Fibonacci's Liber Abaci. Springer, New York.

[7] For an excellent short introduction see Van Egmond W (1994) Abbacus arithmetic. In Grattan-Guinness I (ed) Companion encyclopedia of the history and philosophy of the mathematical sciences. Routledge, London, vol 1, pp 200–209. His indispensible scholarly account of the abbacus tradition is Warren Van Egmond, *Practical Mathematics in the Italian Renaissance: a Catalog of Italian Abbacus Manuscripts and Printed Books to 1600*, Istituto e Museo di Storia della Scienza, Florence, 1980. English translations of abbacus texts are: D E. Smith's translation of the Treviso Arithmetic of 1478 in Frank J. Swetz, *Capitalism and Arithmetic: the New Math of the Fifteenth Century*, Open Court, La Salle IL, 1987 and, more recently, Jens HØyrup, *Jacopo da Firenze's Tractatus Algorismi and Early Italian Abbacus Culture*, Birkhäuser, Basel, 2007.

The abbacus texts and schools are important for the history of computation. First, they improved the Hindu-Arabic number system and calculation methods to meet Western needs, providing a basis for our modern symbolic rule-based methods. Second, the abbacus texts and schools helped to change the conduct of business in Europe by introducing and embedding mathematics into commerce and society. Third, they influenced the development of algebra because the concrete practical problems cried out for general abstract algebraic methods.[8]

Quantification and New Knowledge

Reading the texts it is evident that Recorde offered his readers a vision of knowledge that had these characteristics:

- Knowledge is broadly and practically conceived and includes commerce, land surveying and navigation.
- Knowledge is precise and quantitative.
- Knowledge is firmly based upon sound reasoning that must be open to demonstration and debate.
- Arithmetic and geometry are a foundation for knowledge.

One well-known quotation that reflects Recorde's attitude to the nature of knowledge and authority is this:

> … yet muste you and all men take heed, that … in al mennes workes, you be not abused by their autoritye, but euermore attend to their reasons, and examine them well, euer regarding more what is saide, and how it is proued, then who saieth it: for autoritie often times deceaueth many menne … [9]

Recorde is introducing the Tudor reader not just to mathematical methods but to a new *mathematical frame of mind* and a new idea of what constitutes knowledge.

Recorde and his contemporaries are prominent in all sorts of historical topics of which I will mention two relevant to the history of computation, namely: the origins and development of science, and the rise of Europe.

The origins of the scientific method, and the purpose of science within society, has long exercised scientists, philosophers and historians, who have their own reasons (and audiences) for their investigations. For example, many historians are vexed by the notion of scientific revolution, though the notion continues to be popular among scientists, who commonly use it to magnify the work of Galileo and Newton.

---

[8] Italian mathematicians who are important in the history of algebra, such as Niccolo Tartaglia (c.1499–1557), Girolamo Cardano (1501–1576), Lodovico Ferrari (1522–1565) and Rafael Bombelli (1526–1572), were influenced by these problems.

[9] The quotation is a part of Recorde's comments on Ptolemy in *The Castle of Knowledge 1556*, Renascent Books, Derby, 2009, p. 127. It is carved into the *Robert Recorde Memorial* at the Department of Computer Science, Swansea University, which was designed by John Howes and made by Ieuan Rees in 2001.

The origins of the scientific method is the subject of classic studies by Edgar Zilsel of the period 1300–1600. In 'The sociological roots of science' (1942), he observed the following three things relevant to our topic.[10] First, the principles of causal explanation and methodological experimentation derive from the working practices of craftsmen, artisans, surgeons, instrument makers, surveyors, navigators, engineers and architects; this idea is often referred to as *Zilsel's Thesis*. Second, the mathematical description of nature in the seventeenth century depends heavily on the commercially inspired mathematical work of Pacioli, Recorde, Digges, Tartaglia and others. Third, the historical context is the development of capitalism.

Recorde is not alone, of course; his work belongs to a much larger European movement. But it is worth noting that Recorde initiated practical mathematics in a country that was becoming a world power and in which, in the next century were to appear several great landmarks of science, such as the works of Francis Bacon and of Isaac Newton, and the founding of the Royal Society (in 1660).[11]

Our conception of the modern world is fundamentally European. One feature of modernity is quantification, encompassing the systematic collection of measurements, records and other data and their analysis by computation. The historical study of the rise of Europe has led historians of ideas to enter the scholarly world of Recorde and his contemporaries. For example, Alfred W Crosby wanted to understand the success of European imperialism and the dominance of European based societies. In *The Measure of Reality*[12] he formulated the thesis that this success was due to the development by Europeans of the capacity and the mentality both to organize large collections of people and capital, and to exploit physical reality in order to gain knowledge and power. The crucial factors that determine capacity are the administrative, commercial, navigational, industrial and military skills based on measurement and mathematics. For mentality, a new model of reality is required – a quantitative model. Crosby's thesis and arguments depend heavily on the development of practical mathematics.

To test and develop such historical theories we need substantial new scholarship such as that of Jack Williams.

History of Computing

Studies of science and society suggest that there are deep reasons for the ubiquity and influence of modern computing, especially software. Our current craving for data has a long and intriguing history.

---

[10] Zilsel E (1942) The sociological roots of science. Am J Sociol 47(4):544–562. The article is reprinted along with unpublished material in Edgar Zilsel, *The Social Origins of Modern Science* ed. Diederick Raven, Wolfgang Krohn and Robert S. Cohen, Kluwer, Dordrecht, *c*.2000.

[11] Whilst the significance of Bacon is well established in the orthodox historiography, the influences on Bacon are less clear. Perhaps one can detect an influence of Recorde's writings and connections with Dee, and Dee on Bacon.

[12] Crosby AW (1997) The measure of reality: quantification and Western society, 1250–1600. Cambridge University Press, Cambridge.

At the *technical* heart of the history of computing is data and computation. A computer scientist studies the representation and storage of data; algorithms for transforming data; programming methods for constructing software to represent algorithms; and methods for designing and operating machines and networks of machines to implement software.

The history of computation is founded upon the history of these four topics. And the technical histories of data representations and algorithms span thousands of years, for one only has to study the calculations and reasoning of the Greek mathematicians (such as Euclid, Eratosthenes, ...). As suggested earlier, the evolution of number systems lead us directly to questions about science and society.

However, the history of computing is also about civil and military applications; the creation of vast new businesses and organisations; changes in education and skills, and in social and cultural behaviour. The history of computing cannot be a history of technicalities independent of the world. In search of the origins of computing we encounter technical problems to do with data, representations and algorithms that are close to the world's work and for which quantification, measurement, and data, are the conceptual *sine qua non.*

I think interesting material for the history of computing can be found wherever data is collected and computations are made. In our times, we have become addicted to data, which is used to represent the world in ever more detail and has become a commodity. Today, as in Recorde's time, the ultimate weapon for quantification is an ancient data type, that of money.

I think that the story of data has been taken for granted in the history of computing, leading to a concentration upon the technicalities of software and hardware or upon businesses and markets. It is through the history of data that we can develop new and comprehensive approaches to some problems, such as integrating the history of computing into the history of science and technology and, in particular, of creating a social context for explaining the history of software in our museums.

In my view data is the primary concept of computer science so a big question arises: *What are the origins of our use and dependency on data?* In the long search for an answer we encounter Recorde and his world. I invite you to inform yourselves about this world by studying and working with the fascinating fruits of Jack Williams' research.

Preface

Interest in the life and works of Robert Recorde has persisted at a low and unobtrusive level during the past century, as will be seen from the Select Bibliography appended. Of the two of works that deal with the broad sweep of Recorde's life, Kaplan's unpublished thesis is the most comprehensive. The remaining publications concern themselves with specific aspects of his books. His contribution to the teaching and the understanding of arithmetic, geometry and astronomy received the most attention which is both deserved and necessary, for Recorde's avowed aim in writing his books in English was to facilitate self-education in these three subjects. There is general agreement that this constituted a seminal contribution for which he should be remembered. His activities as a servant of the Crown concerned with the minting of money and with the mining of silver have been examined only patchily. The article by Clarke on Recorde's mining activities in Ireland is now nearly a century old and needs revision in the light of both new evidence and other evidence, available at the time but which was not taken into account. The two articles relating to Mint matters remain authoritative, but both need to be placed in the context of information relating to Recorde's dispute with the Earl of Pembroke, newly uncovered. Easton's papers on Recorde's Arithmetic and Geometry also remain authoritative, but in both cases deliberately forewent any search for possible sources of his material. This deficit will be addressed. The publications that deal with his Astronomy concentrate on the question of whether or not Recorde was the first Englishman to embrace Copernicanism. Where Patterson ventures beyond this remit, she perpetuates the shortcomings of Clarke and adds some of her own. There is far more material of substance in the *Castle of Knowledge*, Recorde's largest and most wide-ranging text, than has been examined to date. Hughes' work on aspects of the *Whetstone of Witte* goes a long way to dispel the notion that the only matter worthy of note in this book is the introduction of the equals sign and underlines the need for a broader debate on the nature of Recorde's algebra.

The general impression left by these publications is that Recorde did a good job of writing sound English texts on arithmetic, geometry and astronomy, but showed little in the way of originality other than by introducing the = sign, arguably embraced Copernicanism and maybe introduced algebra into England. In the absence of firm evidence that he was gaoled for libel, the favoured cause of his imprisonment, perpetuated now for nearly a century and also based only on speculation, was that he was guilty of some form of financial peculation in connection with his activities in Ireland.

There is now documentary evidence concerning the failure of his case of financial irregularities against the Earl of Pembroke and his consignment to prison for inability to pay the fine consequently imposed; this will be presented and analysed. With respect to his activities in Ireland it is clear that failure of the Crown to settle their debts to him bankrupted him but that his family profited from the affair after his death. He was meticulously honest in all his financial dealings.

It is not to be expected that text books intended for self-education at an elementary level would contain evidence of original thought, other than perhaps on matters related to teaching. Nevertheless there is evidence in his texts of a deeper under-standing of the subjects he was treating and of his potential for original contribution, had he had time to indulge. Such evidence is only made available by a detailed scrutiny of these texts, which does not make for easy reading. It is primarily to accommodate such material that this book is presented as a set of self-standing essays. To understand the whole man an understanding of the detail in the essays is needed and no apologies for such an approach are offered. Recorde understood and initiated the detail.

Finally there are the topics that have to date only been referred to in passing in published literature on Robert Recorde. He earned his living as a Physician and addressed himself as such. In common with most of his contemporaries in this profession, little is known of their practical activities. As he was neither a member of the Royal College of Physicians nor of the Company of Barber Surgeons his medical activities are possibly even more obscure than most. Kaplan has given some attention to Recorde's book on urology, primarily to expose and discuss an apparent uncritical acceptance by Recorde of the views of historical authorities on the subject. This topic will be re-examined as also will be the methodology advocated by Recorde for the taking and recording of experiential data. Comments on Recorde's antiquarian activities are sparse and scattered. He had a relatively large collection of English manuscripts on a wide range of subjects whose provenance and relevance to his published work will be examined as also his Anglo-Saxon scholarship. These interests brought him into contact with individuals who are recognised as the founders of English antiquarianism.

His interest in languages also bears on his contributions to the vocabulary of English mathematics. So far there have been only piecemeal assessments of his readers and of their reactions to his works. These readers were widely distributed across English society for over a 100 years, extending well into the initiation and extension of the 'Scientific Revolution'. The largely overlooked role of Recorde's publisher, Reyner Wolfe, in his life will also be evaluated.

Robert Recorde was a committed and devout Protestant. More might have been known about his beliefs had his two theological tracts survived.

Select Bibliography

Baron ME (1966) A note on Robert Recorde and the Dienes Blocks. Math Gaz J Math Assoc 1(374):363–369

Clarke FM (1926) New light on Robert Recorde. Isis vii:50–70

Easton JB (1964) A Tudor Euclid. Scripta Math XXVII(4):339–355

Easton JB (1966) On the date of Robert Recorde's birth. Isis LVII:121

Easton JB (1967) The early editions of Robert Recorde's Grounde of Artes. Isis 58:515–532

Greenwood JW (1981) The closure of the Royal Mint at Bristol. Br Numismatic J 51:107–111

Howson G (1982) A history of mathematics education in England. Cambridge University Press, Cambridge, pp 6–28

Hughes B (1993) Robert Recorde and the first published equation. In: Folkerts M, Hogendijk JP (eds) Vestigia mathematica. Rodopi, Amsterdam, pp 163–175

Johnson FR, Leakey SV (1935) Robert Recorde's mathematical teaching and the anti-Aristotelian movement. Hung Libr Bull 7:59–87

Kaplan EA (1960) Robert Recorde (c. 1510–1558): studies in the life and works of a Tudor scientist. Unpublished PhD dissertation, New York University

Kaplan EA (1963) Robert Recorde and the authorities of uroscopy. Bull Hist Med 37:65–71

Karpinski EA (1912–1913) The Whetstone of Witte (1557). Bibliotheca Math, Series 3, 13:223–228

Lilley S (1957) Robert Recorde and the idea of progress. Renaiss Mod Stud 2:3–37

Lloyd HA (2004) 'Famous in the field of number and measure': Robert Recorde, Renaissance mathematician. Welsh Hist Rev 20:254–282

Patterson LD (1951) Recorde's cosmography, 1556. Isis 42:208–218

Russell JL (1973) The Copernican system in Great Britain. In: Dobrzyck J (ed) The reception of Copernicus' heliocentric theory. D. Reidel, Dordrecht, pp 189–191

Williams J (1995) Mathematics and the alloying of coinage 1202–1700: part II. The english dimension. Ann Sci 52(3):235–264

Acknowledgements

The wide range of Robert Record's abilities, interests and activities, meant that a correspondingly wide range of sources had to be consulted. I have had to depend extensively on the help of many librarians from many locations, off-line and on-line. Time and help have always been gladly given and are gratefully accepted. In Oxford I found succour in the Bodleian Library perhaps most appropriately in Duke Humfrey, at the History Faculty Library, the Radcliffe Science Library, the History of Science Library, the Taylorian Institute, the Sackler Library, the Law Library and at the Libraries of All Souls, Christchurch, Merton and Magdalen Colleges. The University Library of Cambridge and the Pepys Library there provided effective help at a distance, as also did the Library of the University of Columbia N.Y. As always, the resources of the British Library and the Public Records Office were invaluable. The availability of the service provided by Early English Books On-Line greatly eased problems associated with comparison of multiple editions of Robert Record's publications. I am grateful to Elizabeth and Gordon Roberts (TGR Renascent Books) for permission to reproduce an excerpt from one of their facsimile editions of Robert Recorde's books. The discovery by the late W. Gwyn Thomas of the manuscript relating to the trial of Recorde has proved critical. Finally I want to acknowledge the patience and good humour with which my family have accommodated to the collateral damage that research and preparation of this publication has inflicted on them over the past decade.

Contents

Part III   Finale

Abbreviations

APC   Acts of the Privy Council.
BL   British Library.
CPR   Calendar of the Patent Rolls.
CPRI Calendar of the Patent Rolls (Ireland).
CSP   Calendar of State Papers (Domestic); Calendar of State Papers (Foreign).
HMC Historical Manuscripts Commission.
LP   Letters Patent.
PRO   Public Records Office.
STC   Short Title Catalogue (Revised).

Chapter 1
A Chronology

Abstract Son of a respected merchant, Robert Recorde was born in the small port of Tenby, Pembrokeshire, circa 1510. Following graduation at Oxford, he obtained a license to practise medicine. This he did for 12 years and was made a Doctor of Physicke by Cambridge University in 1545. By this time he had begun to move in circles close to the Crown and in 1549 received the first of a number of Crown appointments involving him successively as iron-founder, comptroller of three Royal Mints and extraction metallurgist. Starting in 1543, over a period of some 15 years he produced a succession of books written in English, one on Urology and four on mathematical topics. These latter formed the foundation of the English school of practical mathematics whose influence extended well into the next century. His interests as an antiquary made him one of a select band of intellectuals who saved collections of manuscripts by English authors from potential destruction during the Reformation. Fluent in Greek and Latin he was also an Anglo-Saxon scholar. His introduction of the mathematical sign for equality is well recognised: he also introduced a sizeable mathematical vocabulary still in current use. His theological texts have not survived. He died in a debtor's prison in 1558 following imposition of a massive fine for libelling William Herbert, Earl of Pembroke.

One of the defining show trials of the reign of Edward VI was that of Stephen Gardiner, Bishop of Winchester. Gardiner was the leading English religious conservative of his time. As such he urged Protector Somerset, privately, to avoid religious innovation during the minority of Edward VI. He also made his views increasingly public, to the embarrassment of the regime. To resolve the situation he was asked by the Privy Council to preach a sermon endorsing the religious policy of the regime. He delivered his sermon before King and Court on 29 June 1548, but stopped short of compliance with his instructions on a number of issues. He was re-imprisoned, during which time further unsuccessful attempts were made to bring him to heel. Gardiner was then brought to trial at Lambeth on 15 December 1550. Depositions relating to the content of the sermon of 1548 were made by members of the Privy

J. Williams, *Robert Recorde: Tudor Polymath, Expositor and Practitioner of Computation*, History of Computing, DOI 10.1007/978-0-85729-862-1_1,
© Springer-Verlag London Limited 2011

Council and their officials, members of the king's court and divines. Twelfth in the list of depositions was one by a 'Dr Robert Record, doctor of physicke of the age of 38 years or thereabouts.' A boy from Tenby had travelled a long way geographically, socially and intellectually in his 38 years.

We do not know for certain that Robert was born in Tenby, but it is highly likely that he was, for his family had been resident there for some time. The genealogy of the Recorde family of St. Johns by Tenby, found in Dwnn's Heraldic Visitations, is based on evidence given in 1597 by persons unspecified, but presumably by family members still living there.[1] The earliest member of the family noted was Roger Record of East Wel in Kent who is stated to have had one surviving son Thomas Recorde, who married twice. Thomas had no children by Joan, daughter and co-heiress of Thomas Ysteven of Tenby Gent., but by Ros Recorde, daughter of Thomas Johns of Machynlleth ap Sion he had two sons, Richard Recorde of Tenby Gent. and Robert Record Doctor of ffysig. Richard married Elizabeth, daughter of William Baenam of Tenby, by whom he had a son and heir named Robert, presumably after his uncle, and several other offspring. Dr. Robert Record did not marry and Robert his nephew was to become his heir. The family was reportedly armigerous 'Hi bereth sable and argent quarterly by the name off Record of East Well in Cent', but it has not proved possible to confirm this claim. Thomas Steven, father of Recorde's first wife Joan was a person of standing in Tenby, having been a bailiff in 1462 and Mayor in 1473, 1478 and 1484. Thomas Recorde himself became a bailiff in 1495 and mayor in 1519. William Beynon, father in law to Thomas Recorde's son Richard was mayor the year before Thomas and again in 1527. Richard was a bailiff in 1536 and became mayor in 1559.[2] The Recorde family was thus deeply bedded in the community of Tenby from well before Robert Recorde's birth and became even more so after his death.

Supporting genealogical information comes from Robert Recorde's will. This is given in full in Chap. 14. Probate was dated 18 June 1558. Robert Recorde predeceased both his mother, who being widowed had married again, and also his older brother and only sibling Richard. His nephew, also called Robert was to profit greatly from his uncle's estate, eventually. His nieces Alice and Rose also received minor bequests. His brother and nephew were named as executors.

Nothing is known of his youth. How he obtained an education adequate to enter Oxford is a matter for speculation rather than one of record. There is no evidence of the existence of lay schools in the vicinity of Tenby and its church was not collegiate, in the formal sense of that designation. There is the possibility of a modest chantry school being available. The Church's monopoly of school-keeping was challenged progressively in the larger towns during the fifteenth century as a result of the increasing demands of commerce for such competences. 'In time, as a result, founders

---

[1] Dwnn L (1846) In: Sir Meyrick SR (ed) Heraldic visitations of wales and part of the Marches, 2 vols. William Rees, Llandovery, I, 68

[2] Hore HF (1853) Mayors and Bailiffs of Tenby. In: Archaelogia cambrensis, Second series, pp 114–126, 117–119

of chantries, hospitals, almshouses, began to make teaching the duties of clerks or chaplains of their foundations.'[3]

The Church of St. Mary the Virgin, Tenby, was one of the largest parish churches in Wales. It doubled in size during the fifteenth century following the increased affluence of the town resulting from its maritime trading activities. The last additions of the century were those of a West door with having a large cruciform porch with windows inserted into the structure and a 'college', erected close by. Although in ruinous condition, substantial portions of these two structures were still in existence at the beginning of the nineteenth century, when they were examined and sketched by Charles Norris. In 1657/1658 payment for repair of the church windows and those of the 'Schoolehouse' was recorded. In 1831 the porch was completely demolished despite a public protest by Norris that 'the Corporation of Tenby lately destroyed a venerable edifice, occasionally used as a private school, standing in the churchyard'. When the rubble from this demolition was being cleared a piece of stone was recovered bearing the inscription

M.... cum collegio annexo Fundavit
… et Brigittae anno 1496 retribuat ei

It was mounted inside the church in St. Thomas Chapel in 1868. This inscription had been recorded in its original position in the West porch by Norris during in his earlier inspection and sketching of the Church. Two doorways of the 'college', a two storied building, still survive. The Latin inscription 'Bessed be God in his gifts' is to be found on their arches, echoing those on the existing West doorway and also on the original West Porch doorway. The function of the 'college' can only be guessed at, but it may have provided housing for the chantry priests who served the three chantry chapels.[4] It is not known whether or not the original terms of their endowment included some stipulation regarding teaching. All three chapels formed part of the fifteenth century expansion and therefore were probably funded by municipal monies. The interaction between town and church was strong. One of the chantry endowments payments was channelled through the Mayor. From 1484, together with the Corporation he also had oversight of the two Hospitals of St. John and of Magdalen with their almshouses that were associated with the Church. If the Corporation wanted the chantry priest to teach, then it seems highly likely they would get their way. Thomas Recorde became Mayor in 1519.

It is possible therefore that Robert's early education was begun in either or other or both the West porch and the 'College'. Wherever he was taught, teaching would have comprised Latin grammar and probably writing, but not arithmetic. However, whether attending school or church services, the young Recorde could not have avoided seeing the date of the foundation of porch and college on the inscription,

---

[3] Ibid., pp 19–32

[4] Gwyn Thomas W, The architectural history of St. Mary's Church, Tenby, Archaeologia cambrensis, vol CXV,134–165,161–2,165;

Laws E, Edwards EH (1807) Church book of St. Mary the Virgin, Tenby. J Leach. Tenby 1807, 16

written in Hindu-Arabic numbers. This is an extremely early example of the public use of such symbols and highly unusual.[5] If his father or his business associates kept accounts, unless they were from the mainland of Europe they would have used Roman numerals. With a mind as inquisitive as Robert Recorde's would eventually prove to be, the strangeness of the inscription could not have failed to have had an impact on him. The practical value of such symbols in this specific application would have been obvious. Twelve Roman numbers would have been needed to express the date in comparison with the four actually used. Perhaps this would have been explained to Robert. But whose idea and competence was it that led to the innovation? Was it a priest or a merchant, native or foreign? Was the same person available to teach the young Recorde? This early experience could have provided the start of Recorde's interest in the subject and initiated the seminal role he played in establishing arithmetic based on the Hindu-Arabic numbers in England. Appropriately, Recorde's modern bust now faces the inscription '1496' which is installed on the wall across the floor of the chapel they jointly inhabit.

Robert probably arrived at Oxford about 1525. He was admitted as a B.A. of Oxford on 16 February 1531 and elected a Fellow of All Souls in the same year.[6] Discussing admissions of Welsh students to higher education between 1540 and 1640, Griffiths points out that territorial links assisted in this process. The lands that All Souls College held in South Wales, Montgomeryshire and Denbighshire allowed it to provide scholars' and fellows' places for Welsh students throughout the period.[7] Income from these properties made it possible for students at the College who were in receipt of maintenance to be relatively generously funded,[8] and to have to take in few if any fee paying students.[9] The holding at St. Clears provided an income of about £40 a year during the sixteenth century.[10] It was situated only about 6 miles from Tenby, albeit in Carmarthenshire rather than Pembrokeshire. There were strong links between Town and Church in Tenby extending back to 1484, as already noted, so it might be expected that Thomas Recorde, Robert's father, mayor of Tenby in

---

[5] It is unusual in more than one way. The Fig. 4 is given in its modern form. This form had become increasingly common in lay circles in Western Europe since its introduction in Northern Italy in the early fourteenth-century and had become the dominant form there by the end of the fifteenth-century. This was not the case in England where, when Hindu-Arabic numbers were used by astronomers, astrologers etc., the form that the figure 'four' took approximated to that of a vertically truncated eight or loop, as found in the early thirteenth-century manuscript, Sacrobosco's *Algorismus*. This form is still found in the astronomical manuscripts that Lewis of Caerleon wrote at the beginning of the Tudor era. The mason who prepared the inscription must either have been trained on the Continent or been instructed to follow their practice.

[6] Wood AA (1820) Fasti oxoniensis, London, I, p 84
    Foster J (1891) Alumni oxoniensis: the members of the University of Oxford 1500–1714, Parker and co., London, III, p 1242

[7] Griffith WP (1996) Learning, law and religion. Higher education and Welsh Society c. 1540–1640. University of Wales Press, Cardiff, p 42

[8] Ibid., pp 64–65

[9] Ibid., pp 203

[10] Ibid., p 202

1515 would have known of the links between the priory at St. Clears and All Souls College. It seems that Recorde undertook medical studies whilst at All Souls, but there are no records at Oxford of this activity. The qualification at Oxford for a licence to practice was proof that two dissections had been carried out and that three cures had been effected.[11] It appears that this qualification took Recorde about 2 years to achieve, but there is no firm evidence of when he left Oxford. Lewis points out that 'Neither the Linacre lectureship nor the Regius chair had the prestige to secure for Oxford the teaching service of any physician who had once left and established himself in the main stream of English medical life.'[12] She lists Chambre, Linacre, Clement, Wooton, Recorde, Edrych, Caldwell, Forster and Mathew Gwynne as examples of such physicians.

The probable course of Recorde's subsequent academic career has to be deduced from his records at Cambridge.[13] In 1545 he was granted an M.D. by that University as he had had 12 years of medical studies after being granted a licence by Oxford to start such practice, subsequent to his graduation there. He was entitled to become a regular member of the faculty of medicine at Cambridge, provided that before Easter of that year he attended certain disputations. He was excused from taking certain courses required for a bachelor's degree as he was already a doctor and because there were precedents for such a relief.

As will be seen later, he developed antiquarian interests including a facility in Anglo-Saxon. In her examination of the development of interest in medieval history in Tudor England, McKisack says 'In Recorde, the age of Leland and Bale had produced a forerunner of Parker and Jocelyn'.[14] Later, her analysis leads her to conclude that there was a bias of eminence in this field towards Cambridge, which suggest that Recorde would have been more likely to have been stimulated in such interests at Cambridge than elsewhere.[15] Neither Cambridge nor Oxford showed much formal interest in the mathematical studies which occupied Recorde so much and so early on in his non-academic life. However, it is clear that individuals such as Cheke and Thomas Smith were likely to have been familiar with the form

---

[11] McConica J (ed) (1986) The history of the University of Oxford: the collegiate university. Clarendon Press, Oxford, pp 165, 217

[12] Lewis G, ibid., '4.2 The Faculty of Medicine', 213–256, 238

[13] Venn J (ed) (1910) Grace book. Containing the records of the University of Cambridge 1542–1589. Cambridge, p 27

[14] Professor M. McKisack, in her book *Medieval History in the Tudor Age*, Clarendon (Oxford 1971), 25., gives an overview of the development of antiquarian interests in England during this period. Mathew Parker, Archbishop of Canterbury, and his secretary John Jocelyn, during the reign of Elizabeth were the prime movers in the recovery of many of the historical documents dispersed during the dissolution of the monasteries. John Leland and Bishop Bale's activities in cataloguing such holdings are well documented. Recorde's activities in these areas will be dealt with at greater length in Chap. 12.

[15] Ibid., 6. Sir John Cheke, as he was to become had a distinguished career at Cambridge before becoming tutor to Edward VI. He presented a copy of Recorde's 'The Pathway to Knowledg' to the king. He was also given charge of Leland's books and papers after the latter's death in 1552.

of Hindu-Arabic numbers if not with their elementary uses.[16] This subject was of course the subject of Recorde's first book, *The Grounde of Artes* that was published in 1543, in London, by Reyner Wolfe. Its dedication, couched in the fulsome fashion of the time, was to Richard Whalley the agent of Thomas Cromwell for dissolution of the northern monasteries. It has been suggested that Recorde might have been tutor to some of Whalley's 13 children for a time, but again no direct evidence for such an activity has been found.

Unsurprisingly Recorde's next book *The Urinal of Physicke*, dated 8 November 1547, dealt with a medical subject. It was to be his only foray in the field but the foreword clearly establishes his presence 'At my house in London'. London was to be his home until he died. The dedication of the book to the Wardens & companies of the Surgians of London suggests that he was well embedded in the medical life of the capital by this time. It was reprinted six times during the following 100 years.

Further evidence of his emergence into a wider ambience is given by Edward Underhill, the 'hot gospellor' in his autobiographical memoirs.[17] In 1548 a man called Alen had been making prophesies that were embarrassing the Protectorate. Underhill, presumably in his official capacity as a gentleman pensioner, had detained Alen together with his 'books of conjouracions' and taken him to Protector Somerset at Syon. There he was instructed to take Alen to the Tower, where he was examined by Sir John Markham 'unto whom he did affirme thatt he knew more of the science of astronomaye than alle the universyties of Oxforde and cambridge; wherupon he sent for my frende, before spoken off, doctor Recorde, who examined hym, and he knewe nott the rules of astronomaye, but was a very unlearned asse, and a sorcerer, for the wiche he was worthy hanginge sayde mr. Recorde'. The date of this interlude is not known exactly, but it must have been in late 1548 or early 1549.

Recorde's first Crown post was as comptroller of the newly established mint at Durham House (London) appointed together with John Bowes as under-treasurer and John Maire as assay master on 29 January 1549.[18] This appointment was swiftly followed by a similar posting to the Bristol mint from which Sir William Sharrington had been ejected in disgrace.[19] The subsequent events at Bristol form an important and integral part of Recorde's downfall, which will be discussed in detail later. Suffice it to say for the present that it was intimately connected with Somerset's

---

[16] Sir Thomas Smith became Principal Secretary to Edward VI following a brilliant academic career at Cambridge University, of which he was Vice-Chancellor. The evidence for his numeracy is to be found in several of his publications as summarised by Williams in 'Mathematics and the Alloying of Coinage 1202–1700: Part II', *Annals of Science, 52 (1995), 235–263,249–250.* Smith was also the author of *A Discourse on the Commonweal of England* which will be referred to in Chap. 5.

[17] Underhill E (1854) Autobiographical Anecdotes of Edward Underhill esq., one of the Band of Gentleman Pensioners. In: Gough J (ed) Narratives of the Days of the Reformation, chiefly from the manuscripts of John Foxe the Martyrologist, Old Series 77. Camden Society, London, 132–177, 173

[18] Calendar of Patent Rolls [CPR.], 1548–1549, 303–304

[19] Ibid., 304

fall from power. By the end of October 1549 Recorde was back in London under some form of restraint. At the same time that he was comptroller at the two mints, he was also involved in some form of supervisory capacity, with iron mining and smelting operations at Pentyrch (near Cardiff). Here he fell foul of William Herbert, Earl of Pembroke. These events were also to be part of the story of his ultimate imprisonment. Their consequences wound their wearisome way throughout the year of 1550, to no happy ending as far as Recorde was concerned although he was paid as comptroller until October of that year. As already noted he was still sufficiently well regarded by the Establishment to testify against Gardiner in December 1550.

On 27 May 1551, Recorde was appointed Surveyor of all the king's mines, newly found in Ireland, with commission to rule their affairs. He was also appointed Surveyor of the newly erected mint in Dublin.[20] He left England during June and is unlikely to have returned before his financial assessment of the mining operations at Clonmines in February 1552. Whether he stayed on in Ireland beyond this date to complete his more detailed accounts, which he finally presented to the Privy Council in May 1553, is not clear. *The Pathway to Knowledg*, Recorde's book on elementary geometry was published on 28 January 1552. It seems very unlikely that he was available, in person, in England, to check its proofs before printing. On 27 February 1553, the Privy Council instructed him to proceed to his account and to cease his commission.[21] This he did some time between the closure date for the account, 12 April 1553 and the death of the King in the following July.

How Recorde occupied himself in the interim is not known. His second edition of *The Grounde of Artes*, was published at the close of 1552, and being dedicated to Edward VI must have been completed before the king's death. It seems reasonable to assume that he spent more time with the preparation for printing of this more comprehensive arithmetic than he did with his earlier geometrical text, and was responsible for the corrigenda of the 1552 edition which latter should be regarded as the prototype for the 1558 edition.

By the first half of 1553 he had become involved with some of the maritime adventures of the time. In 1576, a Philip Jones reported on a voyage to the North-West passage that he had undertaken 23 years earlier.[22] He claimed that he had 'set out with the encouragement of Mr. Chancellor that first found for us the Musco, and Doctor Recorde's conference in my house and speciallie the noble pillot Pintagio the portugale encouraged me.' He continued that a map of the West India that he had, was in agreement with 'the opinion of Doctor Recorde, Mr. Bastian Cabotta, Harry Estrege, his sonne in lawe, Mr. Chancellor that founded the Muscovia, and noble Pintagio, the Portugal pilott that was with Windam in Guinea'. The voyage was

[20] Calendar of Patent Rolls Ireland [CPRI.], 1503–1578, 275–276

[21] Acts of the Privy Council (APC.), 1552–1554, 225

[22] British Library. Harleian MS no. 167, fos. 106–108

Andrew KR (1984) Trade, plunder and settlement. Cambridge University Press, Cambridge, p 167 n. 2

Nichols G, loc. cit., 150–151

unsuccessful in that it failed to find the passage. Wyndham's expedition to the Guinea set out in August 1553 and Pintagio [Pinteado] died on the return journey. Henry Estrege [Ostrich] died in 1551 and Chancellor in 1556. Andrews' suggestion that Jones' voyage took place around 1553 seems the latest possible date.[23] It follows that Recorde had become involved with mariners and their aspirations before this date.

No written evidence of Recorde's subsequent activities exists until he sent the fateful letter to the Queen in June 1556 that led to his untimely end. Underhill gives a graphic account of the measures that he, Underhill, had to undertake to preserve his anonymity during this period. Assumption of a similarly low profile by Recorde might have changed the course of his history but maybe his adversaries did not afford him the opportunity.

Contact with Cambridge University was briefly rekindled at the beginning of 1556, when together with John Blythe, regius professor of Medicine at Cambridge, he was granted the privilege of *vestro communi*.

Recorde's next book, *The Castle of Knowledge*, dedicated to Queen Mary, contains astronomical data which could not have been included before September 1556. At this time he was about to make his first formal appearance before his judges in the libel case brought against him by the Earl of Pembroke. From comments he made in *The Pathway to Knowledg* about the material he had in preparation for publication, it is clear that some of the material included in *The Castle* had been to hand for some time. But as will be seen later there was also material, newly emerging from publications by European authors, to be found in the text. This material had to be understood and digested before being incorporated into his texts, which would have taken time. This is less true of the content of his last book, *The Whetstone of Witte*, which was published on 12 November 1557 by John Kyngston rather than his previous publisher and associate Reyner Wolfe. By this time, Recorde must have been in prison. In his dedication to '...the right worshipfull, the governors, consulles, and the reste of the companie of venturers into Moscvia..' he continued to be reasonably cheerful about his condition, but as he did not receive judgement in his case until 10 February of that year perhaps he had completed his dedication prior to that date. He was still promising his dedicatees a further book on navigation.

Whilst he appeared resigned to his incarceration there are no indications that he expected to end his days in prison. The Kings Bench prison at that time was an interesting institution. It was not run by the State but was farmed out and run for profit. Prisoners paid the marshal for their upkeep. Under the so-called Rules they could pay to stay in private quarters outside the prison itself. This might explain the various disbursements to prison officers listed in Recorde's will. Recorde was not the first English writer of a mathematical text to have spent time in this prison. Cuthbert Tunstall, his English arithmetical predecessor, was there for a short time in 1552. Death from cholera was common in the Kings Bench prison, as in most other such establishments but in 1558 a particularly severe epidemic of cause unknown,

---

[23] Andrew KR, loc. cit. p 167 n. 2

swept the country. The exact date of his death is not known nor his place of burial. In the next century John Aubrey discussing Recorde as one of his 'Brief Lives', asked rhetorically where he had been buried but was not answered.

A post-obit commentary which comes closest to an obituary for Recorde was provided by a fellow Physician and successful author, William Bullein. In the Preface to the second book of his '*Bulwarke of Defence....*', published in 1562 but compiled earlier whilst he was in prison, Bullein lists some medical worthies, the last of whom was Robert Recorde to whom he devoted most space. 'How well was he seen in tongues, Learned in Artes and in Sciences, natural and moral. A father in Physicke whose learning gave liberty to the ignorant with his Whetstone of Witte and Castle of Knowledge and finally giving place to eliding nature, died himself in bondage or prison. By which death he was delivered and made free, and yet liveth in the happy land amongst the Laureate learned, his name was Dr. Recorde.' If this list of abilities is augmented with those attributed to him a few years before Recorde's death by Edward Underhill – '...singularly sene in all of the seven sciences, and a great divine ...', together they give some feel for the regard in which Recorde was held across the breadth of his many 'lives'.

The contents of these lives, in Natural and Moral Sciences, as Physician, Divine, Mathematician, Astronomer and Antiquarian will be looked at in greater detail, separately, in the following chapters in the sequence indicated. Robert Recorde is most widely known for having written the first printed Arithmetic in English, introduced algebra and the sign '=' into the English vocabulary and, more doubtfully, of being the first English Copernican. It may seem a little perverse therefore to deal firstly with his less well chronicled activities as an applied scientist serving both Crown and the private sector and as a Physician of some repute. However these two sets of activities provided him with the means of living. Unfortunately his experiences as a servant of the Crown set him on the path that led to his early demise. Between them these activities provide a continuum against which his scholarly activities have to be viewed.

Part I
'Profite and Commoditie':
The Practitioners

Chapter 2
Introduction

Recorde's intentions in writing his six books are expressed in their 'Dedications' and their 'Readers Preface's. The arguments pursued in the dedications are designed to engage the interest and support of the dedicatee and so vary from book to book. There are however no inconsistencies between the arguments, only variations in their emphases and extents. The common threads are that knowledge and learning are to be highly esteemed and sought, but that persons possessing such attributes are in short supply in England.

Recorde argues for the desirability of learning and knowledge from two standpoints viz. that of 'intrinsic worth' and that of practical value or 'profite and commoditie'. Whilst Recorde might be considered to be an adherent of the Platonic view of the intrinsic values of mathematics, unlike Plato he pursues actively the application of mathematics to worldly purposes.

It is clearly stated in the *De Republica* of Plato that the 'Philosopher Kings' should not study arithmetic for commercial ends as would merchants and stall-holders, but its application for military purposes was proper (525b-c). A similar stance was taken with respect to the study of Geometry (526e-527c) and Astronomy (530b-c).[1] The view that Plato disapproved totally of practical applications for the mathematical sciences would seem to be unduly restrictive but such applications were certainly of little interest to him. So even if Recorde's philosophical justifications for the study of the mathematical sciences follow closely those used to justify the educational scheme for the 'Philosopher Kings', he diverged emphatically from Plato on matters relating to the practical applications of mathematics and the other sciences.

As with Recorde's more philosophical arguments, those relating to practical values are tailored to his audience. Thus in the first edition of The Grounde

---

[1] Plato (1993) Republic (trans: Waterfield R). Oxford University Press, Oxford, pp 255–256, 258–259, 262

J. Williams, *Robert Recorde: Tudor Polymath, Expositor and Practitioner of Computation*, History of Computing, DOI 10.1007/978-0-85729-862-1_2,
© Springer-Verlag London Limited 2011

of Artes, in his opening address to the Scholar, speaking of numbering the Master says

...syth it is the grounde of mens affairs, so that without it no tale can be tolde, no communication without it can be long continued, no bargaynyng without it can onely be ended, nor no busynesse that man hath, justly completed. These commoditees (if there were none other) are sufficient to approve the woorthinesse of numbre. But there are other unnumberable farre passing all these, which declare Numbre to excede all praise. Wherefore in all great workes are, are clerkes so much desyred? Wherefore are auditors so richly feeyd? What c[au]seth geometricians so highly inhaunced? [Why] are astronomers so greatly advaunced? [...] cause that by numbre suche thinges th[..] fynde, which elles shulde far excel m[...] mynde.

In his books, other than the *Urinal of Physicke,* as will be shown later the examples he gives show how practical applications of the mathematical instruction being imparted may arise. Additionally, Recorde did his best to practice what he preached and turned his talents to practical ends provided he could find sponsors for such activities. The Crown provided such sponsorship on three occasions which caused him to be successively iron-founder, manager of three Royal Mints and mining engineer cum extraction metallurgist. If Recorde departed from the stance adopted by Plato with respect to the uses of mathematics, the Tudor monarchs, Henry and Edward did not even remotely accord with Plato's concept of the proper behaviour of a Philosopher King. When the Crown employed Recorde they had profit for the Crown in mind much more than that of commodity. It was this employment that was to lead eventually to Recorde's downfall in 1557, but to the ultimate benefit of his family in the long term. Recorde must have had some selfish elements attached to his acceptance of the appointments for they were well remunerated. Between them they should have given him an income averaged over 4 years of about £150 p.a. or about £40,000 p.a. in current value if they had been paid in full, which they were not!

Discussion of his various appointments is complicated by the fact that there are central records of two of them viz. those relating to the Mints of Durham House and Bristol and to the activities in Ireland, but not to the third viz. that of iron-making at Pentyrch which is only referred to by Recorde in the evidence he offers in his trial. Presentation is further complicated by the fact that the silver mining and extraction activities which Recorde oversaw at Clonmines played no part in his eventual trial. For the sake of continuity of argument therefore, those of his activities that proved relevant to his trial will be dealt with as an entity followed by that of the venture at Clonmines as a separate but critical entity.

Recorde's remaining profit making activity, one that probably provided him with a reasonable and steady source of income was that of Physician, an occupation he shared with many of his contemporary 'scientific' colleagues on the Continent and also in England.

Chapter 3
Robert Recorde and William Herbert, Earl of Pembroke

Abstract William Herbert, Earl of Pembroke and Robert Recorde, Doctor of Physicke seem unlikely and unevenly matched antagonists. Herbert was a courtier, soldier and magnate, Recorde the son of a merchant and an intellectual. Their clash arose from Recorde's Crown appointments. In January 1549, Herbert's men wrecked the iron mill which Recorde had set up for the Crown at Pentyrch near Cardiff, seized property there and pursued Recorde relentlessly for the profits from the operation for 2 years. Later that year Recorde, as comptroller of the Mint at Bristol, refused to hand over its assets to Herbert, and was confined to Court. Part of his next Crown appointment placed Recorde as overall Surveyor of the Dublin Mint, where he suspected Herbert of interference with the intent of diverting profits to his benefit. Matters were brought to a head in 1556 by a letter that Recorde wrote to Queen Mary in which he accused Herbert, by this time Earl of Pembroke, of a range of financial peculations and, as Herbert interpreted it, of traitorous behaviour. At the subsequent trial Pembroke asked for damages of £12,000 but Recorde was fined £1,000 and costs. Portions of Recorde's case were rebutted with the aid of William Cecil. Unable to pay the fine, Recorde was committed to the Kings Bench prison, Southwark, where he died.

On Friday 16 October 1556, the earl of Pembroke presented a bill alleging that Robert Recorde, medicus, activated by malice wrote a scandalous and false letter against the earl on 10 June 1556 making him out to appear a traitor, to have injured the Crown and to be worthy of imprisonment. Further it was implied 'by subtle and false prophesy' that the earl, whose heraldic arms had the symbol of a green dragon, was a painted dragon and an enemy to the Queen whose symbol was a red dragon. This letter had been delivered to William Ryce, a gentleman of the Queen's privy chamber, with the intention that it was made known to the Queen. She read the letter and passed it to Nicholas Heath (Lord Chancellor), Thomas Thirlby (Bishop of Ely), both Privy Councillors and Sir William Petre (a Chief Secretary of the Privy Council) for examination. Recorde was summoned before them on 20 June 1556 and confirmed

J. Williams, *Robert Recorde: Tudor Polymath, Expositor and Practitioner of Computation*, History of Computing, DOI 10.1007/978-0-85729-862-1_3,
© Springer-Verlag London Limited 2011

that he had written the letter and that its contents were true. The earl claimed damages of £12,000.[1] Recorde's letter reads:-

To the righte worshypfull mr Ryce, one of the gentlemen of the queenes majesties pryve chamber.

Sir, I am right sorye that the malice of any man shuld hinder the declaracion of my good wyll, namely where I was so wyllyng to have express'd it, and yet suche ys the chaunce presently. For sithe my last being with yow the earle of Penbroke has commensid agaynste me an accion of m li. albeyt I never hadd to do with hym for myselfe but for kyng Edward onely whome god pardon. But I thynk it had byn better not onely for his maiestie but also for me yf nether of us bothe had knowen that good erle. As long as I was in offyce and myghte answere hym accordynglye he would never attempte any accion agaynste me derectly, although craftely he with his confederate of Northumberland with greate preiudice to the Crowne made wonderful attemptes for my distruccion, obiecting againste me firste treason and then heynous contempte, when yf iustice had been free hym selfe had ben worthier the rewards of bothe. And therefore as long as I was absent from the Court he thoughte yt not good to wake a sleepyng dog, but seying nowe, mere suspecting my repair thether agayne seakythe meanes to stay me some other wayes, least by the libertie of my tong he myghte be as wel knowne to other [of] the Quenes highnes true frynds (absit verbo arrogancia) as he ys unto me. I have wrytten to you earnestly and am wyllyng to answer yt as gladly which may redound to the queens maiestie great commoditie yf the matter be handyld accordyngly, and namely such circumspection usid ne mutuo stanat muli. But to conclude: testes advoco celum sidera ac sapientes nunquam hostile dissidium in Brittania extinctum iri donec occulte inimicite prosapie veri et fucati draconis palam erumpant. Knowyng your faithfulnes to the quenes highnes with libertie of access to the same this much have I wrytten with my owne hand. Quelicet occasio precosior sit amplectanda tamen est. But to thentent yt may not be thought that I seke ayde agaynst the earle I wysshe I might aunswere hym before the pryvie councell, for the matter toucheth the quenes maiestie whom God prosper.

                                       yours fully in that he can doe. R Recorde.

John Wykes, goldsmith and Reginald Wolfe, stationer, both of the City of London, on 20 July 1556 entered into a recognisance of a 1,000 marks to the Crown under the condition of ensuring the personal appearance of Robert Recorde before the Lords of the Privy Council to answer to such matters as are raised.[2] This appearance was on 16 October following when Recorde denied the charge of false prophesy and proceeded to make a series of detailed allegations about Pembroke's behaviour over a period extending from early in 1549 to mid-1553 ascribing to him a number of major financial peculations. The earl's response was to ask that he should not be prevented from his action for the unjustified personal attack in the letter to Ryce. The first hearing for 28 November was deferred initially until 25 January following and then further until 10 February. Judgement was given against Recorde with respect to the letter which was deemed to injure the earl without good reason. Recorde was to pay damages of £1,000 to the earl with costs of £10.

No detailed record of the proceedings has been found. The judgement is merely appended to the information presented to the hearing of October 1556. The only glimmer of light on the judgement is thrown by a letter from Thomas Cornwallis to

---

[1] PRO KB.27/1180 f66. Credit for the discovery of this document goes to the late W. Gwyn Thomas
[2] Acts of the Privy Council [APC.]; 1554–1556, p 310

William Cecil dated 5 March 1557 from Calais.[3] In it Cornwallis passes on to Cecil a message from 'my Lord Lieutenant' expressing the latter's earnest inclination of goodwill towards the former. Pembroke was the Lieutenant/Governor of Calais as of November 1556.[4] The reason for this reaction Pembroke gave as being 'when the matter between him and Record was opened before the judges you, being present, replied against Record, and said that ye were well able to clear his Lordship of many articles against him objected by the same Record. Whereunto I answered that I could well witness the same, for when I was in England, I saw how you attended in Westminster to hear the matter, when it should be called on. This doing of yours is so well accepted that my Lord thinketh himself much in your debt; which (in this time of his credit with the Queens Highness) may stand you in good stead, if you shall have any occasion to use him.'

The exact date that Recorde entered prison is not known, but there is reference to his impending apprehension at the end of the *Whetstone* which was published in 1557. In his dedication of the same book to the Muscovite Company, he promises them shortly a book on navigation in which he will touch on not only the old north navigation i.e. the North-West passage, but also the to the North-East. Although he must by this time have either been in prison, or known that he was destined to be committed there, it did not prevent him from saying that he needs no protector for his works but valued expressions of friendship. Thus 6 months before he actually did die he clearly did not expect to die in prison.

The absence of an account of the trial leaves much room for speculation about the detail. For instance, Recorde states clearly that he only sent the letter because he felt that Pembroke was about to start harassing him again, as he had done when Northumberland was in control. Two such actions have been reported in Recorde's own writings and will be discussed in context.

Pembroke was primarily concerned about possible implications of false prophesy presumably present in those portions of the letter written in Latin. Given Pembroke's limited scholastic abilities, some third party would have had to interpret it for him both literally and legally. The charge of false prophesy was not a trivial matter as there had been statutes against 'fond and fantastical prophecies' in 1549/1550 and again in 1552/1553. Recorde had several books on prophecies in his library as described by Bale.[5] There were six books on the prophecies of Merlin (nos. 293, 294.), another by Geffridus Eglyde (no. 96.) and further one by an unnamed author (no. 481). The Latin text used by Recorde may have been a quotation from one of these sources, known to some scholars, but to few others. However he also had four volumes of a work on arms and insignia by Nicholas Upton (no. 310.), so he was well placed to be 'subtle' as alleged, despite his denial. On the other hand it has to

---

[3]Historical Manuscript Commission, Salisbury MSS 1, no. 576 given in extenso in 'A Collection of State Papers, relating to affairs in the reigns of king Henry VIII, king Edward VI, queen Mary & queen Elizabeth, from 1542 to 1570, transc. from authentik memorials left by W. Cecill lord Burghley, and now remaining at Hatfield House' by Samuel Haynes London 1740. p 203

[4]PRO. SP 11/9 fos.94-7, 104-5.

[5]The subject of Recorde's library is dealt with in Chap. 12.

be remembered that Recorde had testified against the 'prophet Alen' in 1549, when he described him to Somerset as one who '…knew nott the rules of astronomie, but was a very unlearned asse, and a sorcerere, for the wiche he was worthye hanginge…'[6] It seems unlikely that he would have offered himself such a hostage to fortune on this count, given the mass of other accusations he had to make. In his response to the Earl's comments he flatly denied making a prophesy. Nonetheless the Latin sentences that caused offence had one property reminiscent of a sybillic utterance, namely it was capable of a number of interpretations.[7] A guilty mind might have taken the view taken by the Earl. Equally well, Recorde was more than sufficiently expert in Latin to have known of and deliberately used such ambiguity. The weight given to this accusation in the final judgement cannot be assessed in the absence of a documented account of the trial.

Recorde intimated that he was drawing closer to Court circles again which, given his known religious convictions must have meant that despite these latter he had knowledge that was of value to the Crown. It seems most likely that this related to the coinage. The Crown was in even worse financial straits when Mary ascended the throne than was her predecessor Attempts by Edward to rectify the disastrous consequences of debasements of the coinage had taken effect only in part. Those responsible for management and production of the coinage were generally in very bad odour, being generally suspected of having feathered their own nests at the expense of the Country.

As Commissioners for the accounts of 1547 and 1550, Warwick, Pembroke and Mildmay were able to extract back payments in excess of £27,000 from Bowes, accumulated during his period in charge of Tower I Mint. Valuable lessons must surely have been learned by the Commisioners and associated parties on how gains made during the minting of money could be concealed. This tradition of mint officials taking full illicit advantage of their activities was continued into Mary's reign. Thomas Egerton had been made undertreasurer of Tower I Mint. On 9 May 1553 he was warranted to provide up to 40,000 lb of fine silver.[8] This was to be minted into silver groats, half-groats and pence of silver 11 oz along with a range of gold coins according to an indenture of 20 August 1553.[9] Production continued up to November 1555. At about this time a Commission was set up to examine Egerton's accounts relating to the purchase of silver bullion. Reporting on 28 November 1555 they calculated that he had charged 16⅛ d. more per ounce of silver than he had paid,

---

[6]Nichols JG (ed) (1849) Narratives of days of the reformation, Old series 77. Camden Society, Westminster, pp 150–151

[7]Recorde's Latin text is open to differing translations. If it was an original statement much depends on the contemporary meanings of individual words.. If it was a quotation from earlier times, the meaning probably would have been different. Freely translated it might read…. Summoning as witnesses the heavens, the stars, and wise men, never will inimical dissent be extinguished in Britain as long as there is hidden hostility to the true lineage and a false dragon may openly emerge… , but 'fucati draconis' might also be read as painted dragon, possibly even red dragon.

[8]Public Records Office [PRO.] E 163/13/30

[9]PRO. E 101/307/1

thus owing the Crown a little more than £9,000.[10] Egerton's defense was that he had charged no more than the maximum he was permitted by the terms of his warrant, an argument that was not accepted. He was dismissed in December and by 2 February 1556 he was in the Fleet prison. He was summoned before the Barons of the Exchequer in April and again in May. A Commission was set up in June to examine his accounts up to Christmas 1555, which he was able to satisfy financially, but the debt relating to bullion purchase remained unpaid until his death in 1574. Egerton was replaced by Thomas Stanley who, starting in the early part of 1556 and continuing into 1557, oversaw the production of silver coins of 3 oz fine for Ireland.

Central government was clearly still exercised by matters stemming from the debasement of the currency during the two previous reigns and persisting into that of Mary. On 14 June 1556, the Marquis of Winchester Paulet, Lord Paget, the Bishop of Ely, Sir Francis Englefeld and Sir Edmund Peckham were appointed to 'take upon them the charge of the Mynt matters, and considre whiche waies may best be devised for the reformacion of the coynes'.[11] This was buttressed by a licence for the melting of base coin in circulation to determine its real worth, thereby enabling a buy-back value by the crown to be established.[12] The other evidence that was needed to establish a correct strategy for recoinage was the nature and quantity of coin in circulation. Challis and Harrison have argued that this latter could have been provided from a number of sources, one of which is found in the Paget papers and sets out the situation up to July 1556.[13] Whether the issue was taken to completion by the development of options for recoinage during Mary's reign is not known. Minting of coins 11 oz fine from bullion provided by the Treasury was sanctioned on 28 June 1557.[14] However no firm decisions about recoining of the baser currency were taken during Mary's reign. Insecurity about the future of the testoon was noted by Machyn in September 1556[15] and the government found it necessary to issue a proclamation denying the revaluation of the testoon on 22 December 1556.[16] Queen Mary sent a plan for crying down the currency to a small group of her councillors early in 1557, but the whole impetus of any action was lost by the middle of 1557, by which time Philip had left for Spain.[17]

It is against this background of concerns with the probity of mint officials and with the restoration of some order into the value of the currency that the missive to the Queen from Recorde on 10 June has to be viewed. The speed of reaction to it is

---

[10]PRO. E 163/13/30

[11]APC, 1554–1556, p 284

[12]Calendar of Patent Rolls [CPR.] 1555–1557, Philip and Mary iii, p 554

[13]Challis CE, Harrison CJ (1973) A contemporary estimate of the production of silver and gold coinage in England, 1542–1556. Engl Hist Rev LXXXVIII:821–835

[14]Challis CE (1978) The Tudor coinage. Manchester University Press, Manchester, [Challis 1], p 117

[15]Nichols JG (ed) (1848) The diary of Henry Machyn .... . Camden Society, xlii, p 114

[16]Hughes PL, Larkin JF (ed) (1964–1969) Tudor royal proclamations. Yale University Press, Newhaven/London, vol 2, no. 341

[17]Challis and Harrison, loc. cit, pp 823–825

perhaps an indication of sensitivity to some of the topics raised, for Recorde was summoned to appear before the Archbishop of York, the Archbishop of Ely and Sir William Petre only 10 days later. He reaffirmed his accusations and added the details already quoted.

It was in the detailed accusations levelled by Recorde on 16 October that must have lain the knowledge of which Recorde believed Pembroke to be afraid. They form three groups viz. that concerning the iron mines at Pentyrch, the operations at the Bristol mint and the mint operations in Ireland.

The Iron Mill at Pentyrch

Recorde alleged that *at an iron mill at Pentyrch on 5 February 1549, the earl took a barrel of iron worth £10 that was Crown property. The workers there were driven off which lost the king £2,000. Further, at Westminster on 20 March 1550 the Earl persuaded Recorde to let him have the profits of the ironworking which would be allowed for in the accounts to the prejudice of the Crown to the sum of £200.*[18]

Three of the statements in this allegation are supported by evidence from other quarters. A barrel of iron, according to the measures of the time current in the Weald, contained four firkins of about 950 lb each and thus weighing about one and a half tons of iron. The market price for finished product was about £7 a ton so the estimate of £10 quoted is about right. As will be detailed in the next section, on 6 January 1550 in London, Pembroke tried to force Recorde to hand over gold and silver from the Bristol Mint. It is credible that he adopted the same tactic with respect to monies arising from the operations at Pentyrch. Correspondence between Chamberlayne, then the ambassador in the Netherlands and Cecil shows that Pembroke was still pursuing this payment at the end of the year, as will be detailed in the discussion of the winding up of the accounts of the Bristol Mint, also in the next section. The sum sought by Pembroke at this time as part payment of a debt by Recorde from the iron mill operation had shrunk to a little over £68. Recorde's claim of loss of £200 of profitseems entirely credible

The technical nature of the operations at Pentyrch is unclear. Iron mining at the Garth above Pentyrch had proceeded sporadically since Roman times and probably from much earlier. The collocation of haematite, wood, limestone and water-power made it a natural site for iron-founding. Confirmation of the existence of iron mining at Pentyrch at the time of the incident under discussion is found in the documentation of the reward of William Herbert by Henry VIII, as attested by his inner circle, with property to the clear yearly value of 400 marks.[19] This was enacted on 10 July 1547, just after the king's death. Most of the property granted to Herbert in Glamorgan lay in a swathe from just north of Cardiff, cutting westwards through the

---

[18] The transcriptions of extracts from the original document will be given in italics.
[19] CPR, I Edward VI.-Part VII, pp 193–198

Welshry and down to the Vale of Glamorgan. It was Crown property that had descended to Henry from Jasper Tudor. Some of it was already rented by Herbert but now it became his as of right. The new acquisitions were the ironmines of Clonne park and those in the parishes of 'Llantrissan and Pentirgh'. They were held as of a fraction of a knights fee commuted into a yearly payment of £3. 11s. 2d. for the manors of Miskin, Glynrothney and Llantrisant, for those of Clonne, Lanmays and Pentyrch 23s. 5d. and for the forest of Miskin 5s. 5d. Herbert was also granted the reversion of rents and rights in a wide range of other properties including the lease of mining rights in the park of Clonne. This latter had been granted by Henry to his servant William Kendall of Launceston by Henry 8 June 1540, for 21 years at a rent of 26s. 8d. Herbert was also granted reversion of the mines in the parishes of Llantrisant and Pentyrch, but the names of the existing lessees and the rents were not given. Deposits of iron lead and silver were all being actively sought during this decade. The mining rights of Kendall did not extend as far as Pentyrch. William Herbert would thus seem to have been the landlord of Pentyrch at the alleged time of his foray there, but the mining rights remained with an unnamed leaseholder. Leland in his Itinerary of 1536–1539, had visited Castell Mynach on the opposite side of the Taff from Pentyrch, held by Robert Matthews at the time, but mentioned nothing about iron mining or smelting in the vicinity.

The main iron manufacturing industry in England in the 1540s was based in the Weald and was undergoing a transition from the traditional small scale operations in what were known as bloomerys producing wrought iron, to larger scale production using blast-furnaces and based on cast-iron. Output from a bloomery was typically 30 tons a year, but if a blast-furnace was employed an output of 150 tons a year was obtainable.

A substantial iron mining and manufacturing industry was eventually developed at Pentyrch and across the river Taff at Tongwynlais, both being sites of blast furnaces. Interpretation of such evidence that exists for the names of the parties involved in the operations at these two sites has proved contentious. The most recent view is that the Matthews family was initially concerned with the Pentyrch site in the 1560s, only later taking over the Tongwynlais operation from the Sydneys of Robertsbridge, the Wealden ironmasters.[20]

Rees notes that 'Claims for the existence of the furnace at Pentyrch have been made for a date prior to 1557, viz in the time of Miles Mathew of Llandaff who died in that year, but that is based on slender evidence'. Miles Mathew was succeeded by his son William and later by his brother Edmund Mathew of Radyr'.[21] The source of this claim is not given. Riden says 'It is possible that the furnace at Pentyrch was preceeded by a bloomery in the same parish, but no historical or archaeological evidence for earlier smelting in the area has yet been located.' Recorde's allegation

---

[20]Riden P (1992) Early ironworks in the Lower Taff Valley. Morgannwg XXXVI:69–93
   Rees W (1968) Industry before the industrial revolution. University of Wales Press, Cardiff, pp 259–264
[21]Rees W, ibid., p 259, footnote 44

establishes the existence of such an earlier operation. In his discussion of Recorde's mathematical works Gunther says 'He was particularly strong on the subject of mining for gold, silver, copper, tin, lead and other metals, probably as a result of early experience in a Welsh mining district,'[22] but offers no reference for the assertion.

Both Recorde and Chamberlayne, as will be discussed later, called the plant an 'iron mill', but this description is not sufficiently precise to distinguish between a bloomery and its successor, the blast–furnace. Neither is the level of profit quoted of much help in resolving the matter. Crossley has pointed out the problems inherent in deciding what was meant by 'profit' at that date.[23] Speaking of the situation at Robertsbridge in 1546, he quotes a margin of 53% between the price charged for a ton of finished iron and the cost of raw materials and labour.[24] Using Recorde's figure, a profit of £200 would have arisen from the production of about 60 tons of finished iron. This tonnage would have been typical of 2 years production of a traditional bloomery and maybe 5 months output of a blast furnace complex in full production. There was little difference in price between iron produced by the the old and new processes at this time. The operation at Pentyrch could therefore have been a bloomery in full production for several years or the initial phases of a new blast-furnace based plant. The quoted loss to the king of £2,000 presumably refers to a capital investment. A bloomery would only have involved expenditure of small multiples of a 100 lb. Even the blast furnace and associated plant at Robertsbridge had involved only about 1,500 lb worth of investment over its first few years. The options are therefore that Recorde had exaggerated the scale of the operation and that it was to be based on a blast-furnace, which in the event never got into full production because of Pembroke's actions, or that a bloomery had operated at Pentyrch for more than 2 years.

Possibly the most significant historical feature of the allegation was that the barrel of iron was **Crown** property, the it was **Crown** profit that was diverted and that it was the **king's loss** of £2,000. The whole operation would therefore seem to have been sponsored by the Crown and if so it would have been prior to Edward's accession. Given the pressure the Government was under at this time to increase production of iron for military use, the rational approach would have been to select the option achieving the higher output, as had been demonstrated in the Weald by this time. Additionally, the iron ore available in the Pentyrch area was relatively pure haematite, more easily dealt with using the higher temperature process.

The statement by Recorde and the action by Pembroke implies that the former had an overall responsibility for the operations at that site. Such an appointment made by the operators would not necessarily find its way into government records. Was the Sydney family of Robertsbridge the prime mover? Their operations in the Weald might have provided the impetus and could have provided the technical knowledge. Sir William Sydney, one of King Henry's courtiers of longstanding

---

[22] Gunther RT (1923) Early science in Oxford. Oxford, vol I, part 2, 'Mathematics', 13

[23] Crossley DW (1966) Iron-works management. Engl Hist Rev XIX:273–288

[24] Ibid., 279

who had been Chamberlain to the young prince Edward, might either have met or
known of Recorde as a person both in favour with Protector Somerset and also of
intellectual eminence. Protector Somerset was also an iron-master. Was Pembroke's
initial act directed against this faction rather than against Recorde personally?
A further point to bear in mind is that a Garret Harman was still overseer of the
king's mines, having been appointed by Henry in 1543. He might have been irked
by Recorde's appearance on the scene. He appeared later in Ireland at Clonmines
and harassed Recorde there.[25] Overall, there seems little reason to doubt Recorde's
account of events.

Mint Matters

Robert Recorde was involved, to a greater or lesser extent, with the operations of
three Mints during the reign of Edward VI. As Challis has pointed out, the reasons
for the opening of mints at Durham House, Bristol and Dublin are most likely
to do with personal ambitions of major political players during this period from c.
1548 to 1550 when the muddle associated with the 'Great Debasement' and the
potential for making illegal profits from moneying reached its peak.[26]

Durham House

Recorde first appears in the records as comptroller of the mint at Durham House,
with John Bowes as under-treasurer and John Maire as assay master on 29 January
1549.[27] An indenture to mint a 4 oz fine silver coin at 48s. to the pound had been
issued on 2 December 1548, so Recorde's formal involvement may have started
a little earlier.[28] The whole operation at Durham House seems to have made only a
minor contribution to the overall coinage in circulation, Challis having estimated
production as being about £70,000.[29] This episode seems noteworthy only for
the conviction of certain French mint workers for forgery in April 1549 and
also for the rumours that the mint had been instituted primarily for the benefit of
Protector Somerset. Whether the latter or his steward Whalley, to whom Recorde
had dedicated his first book *The Grounde of Artes* in 1543, was responsible for
Recorde's involvement is a matter for conjecture.

---

[25] This will be dealt with in the next Chap. 3. 'The Affair at Clonmines.'
[26] Challis 1, 7–8
[27] CPR, 1548–1549, pp 303–304
[28] Ibid., p 304
[29] Challis 1, 238, 241

The Bristol Mint

Thomas Chamberlayne, Berwicke and Fisher were sent to investigate the alleged irregularities at the Bristol Mint where Sir William Sharrington had been under-treasurer. They reported back to Protector Somerset from Bristol on 6 January 1549 that Sharrington's accounts had been taken up to London on the previous Wednesday and that the carrier, a Mr. Paget should be questioned. They were keeping the Mint operational in the meantime. On the way down to Bristol they had called at Laycock and sequestered Sharrimgton's valuables there.[30] The scale of operations at Bristol was an order of magnitude greater than that at Durham House and the potential financial pickings were commensurably greater, as demonstrated by Sharrington's activities.

Swift action followed. Sharrington was replaced as undertreasurer by Chamberlayne, whilst Recorde was formally appointed as comptroller in place of Roger Wigmore on 29 January 1549. Recorde was to have a salary of £133 6s. 8d. per annum, which was the same as that of Sharrington and a number of other under-Treasurers, and greater than that of Roger Wigmore, which had been £40 per annum.[31]

The sins of Sharrington have been well chronicled. He admitted to defrauding the Crown so as to provide funds for Sir Thomas Seymour who was executed for high treason by his brother the Lord Protector in April 1549. Thomas Seymour was Herbert's friend and brother-in-law. Greater than £4,000 of profits had been diverted. Given this background it is not surprising to find that mint efficiency during Sharringtons period of office was woefully poor in terms of quality of product. Challis points out that from April to September of 1547, Bristol coin was deficient in fineness by 5½d. and in weight by 6½d. per pound. Challis continues, 'Under Sharrington's more sober successor, Thomas Chamberlayne, however, Bristol was transformed into a highly efficient mint, turning out silver coin which was of the exact fineness (or better) and only a fraction over 2d per lb. deficient in weight. Indeed Chamberlayne's success was such that the performance of his mint surpassed that at Tower I at any time during the debasement period'.[32] The efficiency of the mint operation can also be judged by the fact that when the financial accounts for this period were scrutinised in 1551 they were found to be in surplusage to the extent of £218 13s. 2d., an extremely unusual occurrence at this time.[33] The accounts extend from 1 January 1549 to 31 October of that year and show a gross profit of some £43,000 and a net profit of a little more than £5,100.[34]

---

[30] HMC., Salisbury I., p 58, no. 252

[31] Challis CE (1965) Mint officials and moneyers of the Tudor period. Brit Numismatic J XLV:51–76 [Challis 2]

[32] Challis 1, 259

[33] Greenwood JW (1981) The closure of the Tudor mint at Bristol. Br Numismatic J 51:107–111, 108

[34] Challis 1, 63–68

The contributions made by individual members of the Bristol team to this success cannot be delineated with any accuracy. Both Chamberlayne and Recorde led peripatetic existences. Chamberlayne's accounts give Recorde's travelling expenses as £218 7s. 6d. The latter was of course comptroller of the Durham House mint also at this time. However, Chamberlayne was directed to Denmark and Sweden in September 1549.[35] It seems unlikely that he was back in England before March 1550, and could not have been in charge of the mint at Bristol during the turbulent events attending its closure, as described by Recorde in his submission at his trial. It must therefore have been Recorde who was responsible for the accounts. He was clearly responsible for the receipt of Bristol church plate.[36] He was probably also the recipient of church plate worth about £2,000 from Salisbury and Wells, although this was formally ascribed to Chamberlayne.[37]

Recorde's future became further and inextricably involved with that of Sir William Herbert during the tumultuous events of October 1549 which saw the overthrow of Somerset. Herbert's part in this event has been well explored and opinion on it is still divided. Although his presence at Hampton was asked for twice by Somerset, he and Russell brought their forces as far as Andover by 8 October, from where they responded somewhat coolly to Somerset's request, saying that they were staying put. The following day they retreated a little to Pembroke's manor at Wilton. On 11 October, Warwick and his followers arrested Somerset at Windsor, where he had taken the King. Smith, Cecil, Thynne and Whalley were also detained. Herbert and Russell gave their approval to the move. On 14 October, Somerset was taken to the Tower. The evidence deposed by Recorde at his pre-trial in 1557 exposes a sub-plot which, if true, scarcely shows Herbert in a favourable light at this juncture.

In his submission of evidence at his trial, Recorde asserted *that Herbert, hearing at Easton co. Wilts that Recorde had sent treasure from the mint to the King, on 11 October sent his servants to the Mint to bring the treasure to himself. Recorde refused to collaborate, as a result of which he was accused by Herbert of treason at Westminster on 26 October and forced to attend the king on 30 October for 60 days. The moneyers had to stop work which lost the king £10,000.* If this latter figure is taken to mean gross profit, it is a reasonable estimate of loss from 60 days of missed production. Central records confirm that production at the Bristol Mint ceased at the end of October.

Recorde says that *on 12 October, Pembroke sent his servant Roger Wigmore to Bristol with demands for all the treasure and despite Recorde's refusal he took £400 to Herbert. Further, letters from Wilton dated the previous day ordered Recorde to buy silver for the mint at a higher price than he was instructed to do by royal command.* This latter move shows prescience on Herbert's part inasmuch as on 28 October following a warrant was sent to the Master of the Mint with instructions on how the expenses of Herbert incurred during the uprising were to be defrayed.

---

[35]This is discussed in the next chapter.

[36]Greenwood, loc. cit., p 108

[37]Challis 1, 161

He was said to have expended the bulk of his plate and money on the affair and had additionally borrowed a lot of money. He, Herbert, was therefore authorised to submit 2,000 lb weight of bullion in fine silver to the mint which was to be coined into money and delivered back to Herbert with all the profits of coining that would normally have accrued to the Crown.[38] Given the statement that Herbert had spent most of his money and plate on funding his military activities, it is not obvious where he was going to find a further 2,000 lb of bullion. Roger Wigmore was the official displaced by Recorde at Bristol and was presumably still tainted with some of Sharrington's misdemeanours.

Recorde also says that *on 6 January at the house of Sir John York treasurer of the Southwark mint and an adherent of the Warwick faction, Herbert again tried to force Recorde to hand over the gold and silver treasure at the Bristol mint. When Recorde refused, Richard Wigmore was despatched to seize the treasure.* This was the date and venue for the first meeting of Warwick's newly constituted Council, which included Herbert.

Recorde accused Herbert of taking from the Mint at various times for his own use £1,826. As mentioned previously, plate from three Bristol churches totalling some 78 lb weight had been delivered to the mint there for the king's use, signed for by Recorde in August 1549. The plate from Salisbury and Wells had an estimated value of £1,824 12s. 6d. How much of these two sources had been used up by the time of the interruption of minting cannot be estimated, but neither can the holdings of gold and silver treasure from other sources. Its acquisition could have provided Herbert with a major contribution to the bullion that he was shortly to be authorised to turn into coin, at potentially a very handsome profit. There is other corroboration of some of Recorde's description of events. On 21 October 1549, John Rither, Cofferer of the King's Household received £1,318 8s.4d. by the hands of a Richard Recorde and Giles Evernet officers of the Bristol Mint.[39] This money was eventually handed over to Warwick and Herbert by warrant on 21 December 1550.

Two commissions established on 3 February, the day after Warwick assumed Presidency of the new Council, were to affect Recorde's future. Both were directed to Warwick, Sir William Herbert and Sir Walter Mildmay jointly and both were concerned with Mint matters. One instruction was that they were to take and examine the accounts of the various treasurers from the beginning of the debasement of the coinage, the other that they were to survey the mint operations in general, the appropriate powers for action being given in both cases. The overall account for Bristol was approved by the Commission on 11 October 1551 and on 16 October the Privy Council issued a warrant for £218 13s. 2d. to be paid to Chamberlayne as under treasurer for the Bristol mint.[40] This payment was blocked by the Earl of Pembroke, as Herbert had now become.[41] Further light is thrown on the proceedings

---

[38] Harleian MSS. 660, ff. 67, 69

[39] APC, 1547–1550, p 180

[40] APC, 1555–1557, Phillip and Mary iii, p 389

[41] Calendar of State Papers [CSP.] Foreign, 1547–1553, p 199

as a result of the activities of Recorde's nephew, also named Robert, when attempting to recover what he saw as his inheritance.

In 1573, Robert Recorde junior opened a case against Chamberlayne for recovery of his uncle's share of the settled surplusage account for the Bristol Mint. This case has been analysed by Greenwood.[42] Recorde jr. claimed that in the declared account, £153 6s. 8d. was due to Chamberlayne, £137. 6s. 8d. to his uncle, and a range of amounts to other individuals totalling £224. It is to be noted that copies of the detailed accounts have not been found in the official records, so presumably Robert jr. found the material in his uncle's papers. Robert jr. said that his uncle, by agreement with Chamberlayne, had paid the latter sum out of his own pocket before the account had been settled. He stated that Chamberlayne had received two payments from the Crown in settlement, one in the time of Edward VI for £218 3s. 4d. and one in Elizabeth's reign for £165 8s. 4d., which was £230 15s.more than was due to Chamberlayne. The surplusage from the operations of the Bristol Mint were £218 13s 4d. Chamberlayne denied receipt of the second payment and of any agreement with Recorde sr. to pay the other creditors. Greenwood was unable to find any account of the second payment in the records.

Chamberlayne admitted to receipt of the first payment but said that he had paid £65 6s. 8d. to Pembroke as part of a debt of Recorde's to Pembroke. In his second submission to the Court, Recorde jr. diluted his claims about the second payment, but stated that Chamberlayne had not reimbursed Pembroke, but even if he had it was improper and that Chamberlayne should repay the money to Recorde. The verdict of the Court was not found. The only fact apparently accepted by both parties to the dispute was that Pembroke claimed that Robert Recorde sr. owed him £65 6s. 8d. Light is thrown on the origin of this supposed debt by the correspondence between Chamberlayne in Brussels and Cecil in England. Payment of the surplusage was approved by Council on 11 October 1551. On 26 November Chamberlayne wrote to Cecil requesting him 'to be mean for him [Chamberlayne] to Lord Pembroke, that he may enjoy the money of his warrant stayed by his Lordship for the thing which he [Chamberlayne]is as ignorant as the child to-night borne; as he has written to his Lordship long since'.[43] The inference is that Pembroke's reasons for delay had nothing to do with Chamberlayne's activities at Bristol. On 4 December Chamberlayne wrote to Cecil acknowledging receipt of his letter and again requesting him to be mean to Pembroke for the money stayed by him in Mr. Mildmay's hands, for he needs money to cover his attendance on the Queen [Margaret of Hungary, Regent of the Low Countries] as she was 'on the move'. He goes on to say 'If his Lordship will take his bill to answer the thing claimed at all times, so far as it can be proved he received ought of the same iron mill, and so let him enjoy his money, he would be much bound to his Lordship.'[44] The inference is that a claim by Pembroke for some reimbursement associated with profits from an iron mill, had become

---

[42] Greenwood, loc. cit, 108–109

[43] Calendar of State Papers [CSP] Foreign, 1547–1553., p 199

[44] Ibid., p 200

conflated in his Lordship's mind with a payment that had been agreed for Recorde's portion of the surplusage at Bristol which had properly been aggregated with that of Chamberlayne as Under Treasurer. The propriety of the latter releasing Recorde's share to Pembroke seems dubious but Chamberlayne was desperate for money, repeating his request in a letter of 5 December. The matter must have been settled fairly soon after this for no further mention is made of it in the litany of requests for further funds to be found during 1552 in the correspondence of Chamberlayne with Cecil and other members of the Council. It is likely therefore that Pembroke took the opportunity of his appointment as Commissioner to behave as spitefully towards Recorde as he could and bully Chamberlayne and probably Mildmay into furthering his ends with respect to his claims on Pentyrch, which may or may not have been justified.

To summarise, it appears that Recorde's allegations against Pembroke with respect to the Mint at Bristol are well supported by such documentation as has been found which, en passant, supports his account of the affair at Pentyrch.

The Dublin Mint

Proceedings at the Dublin Mint were more complex and more opaque than those at Bristol. In his submission of complaints, Recorde says that *he was appointed Supervisor of the mint by letters patent 27 May 1551 and by authority of Parliament was given oversight of all mint officials; no one was to coin money without Recorde's consent, any letters patent, commissions or signed bills to the contrary notwithstanding. The earl was well acquainted with this but nonetheless, on 4 May 1551, appointed Martin Pirrie to be treasurer of the Dublin mint to coin new money. On 20 September 1552 he appointed Oliver Daubney to mint new money, which was carried out between 1 October 1552 and 20 February 1553.*

The patent to which Recorde refers reads, 'And further for the perfection of the king's "lately erectyd mynt" in Ireland, appointment of the same Recorde as surveyor of that mint; and charge to the treasurer, comptroller and "say maister" and others of the said mint " that from henceforth in all their alleyes, asseyes, myxtures meltynges, blanchinges, sherynges, coynynges and all other workes and doynges" they use his counsel;…'.[45] The Treasurer of Ireland was advised on the same day of the sending of Recorde and Gundelfinger to Ireland ' the saied Recorde to have oversight of the myners and the Myntt there,…'[46] A missive of the Council of the same date to Martein Perie advises him of Recorde's mission, including the latter's remit for the mining activities at Clonmines and goes on to say '…Mr Recorde, to whom he must declare and make privie thastate of the Kinges Majestie's Mintte and standard there, and what money he has delivered by his bargayn, and to accord to githers like trewe and faytheful servants.'[47]

---

[45]CPR 1552, p 144
[46]APC, 1551, p 275
[47]APC 1551, p 276

It is clear from these letters that Pirry was in place at the Dublin mint at the time of Recorde's appointment and it is highly unlikely that this appointment was the cause of Recorde's complaint. The reference must therefore have been to Pirry's next indenture in 1552 and that the complaint must have been that Pembroke appointed Pirry as Treasurer of the Dublin Mint without consulting Recorde in May 1552, not May 1551.

It would seem that the latter had cause for grievance, but not necessarily with Pembroke alone. From the beginning of March 1552 the matter of money and mints in Ireland was continuously on the Agenda of the Privy Council. Recorde was in attendance but was finally heard only on 11 May. Whatever the decision, it was finalised on 13 May and disappeared from future agendas.[48] On 31 May it was noted that Pirry was to be dispatched with orders for '10,000 lb to be coined after 3 oz. Himself to have the coinage of 3,000 lb.'[49] King Edward notes in his Journal on 10 June that new money was to be minted at Dublin using bullion from the Tower and that Pirry was to carry it over.[50] Pirry as under-treasurer, Daubney as comptroller and Williams as assay master were commissioned to coin monies for Ireland on 10 June.[51] Pembroke's name does not appear in any of the documentation cited. There is no record of him having been present at any of the Council meetings involved in the commission, if in fact this body was informed before the event. However the appointments to the Mint could have been made independently of any commission. On 24 June the Council ordered Walter Mildmay to engross the indenture and the warrants for the transfer of fine silver to Ireland and for the purchase of further supplies were authorised.[52] The Lord Chancellor was urged to pass the indenture with expedition on 29 July and the Lord Deputy of Ireland, on the following day was also urged to expedite Pirry's mission. Thus within a few weeks of Recorde having reported to the Privy Council about his remit in Ireland, Pirry was in place as Treasurer of the Dublin Mint, with the skeleton of a new indenture on the table, about which Recorde had not been informed let alone having been consulted. He seems to have been deliberately by-passed, but nonetheless his original commission of 27 May 1551 was allowed to run to its original termination date in 1553.

Who was behind all these machinations? Pembroke and Northumberland were both Commissioners of the Mint with powers to appoint officials.[53] This matter appeared to be of wider concern however, for the matter of Sir John York's successor at the Tower Mint was on the agenda for discussion at a Council meeting some time in March 1552, together with Record's account of the mining situation in Ireland.[54] Northumberland was present at most of the Council meetings at this time.

---

[48] CSP Domestic Series, Edward VI, 1547–1553; nos. 599, 601,602,603,605,613,616,617

[49] Ibid., no. 627

[50] Jordan WK (1966) The chronicle and political papers of King Edward VI. Allen and Unwin, London, pp 130,146

[51] CSP Ireland, 1503–1578, p 127

[52] APC, 1552, pp 88–89

[53] CPR, 1549, p 216

[54] CSP Domestic Series, Edward VI, 1547–1553; no. 599

He was specified by Recorde as a confederate of Pembroke in his, Recorde's ill-treatment, but he was of course dead by the time of Recorde's complaint. The nature and balance of the relationship between Pembroke and Cecil at this juncture is difficult to decode. There are a number of recorded items of correspondence from the former to the latter asking favours of him, all dating from the last 5 months of 1552.[55] One of these concerned Pembroke's 'friend' Sharrington and was, not surprisingly, unsuccessful as it involved Cecil foregoing his wardship of an heir in favour of Sharrington. Another friend on whose behalf aid was sought was Thomas Gresham. The outcome of these supplications were generally successful and on 8 December 1552 Pembroke wrote to Cecil, 'You have by sundry means declared your good will towards me that I am much bound to you, and you will find me your friend on any occasion of trial.'[56] There are no written records of favours being requested of Pembroke by Cecil. At this time, Cecil was one of the two Principal Secretaries with the main functions of private secretary to both Northumberland and Edward.[57]

Another person with connections to Pembroke, Cecil and the monarch was William Thomas, brother in law to Walter Mildmay and clerk to the Privy Council.[58] By his own telling Thomas had been familiar with the Pembroke household for some time and acted as a conduit between Cecil and Pembroke in the matter of favouring Sharrington mentioned above. By some unknown turn of events, Thomas also established a personal relationship with the King, as a result of which he was asked by the latter to inform the monarch on a number of matters, privately and secretly.[59] The first of these opinions related to the reform of the coinage. The views expressed were radical but no more so than those of Thomas Smith in his *Discourse of the Commonweal* or, less overtly, by Recorde in his dedication of the *Pathway* to Edward. Its main significance lies in confirmation of the King's keen interest in matters relating to the coinage. Adair concludes that '… it can hardly be too much to say that such personal authority as the King possessed was wielded in this [reform of the coinage] and possibly also in some other matters under the influence of William Thomas.' This conclusion begs the question of whether there were individuals under whose influence Thomas might have been operating. Mildmay, Pembroke and Cecil are all possibilities, whilst the idea that Thomas may have been an agent of Northumberland has also been advanced.[60]

The monarch may well have decided to take the initiative himself. By 1552, the finances of the Government and consequently those of the King had reached

---

[55] Ibid., nos. 693,695,715,758, 759

[56] Ibid., no. 780

[57] Murphy J (1987) The illusion of decline: the Privy Chamber 1547–1558. In: Starkey D (ed) The English court. Longman, London, pp 119–146, 130

[58] By this time Mildmay was deeply involved with a number of Commissions concerned with the nations finances'

[59] Adair ER (1924) William Thomas. In: Tudor studies. Longman, London, pp 133–160, 141–143

[60] Skidmore C (2007) Edward VI the lost King of England. Weidenfeld and Nicholson, London, pp 227–228

a parlous state. The king was fully aware of the detailed position, in which he took a keen interest. The manner in which events were played out has been described by Murphy and it would seem that for the closing year of Edward's reign the finances of the country were managed by a small group of people close to him.[61] The Privy Chamber appeared to have taken over financial management from the Privy Council. Funds incoming were measured in fractions of a 1,000 lb. Revenue from the proposed coinage in Dublin would have been measured in tens of thousands. Whoever appointed Pirry, it is clear that the King would have been party to it and have approved if not have initiated the appointment. Whether Pembroke would have played any part in this process seems an increasingly remote possibility. Recorde would not be expected to be aware of these arrangements within the Privy Chamber, but Cecil would have.

The formalities of Daubney's appointment to succeed Pirry are also obscure. The date of the latter's death is not known, but if Recorde is correct about Pembroke's role, it must have predated 20 September 1552. On 30 October, Northumberland wrote to Cecil saying that funding for the 'furniture of Pirrie's wife's licence' had been retracted, suggesting that by this time she was widowed. Daubney, as treasurer of the mint, was on 24 November instructed to segregate a debt to Thomas Gresham, from the monies owed to Pirrie's executors.[62] As with Pirry, Daubney's appointment could have been made by Pembroke but he could equally well have been acting in concert with the other interested, approving parties.

The second part of Recorde's complaint was that *Further on 20 June 1553 the earl viewed the account of both Pirrie and Daubney and passed them without reference to Recorde thereby losing the Crown £40,000.*

The strength of the accusations relating to the loss of money to the Crown and the approval of the relevant accounts are difficult to assess for the accounts are missing. The bargain referred to in Recorde's appointment was the farming of the Mint by Pirry in pursuance of an act of the Privy Council that was due to expire in August 1551.[63] Challis believes that coining under this indenture may have continued until Michaelmas 1551.[64] The original indenture allowed coinage operation up to a level that would provide the Crown with a profit of £24,000. There was therefore sufficient margin for Pirry to have continued coining provided he could find the bullion to do so. On 17 August 1551, the King had written to Deputy Croft of Ireland sending 16,000 l. and authorising further coining by Pirry.[65] Deputy Croft and Treasurer Brabazon reported the receipts of profits to the Crown of a little over

[61] Murphy J, loc. cit., Chap. 4, pp 136–140

[62] APC, 1552–1553, p 177

[63] APC, 1551, p 276

[64] Challis II, 112

[65] CSP Ireland, 1503–1578, p 116. The terms of the bargain made with Pirrie by the King were that the latter should receive 13 s. 4d. for each pound weight coined. Challis has estimated that returns to the King up to May 1551 amounted to that derived from the minting of 19,910 lb of coins 4d. fine. To produce a return to the King of £24,000, 36,000 lb would need to be minted, i.e. a further 16,090 lb.

£13,000 up to end of May 1551, but no later. If in fact Pirry did mint a further 16,000 lb, this should have meant further income to the King of a little over £10,000 bringing his receipts up to £24,000. There is no evidence that Pirry ever reported the financial outcome of any such transaction either to Recorde, or to anyone else, but Recorde could have known if it had in fact have taken place. Pirry was in London in November and as there was no further coining until the following year, it is possible that he did not return to Ireland until the following year.

How much money, if any, was coined under the indenture of July 1552 is unclear. The first mention of any intent of further coining was noted by Cecil towards the end of May when he cited Pirry being dispatched, '10,000 lb to be coined after 3 oz. Himself to have the coinage of 3,000 lb'.[66] The first formal proposal was for the coining of 1,500 lb weight of fine silver, which would have yielded coins of face value £10,800. The final indenture of 29 July however had the sum of 1,500 lb erased after being engrossed.[67] A letter from the King to the Justices of Ireland, endorsed by Cecil and sent on 27 December 1552, set them free of a previous restriction and permitted them to coin further bullion to the extent of 8,000 l.[68] Challis quotes an estimate of production as £32,400, the coined equivalent of 4,500 lb. of fine silver. This value of coins would weigh 18,000 lb i.e. the original weight envisaged by Cecil in May together with the permit of December. Additionally, there was also the 3,000 lb granted to Pirry for his own use, of face value £5,400 giving a possible total of £37,800 coins minted. Early in 1553 Pirry's executors were granted 2,000 l. out of the 8,000 l. that was authorised to be coined, followed a little later by a further 6,000 l.[69] Even after taking into account the money the Crown owed Pirry for the purchase of bullion, the total payment of 8,000 l. to his executors out of a total of 21,000 l. coined, suggests that the debt was derived from operations other than those authorised solely after May 1552. The situation could be clarified if only the accounts for the Dublin Mint post-May 1551 were available, but they are not. These are the accounts that Recorde accused Pembroke of passing without reference to himself, Recorde, 2 weeks before the King died. They should have recorded the fate of money coined in Dublin between May 1551 and the end of Edward's reign which may have been of face value between £32,000 and £42,000.

To summarise, there is no documentary evidence that Pembroke appointed Pirry and Daubney as alleged by Recorde, but he certainly had the authority to do so. It seems more likely that responsibility lay within the Privy Chamber and therefore must have been with the full knowledge of the King and a few others including Cecil. Decisions were made without consulting or informing Recorde which, given the increasing authority the King was beginning to exercise, is perhaps not surprising. The former was also not consulted by Pembroke over the accounts of the Dublin

---

[66] CSP Domestic Series, Edward VI, 1547–1553; no. 627

[67] APC, 1552–1553; p 104

[68] CSP Ireland, 1503–1578; no. 416

[69] APC, 1552–1553; pp 208, 229

Mint after May 1551 and although Pembroke was said by Recorde to have passed them, they are also not available, which is suspicious. The sum of money unaccounted for is comparable with that listed by Recorde.

London

Finally, Recorde makes the general accusation that *The earl as Supervisor of Mints defrauded the king in the accounts of officials to the sum of £50,000.* Not only are accounts for Ireland missing for the period when Recorde had an interest, but those for English gold and silver coining from 1 April 1552 to 24 December 1553 also have not been found.[70] Pembroke and Northumberland were appointed as commissioners for mints in 1550 and all coinage produced in England from this time to the end of Edward's reign was to the re-instituted fine standards for both silver and gold. Mention has been made earlier in this Chapter of a document, probably written in mid-1556, that gave an estimate of the production of gold and silver coinage in England from 1542 to 1556. Its contents have been analysed by Challis and Harrison who conclude that for the reign of Edward VI (Mich. 1551-July 1553), £124,179..17s..6d worth of silver coinage and £ 21,513 worth of gold coins were produced.[71] They point out that an estimate of 1559 for the sum value of coinage of gold and silver still current, struck during the reign of Edward VI, was £100,000, leaving a shortfall of some £45,000 coins not in circulation. Was this the discrepancy that was the basis of Recorde's accusation and if Pembroke was not involved who was responsible, how was it explained at the trial and by whom? Elizabeth was aware of the discrepancy attributing it to either hoarding or exporting.[72] There is no evidence that she took any action against anyone on this count.

Warwick and his adherents had consistently shown a great deal of interest in minting activities. Challis has commented on this period 'In a grant to William Dunche, former auditor of the mint, on 20 June 1573 it was claimed that Warwick and his fellow commissioners [Pembroke was one such] recovered £48,000 to the Crown's use. CPR, 1572–1575, no. 102.' It is important to remember in this connection, however that Warwick had no qualms whatsoever in rewarding both himself and his followers to the tune of almost £70,000 out of the proceeds of the mint. BM, Harleian Mss., 660 fos.67, 69.[73] Pembroke's share of the largesse was a licence to coin 2,000 l. of fine silver. In all, licenses to the total value of 22,500 l. of fine silver were issued, but by the time that Recorde came to accuse Pembroke, the recipients were either dead or in disgrace. The proceeds of such private licences

---

[70] Challis I, 306–307

[71] Challis CE, Harrison CJ (1973) A contemporary estimate of the production of silver and gold coinage in England, 1542–1556. EHR. XXXVIII, pp 821–835. also Challis I, p 119

[72] Harleian Miscellany, ed. T.Osbourne (1740), viii, pp 67–69

[73] Ibid., p 103

were not recorded, so their history remains private. Possibly Recorde was aware only of the activity of coining, but not of the aegis under which such activity was authorised. The whole situation is greatly confused, with many fingers potentially in the pie.

The documentary evidence needed to substantiate Recorde's claim or refute the charge of malfeasance by Pembroke is missing. The account of the trial might have provided such material, but it also has not been found. The only information about these proceedings found to date has been that relating to Cecil's intervention, quoted earlier. Most of the evidence for the sequencing of events concerning the activities of the Dublin Mint comes from Cecil's archives. As has been shown, the evidence they contained neither substantiates nor refutes Recorde's assertions. No one was in a better position than Cecil to provide the unrecorded material that rendered him 'well able' to clear Pembroke of many of the articles against him, successfully as it appears. It should be noted that the absolution rendered by Cecil is for 'many', but not all of the charges.

Recorde clearly made what proved to be a fatal misjudgement in accusing Pembroke of malfeasance. His timing for such an action was probably as good as possible, for Pembroke's standing with Queen Mary was still unsure. His timing may also have been dictated by the action against him that he alleges Pembroke had already commenced, the grounds for which are not known. Comments made in some of his later books, show that Recorde was being harassed by persons unnamed prior to his final action. Recorde was passionate about his educational activities and probably about his religious beliefs, but he does not appear to have been an intemperate man. His misjudgement of the probable strength and nature of the opposition he was to face at his trial fell far short of the critical judgements he exercised in assembling his mathematical texts. Had he realised that Cecil would speak against him he might have taken a different course. Recorde's service was devoted to the Crown and his actions consistently intended to protect its interests. If his discomfiture was attributable to actions by the Crown rather than by Pembroke, he would have been much distressed, possibly even broken by the revelation.

Chapter 4
The Affair at Clonmines

Abstract The prospect of an indigenous supply of silver at Clonmines, co. Wexford, would have excited Henry VIII given the parlous state of his finances. Initial results were mildly encouraging. Henry died before any serious action was undertaken, although the necessary legal instruments to do so were set in place. The financial inheritance of Edward VI made the option more attractive, even moreso given the reportedly favourable results of an assay of a sample of ore from the mine. A force of German miners was recruited and sent to Clonmines to start extraction operations on a serious scale in mid 1552. Robert Recorde was commissioned to oversee the operation. He met with continual interference from Garret Harman who had spearheaded the project from its inception and had provided the favourable assay results. After 5 months work, Recorde found that the operation was financially unsound, primarily because the initial assay provided was wildly optimistic. The project was abandoned at major cost to the Crown. Recorde's accounts were passed for payment but his personal financial debt was not honoured by the Crown. This bankrupted him. The debt was acknowledged and discharged to the benefit of Recorde's nephew in 1571.

The debasements of both silver and gold coinages, first by Henry VIII, continued by Edward VI, were the outcome of excess of Crown expenditure over its income. Lack of any substantial supplies, indigenous or otherwise, meant that the precious metals involved had either to be found by melting down existing plate etc., by recoinage or by importation of bullion. Any possibility of discovery of indigenous sources of supply of relevant minerals always attracted unduly optimistic expectations. Such was the case for the possibility of the mines at Clonmines, co. Wexford in Ireland providing a viable source of silver ore, made even more attractive as Ireland was proving both such a sink for money from the Crown and also because it had its own Mint at Dublin.

Chronologically, this episode embodied the last of Robert Recorde's documented Crown commissions. The issue was examined by Clarke in 1926 and has not been studied since other than in a subsidiary role by Greenwood, who showed that some

J. Williams, *Robert Recorde: Tudor Polymath, Expositor and Practitioner of Computation*, History of Computing, DOI 10.1007/978-0-85729-862-1_4,
© Springer-Verlag London Limited 2011

of the evidence Clarke used had been misinterpreted.[1] Clarke's surmise that in the absence of anything to the contrary, debt could have been the reason for Recorde's imprisonment is negated by the evidence of the previous Chapter. Further evidence shows that the roots of the affair ran politically and chronologically deeper than immediately apparent and that the histories of other persons involved in the matter need to be examined if the role of Robert Recorde is to be properly understood.

Recorde's primary adversary in the affair was to be one Garret Harman with Joachim Gundelfinger in a supporting role.[2] Both individuals had a long history of involvement with the Crown and with a number of its Ministers during the turbulent years at the end of the reign of Henry VIII. On the diplomatic front, Henry was at this juncture juggling with labile alliances with or against Francis the king of France, Charles VI duke of Burgundy and Holy Roman Emperor, as well as various Protestant leagues. This necessarily involved a deal of communication and intelligence gathering, in which Harman and Gundelfinger were to play their part.[3]

Joachim Gundelfinger had been associated with Henry VIII since February 1539 when the latter interceded on his behalf to get him an import licence from Antwerp.[4] This must have involved goods for Henry, for treasurer Gostwick's account for 5 May 1540 details payments of two lots of £40 to Gundelfinger, one being of a prest dating from the previous November and the other a reward.[5] Between In May 1541 and 1545, Gundelfinger was active on a number of political commissions involving the King.

In May 1541, he was named as an agent of Duke Philip of the Palatinate, sent to London to try to get an answer as to whether the King wanted Philip to return to the capital, presumably to pursue the possibility of marriage with the Princess Mary.[6] In August 1542, Gundelfinger the King's servant had been instructed to provide funding locally for a Royal commission given to Thomas Seymour.[7,8] Its objective was to determine the availability and costs of fighting men to be provided by the Barons Hedyk and Flekestyne of Nuremberg. Seymour wrote to Henry on 29 December

---

[1] Clarke FM (1926) New light on Roberte Recorde. Isis, vii, pp 50–70
Greenwood JW (1981) The closure of the Tudor mint at Bristol. Brit Numismatic J 51(24):107–111, 109

[2] The spelling of the surname 'Gundelfinger' caused problems for English clerks and hence is to be found also as the variants Goldenfynger, Kindelfinger, Goldingfinger, Gwyndelfinger. Similar difficulties arose with Harman's Christian name which commonly was Garret but is also found as Gerhardt, Gerhardo, Garfrido and Gerardus.

[3] Whilst not the most recent discussion of the topic, the concluding chapters of *Henry VIII* (Penguin Biographies 1972) by J.J. Scarisbrick gives a good feel for the complexities and hectic European diplomatic activities of the time. The frenetic nature of these years domestically have been examined more recently by Robert Hutchinson in *The Last Days of Henry VIII; Conspiracies, Treason and Heresies at the Court of the Dying Tyrant*, (Weidenfeld & Nicholson 2005).

[4] LP, 1539 I, no. 1286

[5] LP, 1540, no. 642

[6] Ibid., 1540–1541, no. 835, 941

[7] LP, 1542, no. 701

[8] This was Sir Thomas Seymour, who was to become Catherine Parr's husband. The letter was written in Italian and contained a wide range of political intelligence.

reporting his arrival in Nuremberg and of the opening negotiations.[9] On the same date Gundelfinger also wrote to the King, regarding another commission from the Monarch. It was transmitted by a Garret Harman, factor to Wriothesley the King's principal secretary, authorising Gundelfinger to bring, in person 'sufficient silver to prove its weight and goodness'. 'Sufficient' apparently meant a sizeable proportion of £1,000. This Gundelfimger had arranged on favourable terms, but had been delayed due to the war between the Emperor and the Duke of Juliers, but he would gladly accompany Seymour on the latter's return to England.[10] Another commission for Gundelfinger was proposed on 14 June 1545. Vaughan writes to Chief Secretary of State Paget from Antwerp that he had failed to get coiners as desired by the Council because they would not leave the mint houses without the Queen' permission. He suggests that the King should write to Gundelfinger to procure some coiners and Vaughan would transmit this request to Nuremberg where there were 'workmen as good as are in the world for that science'.[11]

Garret Harman was a London goldsmith, but had also been operating as a Royal courier and had had correspondence with Gundelfinger. On 3 February 1545 Harman was appointed overseer of the King's mines for a fee of 2s. per day.[12] There are no prior indications that he had any special suitability for the post. The phase of activity in Ireland that was ultimately to involve Robert Recorde as well as Harman started on 14 August 1545, with a letter of recommendation from the Privy Council to the Lord Deputy of Ireland on behalf of Moris Horner, Hans Hardigan and John of Antwerp, whom the King [Henry VIII] had commissioned and dispatched to Ireland to search out mines.[13] The latter were granted a prest of 20 l.[14] for the voyage and a further 18 l. for their prior charges.[15] The results of their activities was reported 1 August the following year by Thomas Agard of Ireland, Garret Harman of the City of London, goldsmith, and one Hans a Dutchman who had been in Ireland searching the mines. Various 'pieces' both for coin and alum were shown, 'found to be fayre' and were appointed to be kept, awaiting the king's pleasure.[16] This pleasure seems to have been realised around 25 September 1546, when the Lord Deputy of Ireland

---

[9] LP, 1542, no. 1246

[10] Ibid., no. 1247

[11] LP, 1545 I, no. 946. Steven Vaughan had been the service of Thomas Cromwell since 1524. He was a mercer who spent a great deal of his time in the Low Countries and gathered political intelligence there for Cromwell. At the end of 1539 he had become resident ambassador in Brussels as well as governor of the Company of Merchant Venturers there. Following the fall of Cromwell he remained in Brussels as a Royal agent mainly concerned with commercial matters. He retired as the Crown's chief financial agent there in 1546 having successfully established a firm position for the English currency in the Antwerp market. One of his rewards in 1544 was to be made under-treasurer of the London mint.

[12] Letters Patent II, no. 282. [LP]

[13] Acts of the Privy Council 1542–1547, no. 229. [APC]

[14] A prest was an advance of money by the Crown in lieu of services to be undertaken.

[15] APC, 1542–1547, no. 231

[16] Ibid., no. 501

and certain members of the Council were authorised to form a Commission for Mines and a prest of £1,000 was ordered for miners to be sent there urgently.[17]

This latter message was carried by Garret Harman, who on 24 February reported to Paget that the message had been delivered. His contact was with Joachim Gundelfinger who, reporting to Henry VIII from Augsberg on 16 March 1547, said that he had obeyed the King and obtained the services of a mining surveyor, a smelter and a man skilled in mining works. He had despatched them to Master Garret with a letter and money, but had heard from the latter, with grief, that they had not kept their promise. Maintenance of the families of these workers coupled with their retention had cost him 156 crowns and he feared that he had incurred the monarch's displeasure.[18]

It seems clear that Gundelfinger and Harman were by that time acquainted with one another, were both foreigners, probably of German origin, had both been servants of Henry VIII and both had dealings with members of his inner councils. Their value is reflected in the account, dated 30 April 1547, at the start of the reign of Edward VI, of 'Declaration of fees paid by royal warrant out of the Exchequer to foreigners', wherein Garfrido Harman is stated as having £36..10s for life, Joachim Guydelfynger £50 for life and Hans Hardinger £40 at pleasure.[19] From this time forward their activities became very closely linked in the Clonmines project. Closeness of interest did not necessarily mean mutual trust and respect. From his letter to Henry of 16 March 1546, Gundelfinger appeared doubtful of Harman's knowledge of mining and extraction metallurgy despite the latter's position as Overseer of the Kings mines. He wished that Gerhart had had time to visit the mines [at Augsburg] to see the methods used and to take samples of ore to England. He offered his own expertise and sent Harman 'a little book about minerals'.[20] This expertise was to be called upon in 1550.

What had happened to the Irish mines in the time between September 1546 when the Commission for Irish mines was first established and 9 April 1550 is not known in any detail. There had been continued activity, for on this latter date a warrant was issued to Frances Agard to retain £342..8..2 for payment by his father to Thomas Barnard, Clerk of Mines of Ireland and others for the operation of mines; Frances was to retain £40 for his father's travel expenses in the last 2 years of his life [he died August 1549] for the trying of mines together with a further £40 to cover his charges for two visits to confer with the Lord Treasurer [Edward Seymore, Lord Protector] about the mines.[21] The Commission of September 1546 would still seem to have been active.

Shortly after Agard's death, just before Protector Somerset was unseated, a prest of £200 was granted to Sir Thomas Chamberlayne [22 September 1549] to fund a

---

[17] LP. II, 1546–1547, no. 156

[18] Ibid., no. 402

[19] APC, 1547–1550, no. 38

[20] LP, 1546, no. 402

[21] APC, 1547–1550, p 426

visit of himself and Joachim Gundelfinger to Sweden and Denmark.[22] Greenwood states that the primary aim of this mission was to create a balance of trade by which Swedish silver would flow in to England. Instructions to both of them as Commissioners to the two countries were issued in a document dated the end of '49'.[23] Their joint commission must have been executed expeditiously, for by 27 March 1550 Chamberlayne's expenses for the undertaking had been paid.[24] On 22 April 1550 Chamberlayn reported back to the Council the disappointing results of the mission[25] and on 12 June following he was appointed ambassador to the Queen Regent of Flanders. However Gundelfinger had additional specified duties to discharge. He had to negotiate the exchange of £1,000's worth of kersies and other cloths with the Governor of Silberbergen in Hungary for annual payments of ducats or gold. As soon as these matters were arranged, he was to return to the Court with as much dispatch as possible to give an account of his proceedings and settle with the Council. Additionally he was to bring with him a closely specified group of mining engineers obtained on as reasonable terms as possible, for his Majesties service. The list, written in French, exhibits a considerable knowledge of mining- and extraction- metallurgy. The request was for:

> One of the most experienced and sworn master-mining engineers. A good metal founder. Two others for making shafts tunnels and ditches of mines, called in German 'Schirpffer'. Two carpenters to descend into the mines and prop each side of such shafts and passages, called in German 'Steyper'. One who thoroughly understood the drainage and carrying off of water. One who understands the assaying of all metals. Two smiths for making the necessary tools for the pioneers and others. Two colliers to work the large coals of the mines. Two who understood the separation of sulphur before melting. Twenty pioneers good, strong and experienced, unmarried if possible. All who have harness and weapons to bring the same with them and each to have a long harquebuse. One who understands the baking of alum. Two with the seeds of pine and deal to sow the same according to the nature of the soil, to increase the forests of both England and Ireland. All these to be German and brought to England by sea from Frankfort, the best way he can.[26]

No obvious candidate for the authorship of the document has emerged. Of the parties eventually concerned, only Gundelfinger is likely to have had the detailed technical knowledge displayed in the letter. He might have been able to advise the writer who was facile in French. Sir John Mason was French Secretary to the Privy Council having been secretary for the French tongue for Henry VIII since 1542. He was to become ambassador to France from which appointment he returned in July 1551. In 1544 he had been appointed Master of the King's Messengers, Couriers and Posts and as such seems very likely to have known both Harman and Gundelfinger.

---

[22] APC, 1547–1550, p 327

[23] Calendar of State Papers Foreign [CSPF], 1547–1553, no. 245, pp 57–58. This document was attached to a letter from Gundelfinger to the Council dated Antwerp 18 October 1550. The attachment was presumably intended to justify Gundelfingers request for help from the Council.

[24] APC, 1547–1550, p 419

[25] APC, 1550–1553, pp 8–9

[26] CSPF, 1547–1553, no. 245

He might well therefore have been involved as the writer of instructions to Gundelfinger, but it is very unlikely that Mason had the technical expertise to have constructed the letter wholly. Thomas Agard might well have had the necessary competence and could have been consulted before he died. This latter supposition has the attraction of providing continuity in the intention to exploit the silver mines at Clonmines. The inclusion in the specification of the German team of 'one who understands the baking of alum', implies that its author was familiar with the contents of the original commission of Henry VIII.

On 9 March 1550 a warrant to pay £100 in prest to Garret Harman for Joachim Gundelfinger was issued, followed on 8 July by a warrant direct to Gundelfinger for £500.[27] The reasons for these disbursements lies somewhere in a letter to the Council from Joachim Kindelfinger in Antwerp on 18 October 1550,[28] to which document the original commission to recruit workers was attached. Whether the disbursement was in aid of the mission to Hungary or for the recruitment of miners is not apparent. If the latter, the subvention was not adequate He says that he arrived on the third of the month with 50 master mine-workers and applied to the Ambassador Sir Thomas Chamberlayne and the King's agent Mr. Dansel to supply his needs.[29] He asked for them to help him to complete his mission as he had brought the men to Antwerp at his own expense. Gundelfinger asked for £400 to be sent as soon as possible.[30]

Garret Harman arrived in Antwerp at 10 p.m. on New Years Day 1550 [1551] and writes to Cecil that he has delivered his letter to Chamberlayne at Brussels.[31] The content of some of this message can be deduced from Harman's report of the answer he received. He was advised not to apply for a passport, presumably for Gundelfinger and his men, as it was not needed, because it was hard to get, because the Lady Regent was not at Brussels and because it might raise suspicions. It was suggested that he hired a ship or two, go to Zeeland with Gundelfinger and his men and set sail for England as soon as possible. Harman asks Cecil to get him a passport otherwise he might be 'stayed'. He, Harman happened to have a sample of ore in his bag which he gave to Gundelfinger and the Burgomaster. The latter immediately assayed it 'and found it so good that there is no doubt that the King shall receive such honest profit as will cause the Council to regret that it had been so long delayed'. Dansel had not arrived with the money which grieved him. Two days later he writes again to Cecil with the results of assays that had been repeated two or three times. The Burgomaster was very excited about the matter and said that if there was enough ore the whole realm should thank God for it. <u>One hundred ounces should yield more than eight ounces of silver and half a hundred of good lead.</u> He urged Cecil to show

[27] APC, 1550–1552, pp 181, 233

[28] CSPF, 1547–1553, pp 57–58

[29] By August 1550, Sir Thomas Chamberlayne was in Antwerp as ambassador to the Queen Regent Margaret, Arch Duchess of Savoy and ruler of the Low Countries in her own right.

[30] CSPF, 1547–1553, pp 66–67

[31] By this time Cecil was Secretary to the Privy Council.

the letter to the Council, who would see that he had always spoken the truth, and 'it were pity that men of no experience should meddle in it as they would lose the one half that God had given to them'. Whether Harman had taken this action himself or whether he had been sent by Cecil without the knowledge of the Council, some members of which were sceptical of Harman's claims, is not known. Such sceptics would be justified by the turn of events, even if their judgement was not founded on technical expertise. The sample that Harman had assayed must have contained over 60% galena (Lead Sulphide, PbS), which level of purity would only have been obtained by careful selection of samples. The money referred to by Harman was presumably the warrant to Gundelfinger of 8 July 1550 for £500. This money must have been made available to Gundelfinger, for on 18 May 1551 he was recompensed for his losses sustained as a result of exchange of the money he had received for transport of miners to Ireland.[32] Was the information sent by Harman to Cecil the decisive factor in persuading a reluctant Council to fund the enterprise? The Burgomaster was excited but was Gundelfinger equally enthusiastic? He and his company travelled first to London and on 30 May, Gundelfinger accounted for all the sums of money received by him prior to their despatch from London.[33]

The Council for Ireland must have received some prior intimation of the intention to use foreign expertise at Clonmines. On 20 May the leading members of that Council had written another letter of desperation to the Privy Council about the parlous situation with respect to supplies of bullion for the Irish mint and drawing attention to Clonmines, saying that they had not yet carried out any assays as they were waiting for Lord Deputy Croft's delegation. They thought that English subjects were as capable as strangers for reducing ore to metal.

It was at this juncture that Robert Recorde entered the arena. An entry on the Patent Rolls, dated 27 May 1551 at Leighes, appointed Robert Recorde as surveyor of all the king's mines, newly found in Ireland with commission to rule their affairs including labour and carriage for them. Joachim Gundelfinger captain of the foreign miners there and the other officers, workmen and labourers there are to obey him. Recorde was also appointed as surveyor of the newly erected mint in Dublin, whose treasurer, comptroller and assayer are to heed his counsel in their activities. He also had authority for payment of charges relating to the mines for silver produced there to be coined in the mint at the standard of 4 oz. The mint treasurer was to coin such bullion quickly and not hinder monthly pays.[34] At Greenwich on the following day the instructions for Recorde service were signed by the Privy Council headed by the Duke of Somerset. Additionally, the monthly charge of the officers and miners was declared, the Treasurer of Ireland was advised of the mission of Recorde and Gundelfinger with the accompanying 50 German miners whose rates of pay had been specified and Recorde's terms of office were described. A similar letter was sent to Lord Deputy Croft and the Council of Ireland, but additionally authorising

---

[32] APC, 1550–1502, p 274
[33] Calendar of State Papers. Ireland [CSPI] 1509–1573, p 114
[34] CSPI, 1503–1578?, p 144

the taking of the account of Martin Pirrie (the comptroller) by some 'experte men there, to thende Mr. Recorde .... may know the better his chardge according to his instruction. In other words, Recorde's starting point was to be defined. Pirrie was also informed of Recorde's commission and was told to declare to him the state of the mint, its standard and what money he had delivered in by his bargain. This latter has been discussed in the previous Chapter, but it is important to note the unequivocal nature of Recorde's authority. There was no mention in Recorde's mandates of search for and treatment of alum. This sweeping Commission was given to Recorde despite the way he had been treated by Pembroke over matters at the Bristol Mint and Pentyrch so there must have been a majority in the Privy Council who had regard for his technical skills and probity.

Authorities in Worcestershire and Gloucestershire were to provide 60 quarters of wheat and 100 quarters of malt, whilst ships and carts were to be provided at Bristol.[35] Two days later a warrant was issued to deliver to Robert Recorde 100 l. for the making of the melting house in Ireland. It would appear that this subvention, even if delivered was not enough to defray Recorde's expenses, for nearly a year later Gundelfinger was pursuing him for payment of £62 9s. 4d., the balance of a loan of £391 made to him by Gundelfinger out of a payment of £1,500 the latter had received from the King at Westminster around 8 April.[36] Harman reported later that he had on his way to Ireland passed two ships in one of which was Recorde and his 'Companie' and in the other, Gundelfinger and his miners. They all met up at Waterford at about the last day of June, where Gundelfinger and his crew waited for about 14 days until money came and they could buy their necessities. They arrived at the mines on about 6 July,[37] which date accords with the later request on 1 August 1552 for a check of the book of the German miners from the commencement of their work on 17 July 1551 to date.[38]

The make-up of Recorde's 'Company' is not stated but on 17 April 1551, Captains Randall, Powell, Devenish and Williams together with their bands of soldiers each 100 strong were instructed to go to Ireland,[39] an imprest of £423..14s. being granted to Mr Cavendish for 1 month's pay of the force.[40] The nature of their duties was not specified, but as will be seen later, Randall and Devenish both turned up at Wexford.

Whilst Recorde cannot have been unaware of the part to be played by Gundelfinger, the level of interference by Harman could hardly have been anticipated. On whose authority Harman was there and what his terms of reference were has not been determined, but his original report from Antwerp containing the result of the primary assay of ore was sent to Cecil. Given Harman's previous activities he could have had

---

[35] APC, 1550–1552, pp 275–276

[36] Clarke FM (1926) New light on Robert Recorde. Isis viii: 50–70, 64

[37] Ibid., pp 58–63

[38] CSPI, 1509–1573, p 128

[39] APC, 1551–1552, p 261

[40] Ibid., p 264

direct connections with a number of members of the Privy Council, who themselves seem to have had split opinions on the promise of the affair, which was the reason for Recorde's appointment in the first place. Whatever Harman believed his authorisation to be, it was clearly at variance with Recorde's reiterated mandate. Clarke quoted Harman as being private secretary to the King. No justification for this assertion has been found but Harman does describe himself as 'servant to H.M.' As noted earlier, both Harman and Gundelfinger were paid substantial annual retainers by the Crown but whom they reported to in such roles is not evident and must inevitably clash with Recorde's remit. Agard had reported to Protector Somerset as High Treasurer. The latter had been replaced in this role by Paulet on 3 February 1550 but Somerset still chaired the Council on most if not all of the occasions when the matters relevant to Clonmines were authorised.

By November the Privy Council were beginning to fret about progress. In November 1551, they advised the Lord Deputy that they had written to Recorde asking for specific information about the mines.[41] This letter of 23 November asked to 'certefye hither the hole sume and the effects of his doinges in the mines sence his fyrst entrye thereto, in suche sorte as it may appere what charges hath byn from the begyning and what the gains for lyke tyme be; with also summe particulier declaration of his opinion wherein he thinketh the Kinges Majestie proffit might be better advaunced in the same mines then it is, and whether Englyshe and Iryshe miners might be more profittable than thalmaynes or no.'[42] On 12 January 1552 Gundelfinger wrote to the Privy Council explaining why he had not 'written of their proceedings in the mines' because the building of the workhouse at Ross[co. Wexford] had not proceeded as speedily as he would like. Garret Harman and John Antwerpe were the bearers of this missive.[43] As there had been no production of metal by this time it is not surprising that Recorde's reply was delayed until February 1552.[44] It was damning. He said that the wastes of the Almains in their washings, roastings, meltings and finings were excessive, which suggests that he had some technical knowledge of such matters and that Englishmen and Irishmen could do the work better than the Almains. He hoped to save 2,000 l. yearly until the mines could be sunk deeper with consequent hope of greater gains. At that date the King's charges were above 260 l. a month whilst his gains not more than 40 l., so the King was losing 220 l. a month.

Also at the end of February, the Privy Council received 'A brief Certificat of the proceedings in H. M.'s mines in Irland to be exhibited to the most Hon. Councill by *Garret Herman*, servant to H. M. declaring therby the duetie and faithful hart he owt unto the King his Mr.' It was accompanied by an address by Harman to the Privy Council in which 'he declares the mines to be very rich, profitable and commodious' and complains of the wilfulness, pride, presumption, and covetousness of Doctor Robert Recorde. The brief certificate proved to be a long diatribe against Record but

---

[41] CSPI, 1551, p 119

[42] APC, 1551–1552, p 427

[43] CSPI, 1509–1573, p 121

[44] CSPI, ibid., p 123

does give an insight into the relationship between the two men and the way in which Harman interfered.[45]

The first point in Harmans account that needs to be examined relates to the number of miners needed. Gundelfinger had originally been asked for a closely specified team of 37 operatives. In the event 50 miners were brought, but in practice it transpired that the mining available could occupy the time of only 28 miners working two shifts. It appears that Recorde was importuned by Harman and Gundelfinger to deploy the excess effort elsewhere viz. 'at Blak Rokke at the Nunrye against Waterford where is Knowen to be moche Owre yielding much silver and easy to come' – one wonders why this was not exploited preferentially if this was the case -, and perhaps most significantly 'Also the said Captain and I desired the said Mr. Recorde to survey the Alome mynes and others which ar there well knowen, and to set those men to worke in them which the said Captayn for that purpose had brought thither at the request of King Henry VIII, which mynes would have yelded to the King moche profict, yet the said Mr. Recorde wolde never do it.' It will be recalled that the extended brief to search for silver and alum had been given to Harman, Agard and a Dutchman the results of which, reported by them to the Privy Council in August 1546, were good enough to have the Council of Ireland set up a Commission for Mines, funded by a prest of 1,000 l., with an instruction to send miners there urgently. No instruction from this Commission to Gundelfinger has been found, indeed it is clear that there was opposition from the Irish Council to his involvement. Further it appears that the only request from Henry VIII to Gundelfinger was for three experts which he was unsuccessful in providing by February 1546. It would seem that Harman was trying to re-assert his original broad remit dating from the time that he was appointed Overseer of the King's mines in 1545, possibly reinforced by some more recent support from a Council member or members. This must inevitably have conflicted with Recorde's appointment and terms of reference.

They clashed again over the siting of the melting house and stamp mill for which Recorde had been provided with funds. Harman attributes delays in producing bullion to a dispute between Recorde and Gundelfinger. He seems to assert that Recorde felt they could manage without such facilities, which seems very unlikely and that Recorde queried the Captains competence in these matters in comparison with his own. However, whilst Recorde wanted the melting house to be near to the mines and the Captain did not, they agreed initially on a site near the mines, but then changed their minds and selected a site near Ross, some 8 miles away. This decision was taken jointly by Recorde, Gundelfinger and Harman, the reasoning being that timber, water and other commodities were conveniently obtainable there. Apparently it was 10 September before Recorde approved construction of the facility, a consequence of which was that it was 10 December before any ore could be melted. The house was completed by Christmas Eve and 'on the 12 eve they began to make Silver'. The stamp mill was not operating by the time, unspecified, that Harman had left the site. In his letter of complaint to the Privy Council dated Clomens Ireland

---

[45] CSPI, ibid., p 124

15 May 1552, Gundelfinger says the melting house and stamp mill are ready to produce 300 oz of fine silver per week, providing they have the necessary coal and timber, which they have not had to date. The siting of these facilities at Ross rather than at the mines had aided neither the setting up of these facilities or the provision of fuels when they were operational. This siting also appears to have aggravated the provision of ore to the facility, as it had to be transported across seemingly difficult terrain. A bridge over some water was required by the Captain which Recorde would not provide according to Harman. The latter offered to provide a 'pattern' for a pontoon bridge which Recorde would not sanction. This was presumably at the site of the present Wellington Bridge. The road between the mines needed repairing. The terrain was fairly level apart from a few hills near to Ross. It is to be noted that a site at Wexford, only a little farther away, might have been more easily accessed. A 'hoye' was wanted for transport between Waterford and Clonmines but was not provided. Harman had bought a large lighter which was never used.

Harman's chief objective seems to have been to explain the delay in producing silver but showing no concern for the economics of the process. It will be recalled that the assay of the ore in Antwerp as reported by Harman showed that the sample provided 8 oz of silver and upwards of 50 oz of lead from each 100 oz of ore. This is to be compared with the results of the first smeltings of ore which gave between 254 and 346 lb of lead from 12 hundredweights [1,344 lb] of ore i.e. at best each 100 oz of ore would give about 26 oz of lead, [alternatively 1 cwt of lead from about 4cwts. of ore]. They then tested 200 lb of lead and obtained 14 lb. of not wholly refined silver i.e. about 2 oz of <u>unrefined</u> silver from 26 oz of lead or 100 oz of ore, to be compared with the 8 oz of <u>refined</u> silver found in the Antwerp assay. Recorde and Gundelfinger then went to Dublin to advise Croft of the proceedings, Recorde having sent a sample of the unrefined silver to Dublin for refining. The purity of the unrefined silver must have been low, for Gundelfinger advised Croft that upwards of 4½oz of silver might be expected from each hundredweight of fine lead. Thus to yield 8 oz of silver, 4–6 cwts of ore [448–672 lb] would have to be treated , which has to be compared with the 8 oz silver from 100 oz or ~8 lb of ore promised by the original assay in Antwerp. Even if Harman had mistaken ounces of ore for lbs. of ore in his original report from Antwerp, the latter was still wildly optimistic by a factor of between 60 and 80. Earlier in his missive, somewhat anticipating the new findings, Harman said that it was estimated that 1,000 oz of silver would arise from the 60,000 l. of ore already at the mines. The new estimate of silver content from Gundelfinger would reduce even this vastly reduced expectation to about 600 oz. Despite all these revisions, in his ultimate report to the Council in February 1552, just before the project was stopped, Harman declared the mines to be 'very rich, profitable and commodious.'[46]

Harman also raised the case of the 'trane' or whale oil worth 1,100 ducats that Recorde and his brother Richard took from a French merchant's ship, which act Harman says was condemned by the Lord Deputy, and the Lord Chancellor who

---

[46] CSPI, ibid., p 122

called it plain robbery and commanded him to make restitution. This account does not accord with the records of the affair. On 8 December 1551, the Privy Council sent a letter to The deputy of Ireland to consider a supplication made to him by two Frenchmen that Robert Recorde had 'taken up xl tonnes of oyle of them, and refuseth to pay them but as monny is curraunt there in Ireland'.[47] Presumably the merchants did not want to be paid in debased coins. Croft's response seems to have been to write to the Privy Council on 26 January 1552 asking that Recorde be asked to answer on question touching the mines and to be told that money was to have the same value in Ireland as in England. This latter point was a sore issue with Croft. The Privy Council was leisurely in its response, on 20 August writing to the Lord Mayor of London to summon Recorde and knowledgeable merchants to assess the value of 15 tons of whale oil. A blank warrant was issued on 1 September, and a completed one was issued to Peckham on 13 November for payment of £120 for 15 tons of whale oil, received by Recorde for the King's purposes, at the rate of £8 a ton.[48] A further sum of £34 was paid by the Receiver of Wards on 5 February 1553 to the merchant in full settlement of the account.[49] How the final total of £154, compares with 1,100 ducats demanded cannot be assessed, but 15 tons of whale oil are a lot less that the 40 tons claimed originally. It seems unlikely that Recorde would have been reprimanded by the Privy Council for trying to minimise expenses or that they would have paid the smaller bill from central funds had they disapproved his actions. What does seem clear from Harman's submission is that Gundelfinger did not want to use whale oil for lighting purposes and had asked Harman to get tallow for these ends, although again Gundelfinger does not mention the matter in his submission.

Harman complains that Record refused to subsidise the miners for the repair of foot-ware and also attempted to profiteer in selling them food at a higher price than he paid. The truth of this cannot be tested, certainly food was in very short supply at that time and prices were very variable; there was virtually a famine in the region. However it must be remembered that the terms and conditions of remuneration of all staff of the enterprise had been laid down for Recorde by the Privy Council and he was unlikely to modify them without instructions to do so.

The final complaint was that Recorde deployed 20 men from each of Captains Randall and Devenish bands 'practicing the devises of his own brayne whereof come no proficit to H. M.'.

Gundelfinger's correspondence appears in two parts. The first 'The Captayn's Humble Requests' presumably accompanied Harmans 'brief Certificat'. Apart from His complaints about Recorde were non-specific apart from the £62 9s. 4d. that he was owed by Recorde, his complaints were non-specific e.g. he grumbled that Recorde was arrogant with respect to technical matters where Gundelfinger thought that his own role was to be that of teacher rather than pupil. Gundelfinger seemed to be more concerned with other money troubles as he requested letters into Germany

---

[47] APC, 1551–1552, p 435

[48] APC, 1552–1554, pp 115,120, 167

[49] Ibid., p 215

allowing him to recover an unstated sum of the King's money that had been taken by one of Gundelfinger's chief officers who had fled into that country. The second letter, dated 15 May 1552 was apparently in response to letters he says he had received from 'Mr. Gerot Harmon whereby I Undestand that yor Honors are displeased that silver was not made here. I certifie yor Honors that the fault was not in us the Duchemen, as Mr. Gerot and Mr John of Antwerp may certifie yor Honors of the estate of us Duchmen being here, how we lack vittals all thoge we pay dowble the price for such we receive.'[50] He complains about resulting sickness and some deaths as well as delays in payment. He projects an output of 300 oz of silver per week now that the melting house is operational, providing fuel is available. He mentions again the £62 9s. 4d. that Recorde owes him. There is no record of such a complaint to Harman by the Privy Council in the APC, CSP(Domestic) or the CSP(Ireland) records. Was Harman reporting to Gundelfinger the reaction of the Privy Council to the presentation made by Harman in February?[51]

Recorde was scheduled to appear before the Privy Council in March 1552, but was not called.[52] On 9 May, he presented an account of the total sums of the charges of the mines since 13 April 1551.[53] A few days later Gundelfinger repeated his complaints about the state of health of his men and about Recorde's behaviour as already detailed.[54]

The response of the Privy Council to the reports of the three protagonists was measured. It was however considered to be of sufficient importance to merit an entry in the King's diary. In May 1552, Edward notes that it was determined that Martin Pirry together with Treasurer Brabazon, or another, should go to the mines to see what profit might be taken of the ore that the Almains had dug and to determine whether or not to continue. If the latter was the case, the Almains were to be discharged. The following month it is noted that, as Pirry is staying behind to accompany bullion to the Irish mint, William Williams the assay master of the Dublin mint was to take his place.[55] This instruction was repeated in a missive to Williams and treasurer Brabazon, signed by the King and Cecil on 27 June, in a slightly different form. They were to view and survey Clonmines and all the mines where the Almains had worked, to take full account of the extraction and refining activities and to search for alum mines. On 1 August a book check of the miners from commencement of their work on 17 July to date, was reported. An account for the costs of coal and wood delivered to Ross for the smelting operations was delivered on 23 August. Sometime that month Sir Thomas Luttrell, Justice, and William Williams discharged their instructions.

[50] CSPI, 1552, 122

[51] CSPI. IV, p 122

[52] CSP Domestic, 1547–1553, pp 230–231. nos. 599, 601

[53] Ibid., p 235, no. 616

[54] CSPI, 1551, p 127

[55] Jordan WK (1966) The chronicle and political papers of King Edward VI. Allen and Unwin, London, pp 123, 130

It was not until the beginning of 1553 that the affair was formally terminated. On 28 January, North, Mason, Croft and Bowes [Master of the Rolls] were instructed by the Privy Council to examine the books and other accounts relating to the mines in Ireland, and to examine Gundelfinger and Recorde on the matter as well as on the dispute between them.[56] Their findings were declared on 2 February. Given the speed with which the remit was discharged, the data they needed must have already been to hand. The King's profit was stated to be only 474 l, and his charges 3,478 l. 15s. 1d. not including 2,000 l. paid to the Almains before 'they wrought 1 day'.[57] Taking silver to be worth £2 per lb, £474 profit could have arisen from about 240 lb or 3,000 oz of fine silver. According to Gundelfinger's projections, this would have been 10 weeks production i.e. from January to mid-March. Attached to this declaration was an extract from the daily book check of the Almains who were shown to have been absent from their work from 1 August 1552 to 27 February 1553, at which time they were discharged. It appears that Lutrell and Williams had stopped work at the mines on 1 August but had not discharged the miners as they were instructed to do in the first, but not the second, Royal mandate. On 25 February the Privy Council instructed Robert Recorde to proceed to his account and to cease his Commission.[58] A letter to the Lord Justice and Council of Ireland, dated 2 March, asked them for a certificate of the money disbursed about the affairs of the mines there since Recorde and Gundelfinger had charge, as well as before Andrew Wyse, Vice-Treasurer 'came last hither as since his coming hither'[59] The response, dated 12 April, says that 6,665 l. 15s. 3½d. was expended for the 2 years from 13 April 1551 to 12 April 1553.[60] Between this request and response, the Privy Council had written to Andrew Wyse on 7 March that he was to defray the sum of 4,000 l. he had been appointed to receive towards the discharge of the miners in Ireland, as was prescribed to him by Mr. Comptroller [who was by this time Sir Richard Cotton], Mr. North [Chancellor of the Court of Augmentations], Mr. Mason [French Secretary] and the lord Deputy [Croft]. On March 18 the Council of Ireland was instructed to discharge the English and Irish miners as well as the Almains giving them some money, and also to take receipt of the King's property.[61]

Recorde presented his final account some time after 11 April 1553.[62] This latter is the terminal date for the account which started 20 April 1551. The exact dates of the preparation and presentation of these accounts is not known as the original was said to be lost and we are left with only a copy of the account of Thomas Jenyson, the King's auditor at the time, prepared from books submitted by Recorde during

[56] APC, 1552–1554, p 210

[57] CPSI, 1509–1573, p 130

[58] APC, 1552–1554, p 225

[59] Ibid., p 227

[60] CSPI, 1509–1573, p 130

[61] APC, 1552–1554, pp 233, 239

[62] Hore PH (1901) History of the town and county of Wexforde., II, pp 244–246;
PRO., Miscellania Q.R., 923/2, 1551–1553

the seventh year of Edward VI, and therefore prior to the latter's death in July 1553. This account was attested to agree with the original submission by L. Genyson, Auditor at the behest of Robert Recorde, nephew and executor of Robert Recorde. In 1573 the former was vigorously pursuing money possibly owed to the latter by Sir Thomas Chamberlayne and dating from 1549.

The account runs for 25 months and 22 days, counting 28 days to the month, starting 20 April 1551 and ending 11 April 1553. Recorde was therefore being meticulous in accounting for his own remuneration. It appears that Recorde was responsible for payment of all the labour associated with the mines and presumably also at Ross.

Receipts were

£664 6s. 8d. From the Treasurers and Chamberlains of the Exchequer as certified by Thomas ffelton.

£500. From Sir William Brabazon, late Vice Treasurer of Ireland as by his account.

£120. From Sir Edmond Perkham late High Treasurer of the mines certified by Henrie Coddenham, Auditor of the Mines.

£2,434 9s. From Andrew Wyse Vice Treasurer and treasurer at warres in Ireland, certified by Valentine Browne Auditor of that account.

£266 From Richard Wattes in charge of Victualles for Ireland, for grain delivered by him.

£3,894 15s. 8d. Total Receipts.

Allowances claimed by Recorde were grouped under three heads,

Provisions and necessary charges:-

£522 8s. 8d. Emptions and provisions.
£64 17s. Freights to and from Ross.
£56 16s. 2d. Land and water carriages.
£658 1s.10d. Total.[This should be £664 1s. 10d]

Wages of:-

£924 3s. 8½d. Miners and other officers apperteining to the mines.
£549 16s. Robert Recorde for his own dietts at 13s. 4d. perdiem, two clerks at 12d the piece per diem and 4 men at 8d. the pece per diem amounts to £25 4s. the month appointed by order of the Council.
£488 4s. 3d. Artificers and labourers.
£2,057 3s. 6½d. Total. [This should be £1,962 3s 11½d.]
Diverse and sundry extraordinary payments.
£98 0s. 8d.

Prests delivered:-

£31 10s. Garret Harman.
£5 5s. 11d. Hamice Hardigan.
£2,299 14s. 11d. Joachim Gundelfinger, Captain of the Almaine miners as appears in his account.

£2,266 5s. 10d. Total. [This should read £2,266 10s. 10d.]

£5,038 17s. 3½d. Sum of all the Allowances. [This should be £4,989 17s. 3½d.]

£1,054 1s. 7½d. Supplussage on the trial of the books.[This should read £1,005 1s. 7½d.]

Reconciliation of the figures quoted for receipts is possible only for that of £120 attributed to Peckham, which coincides with that he was ordered to pay the French merchant for whale oil. A number of comments need to be made about the disbursements. The statements about freight charges both by land and sea are quite substantial and must therefore moderate appreciably any validity of Harman's complaints about lack of funding for this activity. The wages of Robert Recorde and his secretariat are quite high compared with those of the miners, but they were as fixed by the Council in London. The sum given for the 25 months and 22 days for the whole corresponds to charges for only 21 months and 22 days i.e. 4 months less. It seems reasonable to assume that 3 months of this difference arises from the fact that the operations for which Recorde had a responsibility only started in July. If this was the case the residual difference would result from a termination of activities in March 1553 when Recorde's commission was terminated. Recorde thus appears to have been meticulous in accounting for his own remunerations. He was clearly responsible for payment of the operational staff at Clonmines and presumably at Ross also, the rates for which would also have been laid down by the Privy Council.

The prests he delivered bear closer examination. All three of them are found as dues of the Crown recorded shortly after the accession of Edward VI. The prest of £36 10s. to Harman was presumably as a result of his appointment as Overseer of the King's Mines at a fee of 2s. per diem. The prest of £31..10s. might therefore represent payment for the part of a year that he spent in Ireland from July 1551 up to his presentation of a report to the Privy Council in February 1552 or possibly to 9 May when Recorde made his presentation. This interpretation presupposes that he still held some official position in the affairs relating to the Mines, which supposition has not been established. The most interesting feature of Hardigan's prest is that he is still involved, having been concerned with the affair as long as John of Antwerp and almost as long as Harman. The prest of Joachim Gundelfinger constitutes over 40% of the whole disbursement. He clearly kept his own accounts but his itemised expenditures have not been found. As there is no evidence for his continued pursuit of Recorde for the latter's debt to him, it may have been subsumed in the overall sum.

The overall picture is that Recorde's accounts and his management of the affair were accepted by the Privy Council. References to Harman, Hardigan, Gundelfinger and John of Antwerp are not found in central records after this date. Recorde appears only in connection with his mathematical textbooks and in the context that eventually led to his death in a debtor prison.

There remains the problem of reconciling the three accounts of the money spent on the project. This is difficult because the accounting periods differ, as does the detail given. The total for the King's profit was stated by Croft et al., to be only 474 1. and his charges 3,478 1. 15s. 1d., not including 2,000 1. paid to the Almains before 'they wrought 1 day' i.e. a total of roughly £5,479 compared with the corrected figure of £4,988 of Recorde. The total given by Wyse for the same period as Recorde was

£6,666. Wyse had the financial authority to pay off the German miners and those of other nationalities after Recorde's commission had ceased. He was enabled to do so to the extent of 4,000 l. Such terminal expenses would not have appeared in Recorde's accounts. This largest figure is probably the best estimate of the total costs of producing less than £500 of fine silver, a very costly affair that had been truly assessed as such by Robert Recorde.

Was it a confidence trick perpetrated by Harman from the very beginning? The results of the assay he produced from Antwerp were highly suspicious. However, as the gift to him of a book about minerals by Gundelfinger suggests that he might have been incapable of critical judgement. He must have persuaded Crown officers with access to funds and having even less judgement than himself that it was a good venture. There are a number of such people who remained in contact with the affair as long as Harman. Paget was in place for much of the time and his close acquaintance Mason was also a Privy Counsellor. William Cecil would have also been aware of the proceedings from inception to closure but leaves little trace of involvement and would not be expected to have done so. Recorde was first involved with matters related to coinage with his activities at the Durham House and Bristol Mints when Somerset came into power, possibly at the instigation of Whalley who was steward to Somerset. Recorde's competence and probity as comptroller of these Mints at a time when the opposite of such virtues were the norm, might have been attested by Thomas Chamberlayne. The latter could have recommended him to members of the Privy Council and possibly to the Council of Ireland, who were somewhat suspicious of the claims being made – justifiably so as it emerged.

The affair did not die in the minds of Englishmen with the death of Recorde in 1558: nor did the interest of the Crown in exploitation of its metalliferous minerals primarily in England and Wales wane, although chiefly concerned with non-ferrous metals other than silver. During the 1560s, first the Royal Company of Mines and then the Company of Mineral and Battery Works were set up to this end. William Humfrey, who for a time was assay master of the Tower Mint was closely involved in setting with the first company and, becoming dissatisfied with it, was the prime mover in persuading Cecil to obtain the patents necessary to set up the second operation. During the course of this debate in 1565, Humfreys quotes the proceeding at Clonmines as an example of exaggerated claims being made for the mineral wealth of Ireland. He speaks of the dispute between Mr. Recorde and Gundelfinger. He said that despite the reputation of the mines for great riches the man currently farming it would be glad to be rid of it. Further, 'By the report of Mr. Record on the general trial, one with another did not yield 2 oz of [silver] per cwt. After which rate the silver would not defray the charges [of manufacture].'[63] The results reported are most likely to be those of the assay carried out at Dublin at Recorde's request by William Williams who was the assay master at Dublin at the time. The latter was also to be a shareholder in the new Company and was also assistant assay master at the Tower mint under a number of Mintmasters, including Humphreys. Indeed the list of Shareholders in both Companies is of interest. Whilst the controlling interest

---

[63] Donald MB (1961) Elizabethan monopolies. Oliver and Boyd, London, pp 11, 150

in both cases was retained in English hands, the remainder was held by German financiers and technical experts imported to carry out the operations. There was again resentment at the importation of foreigners. Government ministers were to the fore as shareholders. William Cecil, the Earls of Pembroke and Leicester held a number of shares each, whilst the holders of single shares included a number of goldsmiths, or people associated with minting. Cecil appears to have taken the lead amongst the Privy Councillors, but Nicholas Wooton was also involved in initial moves. It is also of interest to see Francis Agard, the son of Thomas Agard involved. The chief German participant in the first Company was Daniel Hochstetter, for the Augsburg area, but the foreign input into the second operation came from Christopher Schütz from Annaberg in Saxony near the border with Bohemia.[64] The two companies were moderately successful both in their individual operations and in their collaboration. Hamilton attributes much of the success of the principle of the granting of Royal privileges in the form of monopolies for mining and fabrication of copper and brass, to the skill and statesmanship of Cecil.[65]

Closure of the Clonmines affair had to wait until well into the reign of Elizabeth. Probate of Robert Recorde's will was granted in November 1570, for the second time, to his nephew Robert. The successful pursuit of his uncle's debtors by Robert junior for settlement of Robert senior's accounts for the operations at Clonmines was concluded on 12 April 1571. Robert junior was granted the lease for 21 years of Crown lands in Cambridgeshire, Sussex, Caernarvonshire and Pembrokeshire having rents totalling a little more than £25 p.a. 'In consideration that he will relinquish his claim to £1,054 19½d. owed to his said uncle by Edward VI, as appears by certificate of Thomas Jenyson, Auditor of Ireland, remaining with the Queen's Remembrancer at Westminster.'[66] For the Crown to have acknowledged the debt means that Robert senior's probity was accepted. It can be argued that the long time lapse was as much due to delay in probate as to dilatoriness on the part of the Crown. Whether the delay in granting probate was due to bureaucracy or to lack of pressure from the executors, his brother Richard and nephew Robert, is not known. Had the debt been acknowledged earlier would Robert senior have used it to pay his debt to Pembroke? The latter died in February 1570. Presumably as the fine was effectively a result of libel, its levy would disappear with Pembroke's decease, leaving the family to benefit. Did Pembroke block the granting of payment to Recorde out of malice? As this would have cost him money it seems unlikely. These conjectures all beg the question of how Robert senior had managed to accumulate such a large sum of money which enabled him to pay the accrued expenses out of his own pocket in the first place. His accounting for Crown monies both at Clonmines and earlier at the Bristol Mint seems to have been beyond reproach. Indeed he was owed further sums of money by Sir Thomas Chamberlayne, if his nephew's claims were to be believed.[67]

[64] Ibid., p 33

[65] Hamilton H (1926) The English brass and copper industries to 1800. Longmans, Green and Co. Ltd., London, Chap. I, pp 1–44

[66] CPR, 1569–1572, no. 2097, p 261

[67] Greenwood JW, loc. cit., 108–109

Chapter 5
The Physician

Abstract Three professions were open to graduates in the sixteenth century, the church, law and that of the Physician. Robert Recorde chose the latter, was licensed to practice by Oxford University, did so for 12 years and so was made a Doctor of Physicke by Cambridge University in 1545, being one of only eight granted in that decade. A determined attempt to regulate the practice of medicine was initiated by Henry VIII and the Royal College of Physicians (1514) and the Barber –Surgeons Company (1540) were established by Royal Charters. Recorde belonged to neither bodies, but practised unhindered by the former during his lifetime and dedicated his one medical text to the latter. This text was devoted to Urology, the main diagnostic tool available. He correctly anticipated criticism from entrenched interests to his use of the vernacular and mounted a vituperative pre-emptive strike on such people in his Preface. The content of the book was based on the work by Actuarius, a thirteenth century doctor, which was the standard text used on the Continent. Recorde's objective was to present the work as a single book, ordered so that it could be more readily understood and remembered. It was for this labour he was remembered in the Preface to the book when it was published again 130 years later. This ordered approach to instruction typified all his writings.

Death and disease were never out of sight in the world into which Robert Recorde was born. Their presence was heavily underlined at not infrequent intervals by epidemics of illnesses that have been identified, with varying degrees of certainty, as the bubonic plague, dysentery, sweating sickness, cholera, pneumonia and others unidentified which increased mortality rates by factors of 2 or 3.[1] There are no

---

[1] Slack P (1979) Mortality crises and epidemic disease in England 1485–1610. In: Webster C (ed) Health, medicine and mortality in the sixteenth century. Cambridge University Press, Cambridge, pp 9–59.

Creighton C (1965) A history of epidemics in Britain. Cambridge University Press, Cambridge, pp 282–334.

J. Williams, *Robert Recorde: Tudor Polymath, Expositor and Practitioner of Computation*, History of Computing, DOI 10.1007/978-0-85729-862-1_5, © Springer-Verlag London Limited 2011

records that permit any assessment of the health situation at Tenby during Recorde's childhood, but his father's first wife had died before bearing him any viable off-spring, of which Robert must have been aware. Once up at Oxford he could not have avoided exposure to the general situation. There had been a succession of outbreaks of the plague in Oxford with which the Colleges had learned to cope by retreat to properties in the surrounding countryside, abandoning their academic functions until it was safe to return. Such happenings would still have been fresh in people's minds in the 1520s.

The medical needs of the populace were ministered to by a very wide spectrum of individuals, male and female, qualified and unqualified. There were the University licensed Physicians, the barber/surgeons who were guild-members, midwives who were loosely linked to the surgeons, apothecaries and then the majority of practitio-ners, unqualified by most standards, who have been classified as the 'empirics'. Speaking of the state of affairs in London around mid-century, Webster and Pelling, say 'As far as the vast majority of the population of London was concerned, regular contact with official medicine would have been limited to the lower echelons of the Barber-Surgeons' company, to poorer apothecaries and to midwives. Most of their medical assistance came from the internal resources of the family, from neighbours, priests or finally from unlicensed male practitioners or wise women having no for-mal authorisation to practice medicine from either ecclesiastical or civil bodies.'[2] The aspirations of the Act of 1512, which was the opening salvo in the attempt of central government to regulate the practice of medicine by process of authorisation initially to be managed by the Church, were far from being achieved.

A little more is known about the medical circumstances prevalent at the time and place of Recorde's death than at the time and place of his birth and childhood. He himself subdivided the life of man as being childhood up to 14 years, youth up to 25, manhood from 35 to 50 and as age beyond 50, so himself dying whilst still in his manhood. The year of his death was the year of an epidemic in London. Strype in his annals wrote 'What diseases and sicknesses everywhere prevailed. In summer 1557 they raged horribly throughout the land affecting all sorts of people, but par-ticularly gentlemen and men of great wealth...... . In 1558, in the summer about August, the same fevers raged again in such a manner, as never plague or pestilence, I think saith my author, killed a greater number. If the people of the realm had been divided into four parts, certainly three of those four parts should have been found sick .... In some shires no gentleman almost escaped, but either himself, or his wife, or both, were dangerously sick and many died, so that divers places were left void of ancient justices an men of worship to govern the country.' One verifiable effect was certainly that parish records of death are typically absent for this year as exemplified by those for the parish of St. Margaret's of Westminster.[3] Nothing is known of the situation across the river in Southwark, which was not one of the most

[2]Pelling M, Webster C, Medical practitioners. In: Webster C (ed) loc. cit., 165–235.
[3]Burke M (ed) (1914) Memorials of St Margaret's Church Westminster. Eyre & Spottiswoods, London.

salubrious places in London at the best of times. It seems likely that Recorde, imprisoned, most likely died of the epidemic which has been speculated to be some form of pneumonia. If this was the case, no matter how good a physician Recorde was he could not have been expected to heal himself although by his choice of profession he had probably given himself as good a prospect for longevity as any.

Choice of professions for a university educated man at the time that Recorde graduated was limited. The situation that existed a decade or so later was described by Sir Thomas Smith in *A Discourse of the Commonweal of This Realm of England*, which was probably written in 1549 but not published until 1581. As its title suggests the discourse was wide ranging, taking place over dinner between a Merchant, a Capper, a Knight, a husbandman and a learned and wise man, the Doctor, who clearly spoke for Smith. The latter had a distinguished career at Cambridge University before becoming a Crown servant; hence he would have been well informed on the topic of the value of learning and of the state of learning at the Universities. He was a few years Recorde's junior. A very negative view of the need for learned men was expressed by the Capper, somewhat less so by the Knight. The Doctor then sets about educating the gathering over the value of the seven liberal sciences with some success, for the Knight then admits. *'I had weened before that there had been no other learning in the world but that these men had that be doctors of divinity or of the law or of physic; how of the first had all his cunning in preaching, the second in matters of the spiritual law and the third in physic and looking of folks' water that be sick. Marry you now tell me now of many other sciences very necessary for every Commonweal which I never heard of before, but either there be few of these doctors that can skill of them or else they disclose but little of their cunning.*

Doctor. Of truth, there be too few of them that can skill of these sciences nowadays, and of those that be, few are esteemed anything the more for their knowledge therein or called to any council. And therefore others, seeing these sciences nothing esteemed or set by, they fall to those sciences that seem in some price – as to divinity, to the law and to physic – though they cannot be perfect in none of these without the knowledge of the sciences above touched'.

Robert Recorde exemplified Smith's thesis by choosing to be a Physician, but his library, as listed by Bale, showed he was interested in legal matters and had written two religious works. He had no wealthy family to support him or to provide him with privileged access to either legal or clerical appointments, so his choice of profession was presumably influenced by its potential to provide him with a living. It would seem to have done so, for none of his other activities are likely to have helped much. Indeed his income as a physician must have provided him with the bulk of the money that he used, unforeseen, probably unplanned to subsidize the Crown's venture at Clonmines. Income from his books would have no doubt been welcome, but small. He must have started to practice as a Physician immediately following his license to do so, granted by Oxford in the early 1530s, for his Doctorate at Cambridge in 1545 was contingent upon him having practiced successfully for 12 years. Where he practiced during this period is not known. His first book on Arithmetic was published in London in 1543 but this does not mean that he was domiciled there. The book was dedicated to Richard Whalley who was a country gentleman from the

Midlands. However he held a Crown appointment in Northern England, which must have necessitated attendance in London. If Whalley was Recorde's patron then the latter might have attended him as a physician in town or country. Raach has shown that at the beginning of the following century the country outside London, Oxford and Cambridge in particular, were quite well served by qualified physicians.[4] It would have been in character for Recorde to not have followed the fashionable move to London by other Oxford graduates. Recorde continued to practice as a Physician into the reign of Queen Mary, though contrary to some gossip no evidence has been found that he was one of her Doctors.

His achievements in practice as a Physician must have been the root cause of such success as he achieved, for Recorde held no privileged position in the nascent profession. Compared with the situation to be found in Italy at the beginning of the sixteenth century, medical education and practice in England was at low ebb. Linacre, who received much of his schooling in this subject in Padua, persuaded Henry VIII to establish the Royal College of Physicians in 1518, the intention being to control the quality of physicians operating in London and its environs by means of a licensing process. Graduates of Oxford and Cambridge, licensed by those bodies to practice medicine, were exempted from such provisions. The number of Fellows of the Royal College were limited, but provided physicians to the King and Court. These Fellows had been educated in the subject at Oxford, Cambridge or abroad, chiefly in Italy. Recorde was never a Fellow nor yet a Royal Physician. In 1540, the King agreed to the setting up of a joint Barber-Surgeons Company whose licensed members were nearly an order of magnitude greater in number than those of the Physicians. Recorde was not a member of that Company either. . He was one of a relatively few products of both Oxford and Cambridge. In the decade 1540–1550 he was one of eight MD's granted by Cambridge, of whom four had been licensed elsewhere. One of these was John Blyth, educated in medicine at Ferrara and shortly to become the first Regius Professor of Physic at Cambridge and brother-in-law of John Cheke, tutor to Edward VI. Contact with Cambridge University was briefly rekindled at the beginning of 1556, when together with John Blythe, Recorde was granted the privilege of *vestro communi*. He was therefore accepted by the academic medical community, but was never formally a part of it. He was never harassed by the Royal College in the way they pursued other medical practitioners they had not licensed. In his book, *The Bulwarke of Defense against all sickness etc.* of 1562, in the preamble to the text already quoted in praise of Recorde, William Bullein places him in the company of Caius, Linacre, Turner, Eliot, Fayre[Phaer], Panel[Paynel], Bord [Boord] and others as a distinguished practitioner of medicine. Perhaps he was a little biased as there is evidence that Recorde was Bullein's tutor. Further evidence regarding Recorde's status as a physician in London, has to be adduced from his own writings, specifically from the preface to the *Urinal of Physicke*.

The division of territory between the Royal Companies was not sharp, but the sequence in which their advice and treatment was to be sought had been laid down by

---

[4]Raach JH (1962) A directory of English country physicians 1604–1640. Dawsons, London.

Galen, to whose doctrines both groups adhered. As Thomas Vicary, many times Master of the Surgeons Company wrote, the instruments of medicine were diet, potion and surgery and were to be applied in that order, each successive type of treatment following failure of its predecessor to cure. The judgement of urines was probably the prime diagnostic method used by physicians but it was to the Warden and Company of the Surgeons in London that Recorde dedicated his book on the subject.

As with all his books, the comments made by Recorde in their Dedications and Prefaces are the main source of information on his personal opinions. Two topics provoked Recorde's comments in his Preface to the *Urinal* viz. anticipated criticisms of his competence to write a book on urology and of his choice of the vernacular to do so.

Anticipation of a criticism to be levelled at his use of the vernacular was justified. In writing *The Urinal of Physicke* in English, Recorde was adding to a small but fairly stable base of about six new titles of medical books, also in the vernacular, published during the first three-quarters of the sixteenth century. Alone amongst his publications, at the start of its Preface Recorde mounts a spirited, almost vituperative attack on those who opposed the use of the vernacular. '*But now as touching mine intent in writing this Treatise in English; though this cause might seem sufficient to satisfy many men that I am an English man and therefore may most easily and plainly write in my native tongue rather than in any other: yet unto them that know the hardness of the matter, this answer should seem unlikely considering that it is harder to translate into such a tongue wherein the Art has not been written before, than to write in those tongues in which the terms of the Art are better expressed. ...... But because there is a common saying in the mouths of many men nowadays, that it is a profaning of learning and a means to bring it into contempt, so to set it forth in the vulgar tongue, that everyman indifferently may read it, and study it: To this I will briefly answer, that this saying is not only against many great learned mens acts and examples, but also against manifest reasons: besides, that it includes a pernicious kind of counsel. For if everything should be put away, or left undone, that evil men may pervert and use to an evil purpose so should we have no good thing remain: meat and drinks must be taken away because many abuse of it. And, because evil men do abuse of both eyes and tongues, shall all men therefore pluck out their eyes and tongues? Because many men do abuse laws and authority shall men expel laws and high powers? Many evil men and heretics have misinterpreted Gods word, yet ought Gods word nevertheless be taught vulgarly to all men. Though the Pope, Cardinals and Monkes have practised to poison men even with the very sacrament of the Supper of the Lord, yet no man will be so mad as to eschew the use of that blessed Sacrament. And yet all this follows if men allow that common saying, above written. Better means it were to set forth publicly all that might do good to the public wealth, and straightly the abuse of them, than to punish good men and good things because evil men offend. It is a like error to that sort of doctrine which condemned wine as an evil thing because that many were made drunk with it.*'

Recorde was not alone in his concern over adverse reactions to printing in the vernacular. There was a history of such criticism of books dealing in particular with matters relating to the subject of 'Physicke' by other, contemporary authors. In his book on urology, Recorde made reference to Thomas Elyot's *The Castle of Helth* (1539). In the preface of the latter Elyot found it necessary to defend his use of the vernacular. 'If physitions be angry that I have wrytten physyke in englyshe, let them remembre that the grekes wrote in greke, the Romayns in latyne, Avicenna and the others in Arabike, which were their owne propre and maternal tongues. And yf they had bene as moche attached with envy and covetyse as some nowe seeme to be they-wolde have devysed somme partycular language with a straunge cypher of fourme of letters wherein they wolde have written their syence, whyche language or letters noo man shoulde have knowen that hadde not professed and practysed physyke'.

The Boke of Children, often described as the first book in English on paediatrics was compiled by Thomas Phaer of Cilgerran from previous writings and published in 1546. In its preface he took issue in no uncertain terms with critics of his choice of the vernacular. '.... how long would they have the people ignorant: why grudge they physic to come forth in English: would they have no man to know but they. Merchants of our lives and deaths that buy our health only of them, and at their prices: no good physician is of that mind. .... What reason is it that we should mutter among a few the thing that was made common to us all.'

'The most excellent works of chirurgie made and set forth by John Vigon ...' was translated into English by Bartholomew Traheron and published in 1543. Despite the title, it contained more material more apposite to 'physic' than to surgery, reflect-ing the lack of distinction between the two pursuits in Italy. Traheron says that he was requested to undertake the translation by an unnamed friend and goes on to comment 'For this cause, I have thought it not unprofitable (let some busie speakers rather than doers babble what they lyste) to bestow some labour and tyme in trans-lating this booke which contayneth so many good remedies for the diseases that communeley and justley happen to us.'

In contrast, Thomas Paynel [1534]and Thomas Phaer [1546], translators into English of the text *Regimen Sanitatis Salerni* which was directed towards defining a healthy style of living, found no need to comment on their use of the vernacular. Thomas Vicary, Master of the Royal Company of surgeons published a book on anat-omy in 1548 in the vernacular and also found no need to defend his use of English. When Vesalius' classic text on anatomy was translated from the earlier Latin version by Thomas Gemini, into English by Traheron's friend Nicholas Udall in 1552, no need to defend the action seems to have existed. In his books on mathematics and astronomy, Recorde refers to the issue only once and then only in passing.

Objections to the use of the vernacular thus appear to centre on texts dealing with matters within the purview of physicians, some of whom must therefore be the prime suspects as objectors. Who they were individually is difficult to fathom. Fellows of the Royal College of Physicians were not prolific publishers in any language. Exceptionally, Caius published a book on the sweating sickness in 1552 that was written in English and he found it necessary to preface his text with an apologia for the use of the vernacular. Opposition to the use of English was thus real even if the writer was so eminent as to become the President of the Royal College of Physicians a few years later. There is no overt evidence of objections within the Universities to the employment of the vernacular. Paynel, Elyot, Traheron, Phaer and Recorde were all University graduates, albeit not all having medical qualifications. However, when the first attempt to revise the curriculum at Oxford and Cambridge was made in 1549 the prescribed texts to be used for mathematics, geometry and cosmography were all written in Latin; no similar guidance for medical texts has been found.

Education of the masses would not have been welcomed by the numerous unqualified practitioners that the Royal College was attempting to monitor and control, but the texts by the authors mentioned were too complex to have been of interest to the general public. They could have appealed only to existing practitioners, supplementing their existing knowledge.

Johnson has written 'About ninety per cent of the scientific works in England were published in the vernacular. So far as I can determine, no other country can claim nearly so high a proportion for the period from 1500 to 1640. Some allowance must of course be made for the Latin books printed on the Continent and offered for sale by the English booksellers, but the fact remains that the English tradition was overwhelmingly on the side of the vernacular.'[5] Medical works seem to have followed this trend, possibly a little more sluggishly than for other subjects, accelerating in its adoption towards the end of the century.

Accepting that physicians were less than enthusiastic about the use of the vernacular for texts on medical matters within their preserve, they may not have felt very strongly about Recorde dedicating his book on urinary diagnostics to the surgeons. They may however, not have felt so relaxed about him urging surgeons to take an interest in educating themselves the subject and carrying out their own diagnoses. Admittedly his objective was to better their chances of effecting a cure, or of avoiding blame if their treatments went wrong because of the existence of an unsuspected pre-condition. He does nod in the direction of the physicians by suggesting that if the surgeon detects from examination of urine that the diet should be corrected, the advice of a physician should be sought. Nonetheless, he was blurring boundaries between the preserves of surgeons and physicians, boundaries which did not exist in contemporary Italian practice.

Recorde returns to this thesis in his Preface where he raises the questions about relationships between physician and patient. Error in the judgement of urines may arise not from the ignorance of the physician but from lack of understanding on the part of the patients about the questions that they should ask him 'and not to looke that the Physician should tell him all things at the first sight, more like a God the man.' Having berated the patient Recorde goes on to criticise some physicians. 'So that if there be any Physician so arrogant that he will not take upon himself to tell all things alone, and will not have the Patient speak, specially not knowing the party before, neither seeing other signes but only of the urine, as I do boldly pronounce that such a man is unworthy to be called a Physician. So shall it be good for all men, not to trust to the judgement of such a one: for by such misuse of this thing not only much harme befalls the patients, so that is has been the occasion of many mens

---

[5]Johnson FR (1968) Astronomical thought in Renaissance England. Octagon Books Inc., New York, p 3 footnote 2.

death, but also very much reproach hath ensued to the whole estate and order of
Physicians, and hath caused that excellent and most necessary art to be condemned,
derided and little set by.' Recorde then returns to the instruction of the potential
patients in their duty to have some understanding of their own urines from the beginning of their illness, preferably even before it started, so that it might even be avoided
by choice of diet. At very least they should be able to ask the physician relevant
questions and provide pertinent diagnostic material. The book was written with
such instruction partially in mind. Recorde casts a jaundiced eye on the jesting and
mocking of physicians by presenting them with samples of beasts urine as if they
came from men and men's urine samples as if they came from women 'For in this
doing they deceive not the Physitian, but themselves. ...... Therefore if you seek the
Patients heath, look that you receive the urine diligently: and that as soon as you can,
present it to the Physician, and be diligent to instruct him in all things that you can
and that he shall not have need to ask. And so no doubt you shall receive great commoditie of that Art, to the health of man, and the glory of God, which have given
such knowledge to man.'

In this harangue, Recorde was advocating an approach to medical treatment typical of Galen's teachings as practised in Italy with patient, physician and surgeon
working together, communicating with one another at all stages. When the patient
was not able to attend the physician it was the function of his messenger to speak for
him. The approach that was to evolve in England over the next century was more
concerned with matters of territorial imperatives, with physicians to the fore in stak-
ing their claims to supremacy, with the consequences that Recorde predicted.

A fore-runner of troubles to come took the form of a polemic from a Salisbury
physician, John Securis, published in 1566 and entitled 'A Detection and queriono-
mie of the daily enormities and abuses committed in physick; ...'. It was dedicated,
in Latin, to the Universities of Oxford and Cambridge and lists the shortcomings of
'the wicked sort of physicians', of surgeons and of apothecaries. He wants all physicians to be educated at and licensed by Universities. As knowledge of Latin is a
prerequisite of learning at these establishments, he sees no need for texts to be published using the vernacular, indeed this would only encourage people like surgeons
to operate beyond their station. Encroachment by surgeons and apothecaries onto
the preserves of the physicians formed the remainder of his complaints. This did not
deter the publication of further editions of the *Urinal of Physicke* in 1567 and 1599,
but there was then a gap until it was republished in 1651, 1665 and finally in 1679.
Three books written separately by practising physicians in 1612, 1625 and 1637
chronicle the state of abuse that the judgement of urines had fallen into. The last of
these, by Thomas Brian M.P. was entitled 'The Pisse-Prophet or Certaine Pissepot
Lectures. Wherein are newly discovered the old fallacies, deceit, and juggling of the
Pisse-pot Science, used by all those (whether Quacks and Empiricks, or other
methodicall Physicians) who pretend knowledge of Diseases, by the Urine, in giving judgement of the same.' It occasioned the collective disapproval of the College
of Physicians. The work was an exposé of the practices of the time illustrated from
Brian's own experience coupled with those of others known to him. His conclusion
in the penultimate chapter is 'That there is no judgement of diseases to be given by

the judgement of Urine alone; that the Physician ought not to give Judgement of the Urine before hee have strictly examined how the sicke partie is affected: ...', and attributes the abuse to 'That covetousnesse in the common people to save their money (because they saw Physicians to view the water at the Patients owne house) caused them to send their waters likewise unto Physicians: And Pride in the Physicians, to shew more skill than they had learned out of their Master *Hippocrates*, made this to become a custome, which is become a very strong Plea.

I could show them how this custome mighte soone be abrogated; but since I have no power to put it into execution, I leave it to them (whose power is sufficient to suppresse it) if their care were correspondent......'.

None of these later writers seem to have read Recorde's book, neither had the College of Physicians taken note of his concerns. Perhaps this is why the book was republished at this time, but without the Dedication to the College of Surgeons. The book was published in 1599 by a Thomas Dawson and the editions of 1651, 1665 and 1679 by a Gertrude Dawson. She explains how she had been persuaded to print the book again by someone who signed himself just 'R R', presumably a physician who felt the need for a re-statement of proper practice. He described Recorde as '... one of the first who laboured to reduce the tractate thereof [the judgement of urines] unto order and method and has been seconded by laborious Fletcher, to whom our English Nation oweth much for their labours: the antiquity and paines of the author hath caused it to be presented again to the Press hoping with judicious men, it shall receive the acceptance is desired & studied.' The printer explains how, after 'the fitting of Record this second time for the Presse', she had been pointed towards two other pieces of work, which she had then appended. The first of these is Securis' text already mentioned and the second a short book by Joseph Pape, translated from Latin and dealing critically with the practices of apothecaries of about the turn of the century. The publisher does not explain why she omitted the original Dedication. Superficially, the various books represent an uncomfortable set of bedfellows, but clearly Recorde's book was considered to still be authoritative on his chosen subject. There is no need to go into the details of his text to understand this status, the reasons are those given by 'R R', it is ordered and methodical. It contrasts sharply with contemporary competitors in the vernacular, which it outlasted by more than a half-century.

The anticipated criticism of his competence to write on the subject, real or imagined, is dealt with immediately in his Dedication. Those who do not have the knowledge to be able to comment he dismisses as 'curious carpers' having 'currish stomachs'. 'Wise, sober and lerned men' he woos by stating that he has consulted widely of authorities both living and dead, Greeks, Latins, Arabs and others. He goes so far as to say '... so nearly have I followed their counsels, that I may rightly call this my writing, rather their worke then myne. ....' In other words if you criticise me, you also take on those authorities that I use. He returns to the subject in the Preface, where he lists some 18 ancient authors and goes on to justify his use of their material 'partly by reason that it [judgement of urines] is written about so dispersedly in their works, and not in any one book peculiarly and sufficiently: and partly because that sundry words used in the same (as in the rest of Physick) are obscure to them that have not been exercised in the knowledge of both the Greeke and Latine tongues.'[6] He refers to these authorities later, individually, in the main text at points appropriate to the subject matter. The ignorant critics could well have been those authors and printers of handbooks on urines that he advises his Dedicatees to avoid as being set forth in great numbers for the advantage of the printer rather than for the health of men. The wise men he refers to were presumably his fellow, University educated physicians, many of whom had been educated in Italy and would be familiar with the authorities referred to but in the original Latin or in the Latin transcription from Greek or possibly Arabic. The authoritative work on Urology used on the Continent was that of John Actuarius, a thirteenth/fourteenth Century author based in Constantinople and written originally in Greek. It was translated by Ambrogio Leone, published first in Venice in 1522 but followed by editions published in Paris in 1522, Switzerland 1541 and all of these could nominally have been available to Recorde. The book continued to be published through the remainder of the century. A cause for writing the book given by Actuarius was that he deplored '… for a reason which I do not know, none of the ancient authors, who nevertheless knew the subject well, treated it fully.' It appears from later text that he had in mind Hippocrates, Galen, Alexander of Trolle and Theophilus. Recorde agreed with this assessment, paid his respects to Actuarius on this count but took a different line in structuring his presentation.

The *Urinal of Physicke* is a book on Urology, not just a self-help book on the diagnostic uses of urines. It sought to present the state of knowledge on the subject at that time, an English counterpart of the work of the thirteenth-century Byzantine physician Actuarius. As it is based on the teachings of Galen as modified by the Salerno school, its medical content is now primarily of only historical interest. What is of continuing relevance is the philosophy underlying the presentational approach used by Recorde, an approach which he was to use consistently in all of his books, but expressed specifically in the '*Urinal*'. 'Because nothynge done disorderly cann be well understanded of the reders, and everything the better order it has ye better it may be understand and also much more easily remembred when the order of it is well and certainly known. I have therefore digested this boke into a certayne and orderly processe, which I shall set forthe to the intent that you may rede, as it were in grosse, one whole booke, and thereby keepe it the better in remembrance.' The first section deals with the nature of urine, the way in which it is generated and the way in which it passes out of the body. The next three sections, which constitute the bulk of the book, deal with the use of urine as a diagnostic tool. The fifth section describes how urines in general may be used for medicinal purposes and the last section with urinary problems. A glossary of seven pages is appended, '… because I was

---

[6]*The Urinal of Physick*, sig.Biii., For the Judgement of urines he refers to Hippocrates, Galen, Aetius, Aeginita, Philotheus, Actuarius, Cornelilexus, Plinius, Constantinus Africanus, Clementius Clementinus, also Avicenna, Egidius, Polidamus et al. For the uses of medicines relating to urine he consulted Disocondes, Quintus Serenus, Columella, Sextus Platonicus et al. Works by Hippocrates, Galen, Aetius and Aegidius were all to be found in the Library of Thomas Cranmer.

forced to use some (though but few) terms in this Book, which be not wel known of the most sort of men, though a great number know then well enough by often talking with their Physicians, I thought it good here to declare some certain of them, for the aid of the most simple sort.' He also provided 'A Universal Table for the Judgements of Urine', which contains the essence of parts two to four inclusive and finally an index to the whole book. Recorde thus deliberately provided access to its contents for readers at several levels of interest. He was quite specific on his intent in this matter of ease of access, closing the 'Universal Table' by saying 'These in general be the things meet to considered in Urine, of which particularly in this Book you may read as much as to this time serveth.'

This Table shows succinctly the methodical way in which Recorde tackles his subject. He first treats the topic of correct sampling procedure, defining when it should be taken and that the vessel used should be of thin, clear, not green glass and of a preferred shape. The time for which it may be kept is discussed and it is insisted that it must be kept and examined as a whole, remaining unstirred whilst being kept neither too hot nor too cold. Examination is to be by moderate illumination. It is refreshing to find a realisation of the importance of standardisation. Next he deals with the factors that may affect the urine of a 'healthful man' and what there to expect under a variety of circumstances before considering what such examination may portend for a sick man i.e. he is establishing his fiduciary points. The variables likely to have an effect on urines include sex, age, complexion (temperament?), climate (country and season) and diet.

Having settled that a patient is truly ill and the foregoing factors have been recorded , the urine may be examined with respect to a range of properties. However, as was his wont, Recorde has to explain the principles underlying the procedures to be described and followed.

'Before I treat of the significance of the parts of Urine, I think it good to instruct you of the general qualities which cause all alteration in urine: whereby you shall perceive not only what each urine doth betoken (as I shall anon set forth) but also if you mark well this chapter [vii] you shall see the cause why every urine doth so signify.

You shall understand therefore that there be four chief and only qualities whereof all things that are both in the Sea and the Earth are made: as man and beast, fish and fowl, trees herbs stones and metals. These four qualities are heat and cold, moistness and dryness: and these four continuing duly tempered (as nature ordered them first in every perfect body) be the cause of continual health. But if they be altered wrongly the do they cause diseases diversely, according to the diversity of the alterations. And as they do cause diseases, so they do change the odour, substance and other parts of the urine, whereby we may conjecture the cause of the disease; and so consequently the disease itself, though sometimes it declareth the disease itself and not the cause thereof.' He goes on to discuss how the active qualities, heat and cold and the passive qualities, dryness and moisture, may interact with one another to produce a compound overall result. These interactions are described in qualitative terms with no attempt to quantify them. Only then does Recorde turn to actual judgement of urines. He specifies some 13 main parameters that have to be considered, certain of which e.g. colours are subdivided further. Diagnosis and possible treatments are only soundly based if these parameters have been evaluated and the factors other than illness, mentioned previously, have been taken out of the equation.

The rigour and complexity of this procedure makes it unlikely to have appealed to the vast majority of practitioners. Short cuts would have been taken and abuses practised, as has already been noted. Nonetheless, Recorde had set a standard English text for Urology derived from Galenic doctrines that survived in print for over 130 years, almost as long as did his book on arithmetic. The latter however remained basically correct; the *Urinal of Physick* did not, becoming increasingly incorrect along with much other Galenic doctrine.

Part II
Intrinsic Worth

Chapter 6
Introduction

Abstract Robert Recorde's views on the intrinsic worth of learning are scattered throughout the Dedications and Prefaces to his books. His views accord with those of Plato on counts other than the practical application of mathematics, of which Recorde was strongly in favour. He justifies the values of Arithmetic, Geometry and Astronomy individually with slightly differing emphases, but concludes that learning these subjects is basically good for you. He then sets out his intentions to publish a series of texts, written in the vernacular, that were designed to remedy the parlous state into which the teaching of mathematical subjects had descended in England by the beginning of the Tudor period. The task was great. Overall the situation with respect to the state of mathematics in England at the end of the fifteenth century in academia did not differ greatly from that of the rest of Western Europe, but that of practical arithmetic lagged behind the most advanced uses of mathematics on the mainland by some 300 years. In particular, a transition from the additive system of arithmetic using Roman numerals to the place-value system using Hindu-Arabic numerals had to be made and this is where Robert Recorde arguably made his major contribution.

Whilst Recorde might have parted company with Plato on the latter's somewhat negative views on the uses of arithmetic and geometry for practical purposes and whilst Recorde always had the practical uses of learning in mind, they had common ground in their veneration of learning for learning's sake. The philosophical arguments advanced in support of intrinsic value of learning by Recorde, whilst first broached in the *Urinal of Physick,* are found in most concentrated form in the Dedications of The Pathway to Knowledg of early 1552, in the second edition of The Grounde of Artes of later the same year and lastly in the Reader's Preface of The Whetstone of Witte. The dedications of the first two books were to Edward VI, whose education Recorde clearly felt was sufficiently advanced for him to appreciate such arguments.

In the *Pathway*, Recorde opens the dedication saying.... 'It is not unknowen to youre majestie, most sovereigne lorde, what great disceptacion hath been amongest

J. Williams, *Robert Recorde: Tudor Polymath, Expositor and Practitioner
of Computation*, History of Computing, DOI 10.1007/978-0-85729-862-1_6,
© Springer-Verlag London Limited 2011

the wyttie men of all nacions for the exacte knowledge of true felicitie, both what it is and wherein it consisteth...'. The baser sorts of 'felicitie', together with those of power alone and power coupled with worldly wisdom are then described and dismissed. Examples are taken from the histories of Julius Caesar, Philip, Alexander and Aristotle, accompanied by quotations in Greek, duly translated for the benefit of those less able than the King, who had read the works of both Socrates and Plato. The examples of Solomon and David are also quoted. Recorde finishes this part of the harangue several pages later saying 'Wherefore I may justly conclude, that true felicitie doth consist in wisdome and virtu.'

This description of 'felicitie' is also that of Plato as interpreted by contemporary of Recorde, Sir Thomas Elyot, in his book of 1533, *Of the Knowledg which maketh a Wise Man*. This book relates a dialogue between Plato and Aristippus inspired by the account of Plato's life given by Diogenes Laertius. Whether Recorde derived his philosophy from Elyot, Diogenes Laertius or from Plato's original works likewise has to be a matter for speculation. Whilst Recorde quotes from Aristotle in Greek in a number of places, he does not do so for Plato, which suggests that he was using secondary sources. These latter were becoming increasingly available, particularly in the University of Cambridge during the first half of the sixteenth century.

The link between virtue and wisdom is explored further in the Dedication of the expanded second edition of *The Grounde of Artes*, that followed the Pathway chronologically and in which the history of the Celts in Britain is traced. One of the Celtic kings, Sarron, is credited with the setting up of schools in which were taught the liberal arts of Grammar, Logic, Rhetoric, Arithmetic, Geometry, Astronomy and Music. Recorde continues 'And as these Sciences did increase, so did Virtue increase thereby. Again as these Sciences did decay, so virtue lost her estimacion and consequently was less in use: whereof to make a full declaration, were a thing meet for a Prince to hear, but it would require a peculiar treatise.' The source of this information was given as Berosus i.e. the psuedo-Berosus of Annius of Viterbo, published in 1498. This was a source used by John Bale in some of his works and also by Holinshed in his *Chronicles*.

When Recorde called his first book *The Grounde of Artes*, he clearly meant just that. He saw a firm foundation in arithmetic as a basic requirement if full advantage was to be taken of learning and knowledge across the spectrum of liberal studies. 'Whereof at this present I count it sufficient lightly to have touched on the matter in general words, and to say no more of the particularity thereof, but only touching one of those Sciences that is Arithmetic, by which not only just partition of lands made, but also buying and selling, all assizes, weights and measures were devised, and all records and accounts driven: yea by proportion of it were the true order of Justice limited (as Aristotle in his Ethics doth declare), and the degrees of estates in the commonwealth established. ....Wherefore I may well say, that seeing Arithmetic in so many ways needful unto the first planting of a commonwealth, it must needs be as much required to the preservation of it also; ...'.

His commitment to the pivotal role of arithmetic remains consistent throughout his publications but the coherence of his supporting arguments leaves something to be desired. As with his approach to justifying the concept of the inherent value of learning, this patchiness arises from Recorde's tailoring of his arguments to his putative audience. Thus in the dedication of the first edition of *The Grounde of Artes* to Richard Whalley, the strongest general argument put forward in favour of arithmetic is that the ability to count is that which most clearly distinguishes man from beast. His supporting evidence is versified, the first of a number of slowly improving poetic excursions. This particular thesis is found in the Arithmetic of Isidore of Seville (570–636) and Roger Bacon in his Opus Maius says ..... 'take from the world computation and blind ignorance enfolds all things, and men cannot differ from other animals, which are ignorant of the method of calculation.'

Recorde provides his most complete justification for the eminence of Arithmetic amongst the liberal arts in his Preface to his last work, The Whetstone of Witte in which he depart from 'numbering' using Hindu-Arabic numbers and moves on to 'The Arte of Cossike Numbers' in which he incorporates 'The Rule of equation, commonly called algebers rule.' This book was dedicated to the Governors of the Muscovy Company, which included leading merchants of the City of London and politicians. In one portion of this Preface he quotes from the seventh book of the De Republica of Plato. Almost immediately before this quotation, (which will be returned to later,) the following text is found in that section dealing with the education of Philosopher Kings[1]

'...But then, what sort of intellectual pursuits are we after? I mean, all the professions seemed servile somehow.'
'Of course they did. But if we exclude cultural studies, physical exercise, and the professional occupations, what else is there that anyone could study?'
'Well', I said, 'if they're all we can take, then let's take something which applies to them all.'
'What?'
'For example, there's that everyday thing - one of the first things everyone has to learn - which all the modes of expertise, thinking, and knowledge make use of.'
'What are you getting at?' he asked
'It's nothing special, I said. 'It's the ability to distinguish one, two, and three - in short I'm talking about number and counting. I mean, isn't it the case that every branch of expertise and knowledge is bound to have some involvement with numbers and with counting?'
'Definitely', he said.
'Even military expertise?' I asked.
'Yes, certainly. It has to,' he said.[2]

It is highly likely that Recorde had read this argument. Certainly he expanded upon its thesis in the Preface to the *Whetstone of Witte*, the major part of which is a paean in praise of number. There he speaks firstly of the infinity of number and therefore of the infinity of 'commodities' that it affords. He continues, saying that there can be no certainty in other things without number, no infallible knowledge other than that borrowed from the 'Mathematical Artes'. The place of certain special numbers in 'Divinitie' is referred to. With respect to the 'Lawe', neither of the two kinds of Justice, distributive and commutative can be properly carried out unless there is a knowledge of and understanding of the difference between geometric and arithmetic proportion. The practice of 'Physike', amongst other things needs number

---

[1] The translation used is that of R. Waterfield,' Plato's *Republic*'(OUP. 1993). The line numbers refer to the original document he used.

[2] Waterfield R, pp 251–252. lines 522b and c

for the taking and use of the pulse and for the proper mixing of medicines. The basic need for numbers in 'Astronomie' is manifest. Recorde finishes the general part of the argument by quoting Plato twice, first from 'his Booke De Summo bono. Take awaie arithmetike, with measure and weightes, from all other artes, and the reste that remaineth is but base, and of no estimation. where although Plato dooe name three thinges in apperaunce, thet is Number, Measure and Weighte. What are Measure and Weighte, but number applied to several uses. For Measure is but the nombryng of the partes of lengthe, bredthe or depthe. And so Weighte (as here it is taken) is the nomberyng of the heaviness of any thyng. So that if nomber were withdrawen, no manne could either measure or weigh any quantitie. And therefore it must followe: that nomber only maketh all athres perfecte, and worthie estimation: seeing that without it, all artes are but base and without commendation.'

Recorde then produces his last general argument for the value of arithmetic viz. the sharpening of mens wits. 'It teacheth menne and accustometh them, so certainly to remember things paste: So circumspectly to remember thynges presente: And so providently to foresee thynges that followe: that it may truelie be called the File of witte. Yea it maie aptly be named the Scholehouse of reason.' Then follows the quotation from the seventh book of the *de Republica*, mentioned earlier. 'Thei that be apte of nature to Arithmetik, bee ready and quicke to attain all kinds of learnyng. And they that be dulle witted, and yet be exercised in it, though they gette nothyng els, yet this shall thei all obtain, that thei shall bee moare sharpe witted then thei were before'. Recorde's belief in the status of mathematics in the pantheon of the liberal arts was thus along the lines promoted by Plato and his adherents whose philosophy was very much in vogue in intellectual circles in Tudor England.

His advocacy of Geometry is less strident than that of Arithmetic, although he usually associates them in terms of importance and allies himself with the Greek philosophers generally in the matter. In his Dedication of *The Pathway to Knowledg* to the king he pronounces 'And for humaine knowledge thys wil I boldly say, that whosoever wyll atain true judgement therein, must not only travail yn knowledg of the tungs, but must before al other arts, taste of the mathematical sciences, specially Arithmetike and Geometry, without which it is not possible to attayn full knowledg in any art. Which may be sufficiently by gathered by Aristotle not only in his bookes of demontration (which cannot be understood without Geometry) but also in all his other workes? And before him Plato his master wrote this sentence on his schole house dore.......Let no man entre here (saith he) without knowledg in Geometry.'

Recorde reserves further discussion of the merits of geometry to the Preface of the same book, perhaps because he felt that Edward was sufficiently acquainted with Greek philosophy to need no reminder of the pivotal role that Geometry played in their thinking. He returns to the topic speaking as Geometry, '... yet can no humayne science saie thus, but I onely, that there is no sparke of untruthe in me: but all my doctrine and workes are without any blemishe of error that mans reason cam discerne. And next to me in certaintie are my three systers, Arithmetyke, musyke, and Astronomie, which are also so nere knitte in amitee, that he that loveth the one, cannot despise the other, and in especiall Geometrie. of whiche not only these thre, but all other artes do borrow great ayde, as partly hereafter shall be shewed.' In placing geometry ahead of arithmetic in 'certaintie', Recorde was espousing the idea that if the postulates of Euclidian Geometry were accepted along with the definitions and axioms, then the structure erected upon them was rigorously obtained and therefore 'certaine'.

He deals first with the commodity of Geometry for the 'unlearned sort' before returning to its value to the 'learned professions'. He prays Aristotle's testimony in aid of the crucial use of geometry in Rhetoric, Logic and Philosophy. He reiterates Plato's prior requirement for knowledge of geometry before he would teach anyone. 'And what mervalle if he so much estemed geometrye seeing his opinion was, that Godde was alwaies workinge by Geometrie! Galen and Hippocrates' views on the key value of geometry in Physike are invoked. Divines call on the help of geometry in interpeting the scriptures. Lawyers cannot fairly apportion judgements without and understanding of proportions. Returning to the position of geometry in the spectrum of mathematical sciences, Recorde says 'It shuld be to longe and needlesse also to declare what help all other artes Mathematicall have by geometrie, sith it is the grounde of all theyr certeintie, and no man in them is so doubtful thereof, that he shall nede any persuasion to procure credite thereto. For he cannot reade ij. lines almoste in any mathematicall sciences but that he shall espie the nedefulnes of geometrie.' As with Arithmetike, Recorde's attitude is still very much in line with Platonic thinking about the role of Geometrie.

The justification for the intrinsic virtue of the study of Astronomy is somewhat different, involving theology in large measure. The Preface to the Reader in *The Castle of Knowledge* says 'Yea let him think (as Plato with divers other philosophers dyd trulye affirme) that for this intent were eies geven unto men, that they might with them beholde the heavens: whiche is the theatre of Godes mightye power, and the chief spectakle of al his divine workes.' Recorde then elaborates on this theme of the glory of the Creators work. However he does devote as much space to the ways in which unusual activities in the heavens had portended momentous earthly events '...yet dyd God at the beginning ordaine the starres to be as signes and tokens of times alterations.' The birth of Christ, the Flood and the fall of the Roman Empire are given as examples of such events and there would probably be many more of men knew how to interpret the signs. This was a firmly held belief of Recorde's, for he opens his Epistle to Cardinal Pole with an account of the observation by Dionysius the Areopagite and his companion Apollophanes when in Heliopolis, of the eclipse at the time of the death of Christ. This was deemed a totally unnatural phenomenon. Recorde returns to this topic at the very end of the book, in the Fourth Treatise of the book, where he explains why the eclipse cannot be due to the position of the moon i.e. 'be accompted a natural eclipse'. It was the cited opinion of Dyonisius and Apollophenes that 'it could not happen without some mervailous cause, and a wondreful imitation of natures workes.' This view is also attributed to the author of the Sphere, presumably Sacrobosco. Commenting on the matter the Master says 'With this good clause did he eande his booke, and so wyll we eith the same eande close up our talke. Learnynge this good use in this natural arte, that it leadeth men wonderfully to the knowledge of God, and his highe mysteries, as not only by example of these two philosophers here it doth appear, but by the testimonies of the Scriptures in sundry places.' and in his last statement 'So may wee gather manye argumentes by like maters against the infideles and false christians also: but that frute will I reserve for an other place: and for this presente will only saye, that there was never any good Astronomer that denyed the Majestie and providence of God, though many other denyed bothe:…'. Coupling these concluding remarks with what Kocher has described as ' …an ecstatic paean to the beauty and incorruptibility of the heavens …' given in the Preface, an impression is left that the study of Astronomy was almost an act of worship.[3] Perhaps this view should be tempered a little by the opening verse of the Preface

> If reasons reach transcende the Skye,
> Why shoulde it then to earthe be bounde?
> The witte is wronged and leadde awrye,
> If mynde be maried to the grounde.

which suggests that the pursuit of knowledge per se is still very much in Recorde's scheme of things. However it is difficult to do other than agree with Kocher's overall assessment of the book 'Here faith and reason stood in delicate balance, both strong, neither denying the other, integrated in a personality at peace with itself.'[4]

The philosophical arguments for continuation of the study of arithmetic into the subjects dealt with in *The Whetstone of Witte*, extend no further than that presaged by the title and the accompanying verse and the quotation from Plato's De Republica that has already been given. Arithmetic is good for you!

So what system of arithmetic was Recorde to expound and what foundations were available on which he might build? In England at the start of the Tudor era, arithmetic generally employed Roman numerals. Astronomy/astrology, usually the province of academics, used Hindu-Arabic numerals and sexagesimal, 'astronomical' fractions. During the latter part of the fifteenth century interest in this topic and the status of its practitioners had declined from the eminence they had during the previous century. Nonetheless the written tradition still existed. Whilst Englishmen had been well to the fore in writing about algorithmic arithmetic based on Hindu-Arabic numerals in the thirteenth century, by the following century, the standard text in the English Universities seems to have been that of Sacrobosco which dealt only with integers and ignored fractions. In this respect they seem not to have differed greatly from their Continental colleagues in academe. The practical, commercial advantages that presumably drove the practitioners of arithmetic in Northern Italy to a wholesale embracing of algorithmic methods using Hindu-Arabic numerals seems to have been absent for their English counterparts. Planar geometry remained that of Euclid written primarily either in Latin or Greek. The geometry of the sphere was that of Ptolemy and here the premier transcription until the end of the fifteenth century was that of an Englishman, Richard of Wallingford. Here again however, the most widely disseminated text was the *De Sphera* of Sacrobosco, a greatly simplified presentation of the topic.

---

[3] Kocher PH (1969) Science and religion in Elizabethan England. Octagon Books, New York, p 155
[4] Kocher, ibid., p 154

Thus by the time that Robert Recorde was born, academic mathematics in England was dormant at best and the needs of an arithmetic to meet practical ends were being met by practices that had changed little in 400 years. How this knowledge of practical arithmetic had been handed down is far from clear. There was no obvious formal structure for doing so. Presumably it was achieved literally by practice, which might conceivably have involved the guilds and their 'mystery's'. Overall the situation with respect to the state of mathematics in England at the end of the fifteenth century in academia did not differ greatly from that of the rest of Western Europe, but that of practical arithmetic lagged behind the most advanced uses of mathematics on the mainland by some 300 years. It may be argued that this latter situation had arisen primarily as a result of differing structures of trade and commerce between England and the rest of Europe. The political turmoil that characterized fifteenth century England had also meant that it was more concerned with domestic matters than it might otherwise had been. This situation was to change with the advent of the Tudor dynasty and also with the broadening of geographical horizons and the change in accompanying patterns of trade. A transition from the additive system of arithmetic using Roman numerals to the place-value system using Hindu-Arabic numerals had to be made and this is where Robert Recorde made his major contribution through his English mathematical texts.

Order Between and Within the Books

The order in which Recorde wished his books to be read was given in the *Castle of Knowledge*, his penultimate work.

AN ADMONITION FOR THE ordrely trade of studye in the Authors woorkes, appertayning to the mathematicalles.

The grounde is thought that steddye stay.
Where no foote faileth that well was pyghte
Whereon who walketh by certaine waye,
His payce is like to prosper ryghte.

1. *The* Grounde of Artes *who hath well tredd,*
   And noted well the slyppery slabbes,
   That may him force to slyde or falle,
   He hath a staffe to staye withall.
2. *Then if he trade that* Pathwaye *pure*
   that unto Knowledge leadeth sure:
   He may be bolde tapproache The Gate
3. Of Knowledge *and passe in thereat.*
   Where if with Measure *he doo well treate;*
4. *To* Knowledges Castle *he may soon get.*
   There if he travaile and quaint him well.
5. *The* Treasure of Knowledge *is his eche deale.*
5. *This* Treasure *though that some would have,*
3. *Whiche* Measures *friendshippe do not crave,*

4. Nor walke the Patthe *that leadeth the waye,*
1. Nor in Artes *grounde have made their staye,*
  Th.oughe bragge they maye, and get false fame,
4. In Knowledges *courte thei never came.*

There was a major augmentation of the *Grounde of Artes* between the first edition of 1542 and the second edition of 1552, portions of which included material that required an understanding of elementary geometry. This was presaged in its first edition where he concluded the section on Reduction in the latter book by saying 'Here mought I tell you many thyngs els touching measure: and also howe to reduce straunge measures to oure measures, but because it cannot well bee done without the knowledge of fractions, whiche as yet you have not learned, I will let them passe till an other tyme, when I shall instruct you in Geometry, wherein I shoulde be enforced els to repete the same to often agayn.' Recorde's book on the latter subject, the *Pathway* was in fact published between the first and second editions of the *Grounde of Artes*, so that the presentation of educational material in a logical sequence was preserved. Formally therefore the *Pathway* should be read before the second edition of the *Grounde of Artes* but after the first edition. This detail would have been difficult to incorporate smoothly into a poem so Recorde would seem to have exercised the necessary poetic licence. Knowledge of the geometry of two dimensions was needed in the second edition of the *Grounde* to solve problems concerning land and building.

Prior knowledge of elementary geometry would certainly have been needed to understand the subject matter of the no longer extant *Gate of Knowledge*, the likely content of which is known through its description in the preface to the second half of the *Pathway*.[5] In this preface Recorde outlines first the contents of what he calls the second part of his arithmetic, which in fact embraces the augmentation of the first edition of the *Grounde of Artes* and the *Whetstone of Witte* and then describes an untitled work as including

THE arte of Measuryng by the quadrate geometricall, and the disorders commited in using the same, not only reveled but reformed also (as much as to the instrument pertayneth) by the devise of a newe quadrate newely invented by the author hereof.

THE arte of measuryng by the astronomers staffe, and by the astronomers ryng, and the form of making them both.

THE arte of Makyng of Dials, both for the daye and the nyght, with certayn new forms of fixed dialles for the moon and other for the sterres, which may be sette in glasse windowes, to serve by daie and by night. And howe you may by those dialles know in what degree of the Zodiake not onely the sonne, but also the moone is. And how many howrs old she is. And also by the same dial to know whether any eclipse shal be that moneth, of the sonne or of the moone.

The makyng and use of an instrument, whereby you may not only measure the distance at ones of all the places that you can see togyther, how much eche one is from you, and every one from the other, but also thereby to draw the plotte of any countreie that you shall come in, as justly as maie be, by mannes diligence and labour.

---

[5] Record R (1974) The pathway to knowledg, S.T.C. 20812. Theatrum Orbis Terrarum, Amsterdam, pp a iij (v), [iiij (r)]

This was followed by a description of what eventually constituted the subject matter of the *Castle*. There are a number of examples in this latter book where a prior knowledge of material from the '*Pathway* was invoked.'[6] There are further references to geometrical proof when calculating the relative volumes of earth, sun and sky and the Scholar says 'That is sufficiently proved in Geometry,..... .' and a little later confirms that it is something that he learned in Geometry.[7] The *Pathway* did not deal with three-dimensional figures so the Scholar must have obtained this information on the calculation of volumes from elsewhere. The only other of Recorde's publication possibly available at the time that might have dealt with the subject, was the no longer extant *Gateway to Knowledge*. This work is referred to once explicitly in the *Castle*, 'You shall take a Quadrante (whose composition I have taught amongst other instruments in the gate of knowledge, but which you see here, [illustration] is the forme of the plainest sorte), and by the two syghtes of it you shall marke the height of the Northe starre.... .'.[8] A more sophisticated quadrant is illustrated later, which presumably, although not explicitly stated would have had its origins in the same book.[9]

In setting out the *Whetstone of Witte* knowledge of the *Grounde* was taken for granted. The opening verse states '...*The* grounde of artes *did brede this stone.....* '. It is mentioned again in the Preface 'Many praise it [arithmetic], but fewe dooe practise it: onless it bee for the vulgare practice, concernyng Merchaundes trade. Wherein the desire and hope of gain, maketh many willyng to sustaine some travell. For aide of whom I did sette forth the firste parte of *Arithmetic*.'[10] Recorde regarded cossike numbers as a sub-set of denominate numbers, the other member being 'nombers denominate vulgarely'. The latter he had already dealt with in the *Grounde* and therefore did not repeat himself.[11] All of the mathematical operations used in the *Whetstone* would have been learned from the latter book.

There are references to Geometry in the *Whetstone*. Winding up his treatment of the subject of proportions Recorde says 'For of the Arte of Proportios, dependeth all the subtilties, and fine workes, not only of *Arithmetike*, but also of *Geometrie* besides farther mater that as now I will not touch.' A little later, speaking of superficial figural numbers he says, 'Whereof there bee as many varieties, as there bee diversities of figures in *Geometrie*. As numbers *Triangulaire, Quadrate, Cinkeangled, Siseangled* and so furthe. Also numbers *circulare, diametralle, & like flattes,* all whiche nombers have both lenghte and breadthe: and therefore be named *superficiall nombers.*[12]

---

[6] Record R (1975) The castle of knowledge, S.T.C. 20796. Theatrum Orbis Terrarum, Amsterdam, pp 17,38,47,120,124

[7] Ibid., pp 249–250

[8] Ibid., p 68

[9] Ibid., p 166

[10] Record R (1969) The whetstone of witte, S.T.C. 20820. Theatrum Orbis Terrarum, Amsterdam, p b.iij

[11] Ibid., S.i

[12] Ibid., C. iij

This terminology derives uniquely from the *Pathway*, as also does a reference to Pythagoras' theorem quoted as theorem 33 of that book. Those references to Euclid that Recorde gives when dealing with the extraction of cubic roots, are not however to be found in any of his extant books. 'But to help you in this question, and in all soche like, you shall marke well *Euclide* his saying, in the 18 proposition of his 12 booke, whiche is this '*All Globes bere together triple that proportion, that their diameters doe.*' '[13] Also, on the next page, '... Remembryng that *Cubes* also, as well as Globes, doe beare triple proportion, in comparison of their sides. As you learned before by the 19 proposition of the 8 booke of *Euclide*.'[14] The *Pathway* dealt with planar geometry only, there are no references to these features of Euclid in the *Castle* which leaves only the missing *Gateway* as a potential source for the instruction of the Scholar in such matters.

Recorde intended to continue his development of mathematics based topics in a progressive manner. At the closure of the *Castle* there arises the conversation

'Scholar. In deede I thinke this too harde yet, but of the Progression, retrogradation, and station of the Planetes, and also of the Eclipses of the Sonne and the Moone, I know that Iohn de sacro Bosco dyd write somewhat, and so myght you brieflye nowe do.

Master. His woordes are shorte and therefore obscure, and so should my wordes be. beside that, it is a disordrely forme to put the carte before the horse: I meane to write of the passions of the Planets, before I have sufficiently taught the full ordre of their motion. Therefore I will saye in few wordes, that the reasons of the passions canne not bee taught aptly, before the Theorikes of their motions. ..'. He goes on to set out some aspects of solar and lunar eclipses but deliberately omits certain topics as too difficult.[15]

Similar indications of an intention to continue with further presentations are found in his last book, the *Whetstone*. In the dedication to the Company of Venturers into Moscovia, Recorde promises a book on navigation, which presumably would have extended the coverage of the topic tentatively broached in both the *Gateway* and the *Castle*.[16] At the end of the Preface of the *Whetstone* he says that if it was as well received as his first book on Arithmetic he will '... set out the reste of this arte, moare completely in Englishe, then ever I sawe it in any toungue, hetherto doen. ..'. This theme is elaborated a little, later. 'But of these and many other verie excellent and wonderfulle woorkes of equation, at an other tyme will instructe you farther, if I see your diligence applied well in this.'[17] Had Recorde lived to do this the cause of algebra in England might have been advanced more rapidly than it actually was. Clarification of Cardano's presentation could only but have helped this process.

---

[13] Ibid., Piiij.v

[14] Ibid, R.i.r

[15] Castle, 279–820

[16] Ibid., a.iii

[17] Whetstone, Ll,iij. v

Chapter 7
The Grounde of Artes

Abstract The choice of title was not an idle gesture. Record took great care to make the foundations of his arithmetic as sound and durable as possible, as is apparent from the way in which he introduced the Hindu-Arabic place value system and his meticulous treatment of Numeration, the first topic he deals with in his first Edition of the book. This thoroughness he carries through the remainder of this Edition and into the second Edition that was augmented to deal with the arithmetic of fractions. A rigorous approach was not allowed to disregard the need for the presentation to be made as comprehensible as possible. Practical worked examples were used to illustrate procedures and were ones to which readers could relate, being posed in an English context using English units of measurement. In the first edition that dealt only with integers, the examples were chosen to be of interest to merchants and tradesmen but in the augmented edition examples dealt with issues of significance to the State, without overmuch regard for political correctness. Recorde's treatment of the mathematics of alligation was the most extensive since Fibonacci. Of Continental authors of practical arithmetics, only Gemma Frisius can be demonstrated to have been consulted.

The numeration system which Recorde introduced, expounded and used in his texts was the place value system using Hindu-Arabic numbers. He must have been self-taught, for as will be seen later there is no evidence that texts treating the manipulation of fractions using these symbols were available in the Universities.

Easton's review 'The Early Editions of Robert Recorde's *Ground of Artes*' remains the most comprehensive overview of this topic.[1] Whilst more than 15 editions of the work were published before the end of the sixteenth century, she considered that only three were published before his death and one shortly after that event. The first edition is thought to be that of October 1543.[2] Easton demonstrated that the

---

[1] Easton JB (1967) The early editions of Robert Recorde's Grounde of Artes. Isis 58:515–532.
[2] STC. 20797.5, British Library G16099.

J. Williams, *Robert Recorde: Tudor Polymath, Expositor and Practitioner of Computation*, History of Computing, DOI 10.1007/978-0-85729-862-1_7, © Springer-Verlag London Limited 2011

supposition, based on monetary values quoted in the text that an earlier edition existed, is not supported by the known evidence. However, at the end of the list of contents of the Second Dialogue in the 1543 edition and, indeed in all subsequent editions of this version, is listed 'The commen kyndes of castinge accomptes after marchauntes fashion, and auditors also. Numbrynge by the hande, newley added.' Does 'newely added' mean that this was an addition to an earlier version or that it was a lately added afterthought? It deals only with the representation of numbers by finger positions which treatment Recorde promised to augment, following his treatise on fractions, but never accomplished. There was a second, undated, edition which must have been printed after Recorde received his doctorate in 1545 and close to the time that its dedicatee, Whalley, fell from favour in 1550.[3] These two volumes constitute what Easton designated 'Edition A'.

'Edition B' is 'Edition A' augmented by a comprehensive treatment of fractions and their uses, together with sections on the golden rule, the backer rule, fellowship and alligation. Confusion over the attribution of authorship of the greatly extended portions of Edition B arose because for some time the earliest known version of this edition was dated 1558 i.e. after Recorde's death. Easton's analysis of the revised contents of 'Edition B' demonstrates that such alterations as were made, took place between 1549 and 1552. In his preface to *The Pathway to Knowledg* dated 28 January 1552, Recorde refers to an extension of his published work on Arithmetic which he has ready and, in the preface to Edition B, he writes of his work on Cosmography about to be set forth. This latter was published at the beginning of Mary's reign in 1556. Easton's argument for dating of Edition B to 1552 was substantiated by a note added to her article in press that 'The British Museum has recently acquired a 1552 *Ground of Artes* which is the prototype of the 1558 edition.'[4] The additions that Dee made in the editions of 1561 and 1566, which Easton calls 'Edition C', she rightly describes as being of a minor nature. Indeed such contributions that Dee made to the later editions with which he was associated are of a practical nature rather than any alteration of the mathematical insights. Whatever substance may be attached to the *Grounde of Artes*, it is to be found in those editions that are to be attributed solely to Recorde.

Eaton's views on the sequencing of Editions A and B are reinforced by a closer examination of the 1552 edition than she was able to carry out. Further evidence also arises from a facsimile version of the 1543 edition.[5] Both the latter and its earlier version carry an errata sheet, prefaced by a short piece of doggerel that reads

No hede so hedely can be given
But error slyppery will crepe in
For man without errour scarcely can be
So ye errour exceedeth all dylygencye
Paciently therefore I praye you bere,

---

[3] STC. 29798, Bodleian Douce 301.
[4] STC. 20799.3, British Library C113.a.22.
[5] Bodleian Library f. 64.

Those fewe fautes commyted here.
More pleasaunt profyt I give by reding
The grevus grefe by errours offendyng.

There follows a list of 12 errors of which 3 are numeric and all of a very obvious nature. In the Bodleian facsimile copy, all the errors are corrected by hand in the text. They are then incorporated into the body of the text in the 1551 version.

A new set of errors were made in copying the 1551 edition of the *Grounde* to form the first half of the augmented Edition B of 1552. That such copying took place is suggested by the fact that the same illuminated initial letters are used to introduce those topics common to the 1552 and 1551 editions, but both deviate from those of the 1543 version.

The blame for this new set of errors can be laid at the printer's door. What may be taken as an apologia for such defects is found in the 1552 edition, inserted before the opening dialogue between Master and Scholar.

The Printer to the Reader.
To take in woorthe my payne
T'intreate it seemeth vayne:
No man there is I gesse
Wyll slacke such thankfulnes
And where my prynt offended.
I trust I have amended
And if I have omitted ought,
Such faulte of mendyng needed nought.

It was unusual for the printer to address his reader directly and the verse was not included in subsequent editions. It surely signifies something out of the ordinary in the history of the printing of this document. The errors associated with this part of the first book, which deals with the arithmetic of integers, are not numerous, some 13 in total, and of those only 3 are of numerical nature and easily recognised. A further 3 are associated with misalignments of columns or signs and the remainder with textual errors. None were likely to have misled a reader at the stage of understanding of the subject he would have reached at the point where they occur.

The errors associated with the augmentation are more numerous, more serious but also more understandable. The theme of the Second Book was the manipulation of fractions, a topic that would have been totally alien to any English printer. Furthermore, Recorde was attempting to regularise the presentation of compound fractions so as to indicate unambiguously whether they were to be added or multiplied. By far and away the majority of the corrections needed in the second part of the arithmetic were to do with the adjustment of the symbols to meet Recorde's proposed protocol. It is not surprising that the printer erred, for the matter involved needed an understanding of the arguments advanced by Recorde in favour of his proposal. John Dee managed to get it wrong in the first edition of the *Grounde* that he edited following Recorde's death. There was a sprinkling of numerical errors but none of a critical nature. All the errors listed in the corrigenda of the 1552 edition were absorbed into the body of the text in the 1558 edition, which would explain why Easton did not comment on them.

The only person capable of making up the list of 'Fautes escaped' would have been Recorde himself. The 1552 edition was dated October of that year. Recorde had returned from Ireland a few months in advance of that date and so would have been available to act as corrector, but probably not before the main body of the text had been set. Recorde was thus wholly responsible for the main part of the text that was to endure for a century and a half. The next revision of this text of any substance was that of Mellis in 1591.

The general structure of Western European texts expounding the fundamental operations of arithmetic using Hindu-Arabic numbers had remained virtually unchanged since their introduction in the twelfth and thirteenth centuries. Thus the thirteenth century Latin texts of John de Sacrobosco and Alexander de Villedieu, which were still in use into the sixteenth century, spoke of the seven rules of arithmetic and dealt with numeration, addition, subtraction, duplation, multiplication, division, mediation, progression and extraction of roots of both square and cubic varieties. During the sixteenth century, English authors gradually abandoned the separate treatment of duplation and mediation. Tunstall writing in Latin in 1524 treated duplation only briefly, but took two pages to dispose of mediation. *An introduction for to lerne to recken with the penne.* of 1539 has been shown to have been compiled from an anonymous French text believed to have been published between 1530 and 1537 and an anonymous Dutch text of 1508.[6] Both works contain treatments of duplation and mediation. However the author of 'An Introduction', in the section dealing with multiplication, wrote [Diiii] 'As for the multyplycatyon by squares is neyther worth the wrytyng. And where as in other copies is set duplacyon, tryplacyon and quadruplaycyon, all that is superfluous, for so much as it is contayned under the kynde of multyplycacyon: and they that are experte in this feate, may ryghte perceave it.' Neither does the author deal with mediation, although there is no specific mention of its exclusion.

In his 'Dialogue betwene the Maister and the Scholar..' at the start of *The Grounde of Artes*, Recorde wrote [Biiii.v]

> S. ...But how mani thinges are to be learned, to attayn this arte [arithmetic] fully? M. there are reckened commonly, vii, Numeration, Addicion, Subtraction, Multiplication, Division, Progression, and Extraction of radicals: to these some men adde Duplacion, Triplacion, and Mediacion. But as to these last thre, they are contayned under the other vii. For Duplacion and Triplacion, are conteyned under Multiplication, as it shall appear in their place. And Mediacion is conteyned under Division, as I will declare in his place also. Scho. Yet there remain the first vii kyndes of numbryng. M. So there dooeth: How bee it, if I shall speak exactly of partes of numbryng, I must make but v of theym: for Progression is a compound operacion of Addicion, Multiplicacion and Division. And so is Extraction of Radicals. But it is no harme to name theym as kyndes severall, seying they appere to have some severall workyng. For it forceth not so muche to contende for the numbre of theym, as for the dewe knowlage and practisyng of them.

Recorde has here defined what the five basic 'operacions' of arithmetic are and, in effect, the way in which arithmetic was henceforward presented in English texts; a small matter, but evidence of his clear, orderly thinking.

---

[6]Bockstaele P (1960) Notes on the first arithmetics printed in Dutch and English. Isis 51:315–321.

Within the structure so defined, in any treatment of elementary arithmetic there is very little room for the introduction of novel material. At most the author will be able to modify the presentation of material. In this matter Recorde excelled in all his books. One is therefore left with the task of sieving through nuances of the text, looking for clues that might permit some guesses to be made about a deeper understanding of the subject possessed by the author, which it would have been inappropriate for him to discuss in the context of this presentation, or indeed possibly confusing to those readers he had immediately in mind.

Such a situation presents itself at the very opening of *The Grounde of Artes*. Following the Preface and lists of contents of the two dialogues, two lists of figures are given with the admonition 'Before the Introduction of Arithmetik these figures must be learned.' The first set are the 'Figures of numbers', which give the Roman numerals i,v,x,l,C,D and M and their names, individually and in combination, to an extent that would allow representation up to a 1,000 to be derived. The second set is the 'Figures of mony' which give the letter symbols and names for four successively smaller parts of a penny together with those for a penny, shilling and pound. The first set of figures is used immediately following the introductory Dialogue between the Master and Scholar, in the treatment of Numeration. This latter is defined as '..the arte to expresse and rede all summes proposed, and is of two sortes, for either it gathereth the value of a summe proposed, eyther els it expresseth a summe conceyved by fygures and places due'. The Master goes on to explain that there is a difference between the figure and its value, primarily because of the place in which it is set. The Scholar deduces from this that three parameters are needed, the Value, the figure, and the Place. The Master says that a fourth parameter viz. Order has to be added. Only then are the ten Hindu-Arabic numerals introduced, of which nine are signifying figures and one, made like an o, 'called privately a cypher, though all the other sometyme be lykewyse named'. The nine signifying figures from 1 to 9 are then given their values in Roman form and the Scholar exercised in converting between the two sets of symbols.

The place value system is then introduced and exercised at length. The concept of *ternaries or Trinities* is introduced i.e. the practice of breaking down sequences of large numbers into multiples of a 1,000 by the insertion of 'prickes' or lines. This in turn leads into the introduction of *Denominations* defined as follows 'M. It is the laste valewe or name added to to any summe. As when I saie: CCxxii poundes: poundes is the denomination And like wayes in saying: 25 men, Men is the denomination. and so of other. But in this place (that I spake of before) the last number of every ternary is the denomination of it. As of the first ternary, the denominacion is unitees: and of the second ternary, the denomination is thousandes: and of the thyrd ternary, thousand thousands or myllions: of the .iiii, thousand, thousand thousandes, or thosand myllions, and so forth. S. And what shall I call the value of the. iii. figures that may be pronounced before the denominators, as in saiynge: 203000000, that is, CC.iii myllions, I perceive (by your wordes) that myllions is the denomination, but what shall I call the CC.iii. joined before the myllions. M. That is called the numeratour or valewe, and the whole summe which resulteth of them bothe, is called the summe value or nombre.'

This description is more complex than that of Sacrobosco, Tunstall or that found in the anonymous text of 1539 and in that of Gemma Frisius of the same time, where the reader is presented with the Hindu-Arabic numerals, their relationship to their Roman counterparts and to the place value system, for which only the place and figure are the defined parameters. Of the additional parameters introduced by Recorde, that of 'value' occurs naturally but is not singled out for comment by the other authors. Recorde made the introduction of the parameter of Order a noticeably deliberate act and uses it to describe the ascending sequence of places first, second, third, etc. into which the figures are placed. The other authors did this also, but without drawing attention to the nature of the parameter. What Recorde did effectively was to differentiate between Cardinal and Ordinal numbers. Whether this was a vague expression of a distinction that he had in mind is a matter for debate. It has to be remembered that Cardinal numbers by themselves are not capable of creating an arithmetic. Arithmetical operations assume that it is possible to pass from any number to its successor i.e. that there exists a regular sequence of numbers described by Ordinal numbers.

It is possible to view Recorde's introduction of 'denomination' in a similar light. The anonymous author of 1537 does not introduce the term until late in his text when he deals with the topic of Reduction and does not define it. The other authors do not use it. Recorde used the term almost immediately at the end of the section on Addition. The Scholar attempts to add together 848 sheep and 186 other beasts that he met in Cheapside. The Master instructs that only like things should be added together unless there is a 'need to find only the number and care not for the things.' He does however add together sums of pounds, shillings and pence to illustrate his point that like must only be added to like. This need for definition of denomination surfaces again in the matter of algebra as dealt with in Recorde's ultimate book on arithmetic '*The Whetstone of Witte*'.

Casting the subject in the language of sets, Recorde has described numeration in terms of Cardinal numbers which indicate how many members there are in a set, Ordinal numbers which indicate the position of members in a set and Denomination which defines the set, which together adequately describe the basis for an arithmetic. In practice he does this more overtly than the other authors of contemporary texts on arithmetic using Hindu-Arabic numbers, but the impression left is that he was more aware of the formal logical structure within which he wanted to operate.

He was aware of more than just the formal structure of his subject. At the very end of the section on Numeration the Student asks why the values of the Hindu-Arabic numbers were placed backwards from the right-hand to the left. The Master says

'In that thing all men do agree, that the Chaldeys whiche first invented this arte, did set these figures as thei set all their letters, for they wryte backwarde as you tearme it, and so doo they reade. And that may appeare in all Hebrewe, Chaldaye and Arabike bookes. for they be not only written from the ryght hande to the left, and so must be readde, but also the.ryght end of the booke is at the beginninge of it: where as the Greekes, Latines and all nations of Europe, do wryte and reade from the left hand towards the ryghte. And al theyr bokes begyn at the lefte side.' This satisfies the Student. It also shows the breadth of Recorde's reading if not the sources of his information. The nature of these latter becomes more intriguing and frustrating following the Master's next statement, '*It nother satisfieth me nother liketh me well, because I see that the Chaldays and Hebrues do not so use their own numbers. As at an other tyme I wyll declare.*' It suggests that Recorde was familiar with sources, which have not been identified, that understood both Hebrew and, more intriguingly, Chaldean arithmetic.

With respect to Recorde's philosophical attitude towards the status of 'numbers' per se, in Edition A he makes no specific reference to Plato. These are only to be found in the Dedications of Edition B and that of *The Pathway to Knowledg* of the early 1550s, at which time his Platonic view of the abstract nature of numbers was well established. Is it possible that in 1543 he was still struggling with the problem of presenting an arithmetic in terms of abstract numbers, but at the same time preserving a stance that allowed of practical advantage being taken of such an arithmetic, which would necessarily involve working with real quantities i.e. denominations?

Recorde adds nothing significantly novel to the methods described for the basic arithmetical operations of addition, subtraction, multiplication and division on integers. He does however give a wider range of options than other arithmetical texts before offering his preference. For Addition he demonstrates the checking of calculations only using casting out of nines, which the student professes not to understand and for which the Master delays an explanation. As already mentioned, the addition of sums of money of mixed denomination is also introduced and the Master explains the use of the casting out of nines at some length for such cases. Following his initiation into the operation of Subtraction, the Scholar is told that, whilst 'proof' of results by the casting out of nines is possible the preferred method is via the inverse operation of addition. The Scholar sees that now he knows how to subtract, he can apply this inverse method also to checking the result of additions. The operations of Multiplication and Division deal only with operations on integers with checking procedures as for the two previous operations.

Treatments of the basic arithmetical operations are followed by a section on Reduction, which starts in earnest the demonstration of the practical uses of arithmetic. Reduction is defined as the means 'by which all summes of gross denomination, may be turned into summs of more subtil denomination. And contrariwise: all summes of subtyle denomination may be brought to summes of grosser denomination.' To pursue the examples involving coinage meaningfully, Recorde has to give the relative values of a diversity of denominations, which are of interest to historians other than those of mathematics. Challis, although pointing out some deficiencies in Recorde's list of current coinage, found the data of value inasmuch as it provided a picture of the situation pertaining immediately before Henry VIII entered into the process of debasing the currency.[7] In view of Recorde's subsequent involvement with minting it also provide an important baseline for these future activities. A little attention is also given to the values of French and Flemish currency, but he was reluctant to deal with them at length because of their instability.

[7]Challis CE (1978) The Tudor coinage. Manchester University Press, Manchester, pp 221–223.

The position with respect to weights that Recorde reports was also confused. He presumably was quoting the ancient Statutes of England as saying that 'the leaste portion of weight is commonly a grayn, meanyng a grayne of corne or wheat, drye, and gathered out of the myddell of the eare. Of these graynes in tyme passed 32 wayed juste 1 pennye of Troye, and than was but 20 pennies in an ounce: But now are there 46 pennies in an ounce, so there are not fully 14 graynes in one pennye.'[This probably derives from the Assize of Weights and Measures which is commonly attributed to 31 Edward I.]. More serious is the contention that the pennyweight had been reduced to less than half its value as assumed in the Statutes to which Recorde referred. However it had stood at 37½ pence to the ounce throughout the reign of Henry VII, was reduced to 45 to the ounce by Cardinal Wolsey in 1526 and to 48 to the ounce in 1542, where it remained until it was halved at the beginning of 1549.[8] How far this reduction in weight was appreciated by the population at large cannot be judged, but clearly Recorde was sufficiently troubled by it to draw attention to it. The student is then introduced to the pound 'aberdepoyse' having 16 oz to the pound. This is followed by listings of the varieties of multiples of a pound weight, of liquid measures and of volumes of dry goods, which complexities seem to have stunned the Scholar into silence. The Master finally lists the small number of linear measurements in use and also a little about measurements of area. However he concludes this section by saying 'Here mought I tell you many thyngs els touching measure: and also howe to reduce straunge measures to oure measures, but because it cannot well bee done without the knowledge of fractions, whiche as yet you have not learned, I will let them passe till an other tyme, when I shall instruct you in Geometry, wherein I shoulde be enforced els to repete the same to often agayn.' There are thus repeated commitments to future treatments of fractions and also of geometry before the author returns to considerations of measurement of areas.

The exposition continues with descriptions of arithmetic and geometric Progressions. Considerable space is devoted to applications of 'the rule of proportion which for his excellency is called the Golden rule'. This section includes descriptions of the backer rule, the double rule and the rule of fellowship or company. Practical examples of application are taken from a range of subjects, measurement of cloth, provisioning of a military force, payment for work, partnership of a flock of sheep and remuneration of classes of clergy. This latter example treats the division of a total stipend of £2,600 per year between 20 canons and 30 vicars, the canons receiving five times as much as a vicar. This mirrors an example to be found in Gemma Frisius' text where 12 canons and 20 chaplains share 3,000 aurea per annum between them with canons receiving five portions to the and chaplains' four. It seems sufficiently different however to suggest that it was the topic that was of common interest rather than the method of calculation.

The second Dialogue of Edition A gives a clear account of the five kinds of arithmetical operation possible using counters. This is then applied to the casting of merchant's accounts and those of auditors. They are a reasonably comprehensive

---

[8] Challis, loc. cit., 309–316.

description of such historical methods. Finally a description of representation of numbers using fingers is given. The manipulations illustrated derive form those presented in Continental arithmetical texts, not from earlier English sources. Instruction in their use for calculations is promised following the treatise on fractions, but it did not materialise.

The overall impression resulting from examination of Edition A of the *Grounde of Artes*, is that it is a more rigorous and comprehensive treatment of the arithmetic of Hindu-Arabic integers than that of other contemporary texts. This could have been a result of Recorde having a deeper innate understanding of the subject than the other authors and, in view of his avowed objective of dissemination of knowledge, more anxious to be formally correct than the others thereby establishing the appropriate basis for his extension of arithmetic to include cossicke functions eventually. With regard to the description of available options for the operations described Recorde offers the greater choice. He offers both 'prickes' and lines for partioning of sums into thousands, whilst Tunstall offers only the former and Gemma the latter. Whilst the other authors plump for casting out of nines as the method for checking calculations, Recorde offers and generally prefers the use of the inverse operation. His practical examples cover a wider range of subjects. This wider coverage suggests that he may have had all the other texts available for his own self-education and use, although it is only the Anonymous text to which he most likely makes reference in his Dedication. 'And if anyone can object that other bookes have beene written of Arithmetike all readye so sufficiently that I neede not nowe to put penne to the booke, except I wyll condemne other mennes wrytynges: to them I answere, That as I condemne no mans diligence, so I knowe, that no one man can satisfie every manne: and therefore lyke as many doeth esteme greatly other bookes, so I doubt not, but some wyll lyke this my booke above any other Englyshe Arythmetike hytherto written, and namely suche as shall lacke instructers, for whose sake I have so plainely set foorth the examples, as no booke (that I have senne) hath doone hytherto, which thyng shall be great ease to the rude reader.'

Recorde wrote his medical text *The Urinal of Physik* and his text on geometry *The Pathway to Knowledg* before he completed his treatment of arithmetic which Easton has designated Edition B of *The Grounde of Artes*. This latter has in addition to an unchanged main body of Edition A, 'The Second Part of Arithmetike, touching Fractions brieflye sette forthe'. Easton has pointed out that, although this version was not published until 1552, its compilation had probably been completed as early as 1549.[9] The evidence adduced by Easton deserves reconsideration. The new addition starts with the Scholar saying 'All bee it I perceive your manifold business doth so occupye, or rather oppresse you that you can not as yet completelye end that treatise of fractions Arithmeticall, which you have prepared, wherein not only sundrye works of Geometry, Musike and Astronomy be largely set forthe, but also divers conclusions and natural workes, touching mixtures of metalles, and composition of medicines with other strange examples, yet in the meane season I cannot staye

---

[9] Easton, loc. cit., 521.

my earnest desire, but importunately crave of you some briefe preparation towarde the use of fractions, whereby at leaste I may be able to understande the common workes of them, and the vulgare use of those rules, whiche without them cannot well be wrought.' The Master replies, 'If my lesure were as greate as my will is good, you should not neede to use any importunate craving for the attayning of that thing, whereby I might be persuaded that I shall annye wayes profite the common-wealth, or help the honest studies of any good members in the same: wherefore while mine attendaunce will permit mee to walk and talke, I am willing to help you as I maye.' The turbulent events in which Recorde became involved at the time of the overthrow of Somerset have been discussed in Chap. 2. By his own telling, he was forced to attend the king for 60 days from 30 October 1549, consequent upon being charged with treason by the Earl of Pembroke. Recorde was further harassed by Pembroke directly during the first few months of 1550 and indirectly up to the end of that year. It was not until the middle of 1551 that Recorde was back in good odour with the Crown. It seems reasonable therefore to assume that Recorde wrote this portion of Edition B around the end of 1549.

The arguments put forward by Easton for dating the movement of the original dedication to Whalley to that for the 'gentle reader' to some time shortly after the former's imprisonment in 1551 still stand. This is also true for the arguments dating of the Dedication to Edward as probably lying after the publication of the 'Pathway' in early 1552.

It would seem prudent to assume that Edition B was essentially complete by the end of 1551, only requiring addition of the Dedication before it was finally published at the end of the following year. It will be recalled that the Master, in Edition A, had declined to enter into a discussion in depth of the reduction of measurements of area until the Scholar had acquired some knowledge of Geometry. It follows that publication of the 'Pathway' should preceed completion of the work on arithmetic. This sequence of events is presaged in the preface to the 'Pathway', where a list of future books 'shortly to be set forth' starts with 'The seconde parte of Arithmetike, teaching the workyng by fractions with extraction of rootes both square and cubike: and declaring the rules of allegation, with sundrye pleasaunt examples in metalles and other thynges. Also the rule of false position, with dyvers examples not onely vulgar, but some appertaynyng to the rule of Algeber, applied unto quantities partly rationall, and partly surde.' In the event, the work dealing with extraction of roots, with surds and algebra was not published until late in 1557.

Recorde's treatment of fractions is as formal and meticulous as that for integers. Thus he starts by showing the Scholar that his description of a fraction as a 'broken number' is improper. Such a description was common at that time. The Master points out that an integer such as 60 could be broken into a number of parts [factors] which were all integers. 'Wherefore properly a fraction expresseth the partes or parte oneley of an unite, that is to say that the number whiche is the whole or entier summe of any fractions, may not be greater than one; and therefore it followeth that no one fraction alone may be so great that it may make 1 .....'. The Master then goes on to discuss the form of expression of a fraction as rigorously as he had done for integers.

In common with Tunstall, the anonymous author of the 'Introduction', Gemma Frisius and Stifel, Recorde states that a fraction is represented by two numbers, set one over the other with a line drawn between them. The upper number is called the Numerator and the lower one the Denominator. The other authors leave the matter there, but Recorde's Scholar is made to ask why both numbers are not called numerators as they both express numeration of the fraction. This allows the Master to explain that the denominator declares the number of parts into which the unit is divided but it is the numerator that actually expresses the number of such parts. In doing this he places the denominator of a fraction alongside the denomination of an integer as described in the section on Numeration in Edition A, where it was coupled with a numerator to express value.

The Master then elaborates on why there can be no more parts of a thing than it was divided into and that if the numbers are equal it is better to call the sum a whole. He then explores the situation where the numerator exceeds the denominator, which may arise for ease of calculation but avers that should properly be called mixed numbers. By dealing with it at this stage he is anticipating trouble later in making more general calculations. More interestingly and again differently from his contemporaneous authors he interjects the statement 'Nowe must you understande, that as no fraction properly can be greater than 1, so in smalness under one, the nature of fractions doth extend infinitely: as the nature of whole numbers is to increase above one infinitelye, so that not onely one may be divided into infinite fractions or partes, but also everye fraction may be divided into infinite fractions or partes, whiche commonly be called fractions of fractions'. No further elaboration or use is made of the concept but it does again round off Recorde's understanding of the relationship between the arithmetic of integers and fractions and hints at his possession of a greater depth of understanding of the properties of numbers than his contemporaries had.

The diversity in representation of fractions of fractions is seen as a potential hazard to the understanding of the Scholar. This topic is either ignored or confused by other authors, with Stifel as an exception. Recorde suggests that when a succession of fractions are to be added together, all fractions retain the horizontal dividing lines, but when it is intended that they are to multiplied together only the first fraction has the dividing line. Thus $\frac{3}{4}\frac{2}{3}\frac{1}{2}$ means that the fractions are to be added, but if the dividing lines are omitted from the last two fractions it means that they are all to be multiplied together. The Master says that this is but one of a number of approaches one of which is to add 'wordes of distinction' such as $\frac{3}{4}$ of $\frac{2}{3}$ of $\frac{1}{2}$, which is acceptable, whilst others stagger vertically those fractions to be added to one another. Frisius quotes this latter method together with a version of Recorde's proposal, but in practice uses 'words of distinction'. Stifel, quoted by Recorde joins numerators and denominators to be multiplied together by lines. Edition C of the *Grounde of Artes* confuses the issue completely by not using Recorde's proposed format. So much for John Dee's editorial abilities! However the Master makes it very clear what the consequence of confusion between addition and multiplication could be. An English Angel is worth 90d, four ninths of an Angel is 40d, two fifths is 36d, which together make 76d. However four ninth parts of two fifth parts is only 16d. This latter operation is in essence multiplication, but it is not so named although, as will be seen the same example is invoked later to explain the process by encouraging the Scholar to undertake calculations based on subdivision of the 'old Angel'.

At this juncture, Record introduces a major structural change in his presentation compared both with his treatment of the arithmetic of integers and also with that of the 'Introduction' and that of Gemma Frisius and Stifel. The Scholar wishes to proceed directly to the addition and subtraction of fractions but the Master demurs insisting on dealing first with their multiplication, division and reduction, thereby following his practice of dealing with the easier work before moving to that which is more difficult. The reasoning is that addition and subtraction cannot be taught without the use of multiplication and reduction, whilst multiplication and reduction can be taught without recourse to addition and subtraction. Gemma Frisius nods in the direction of this approach by treating reduction i.e the generation of common denominators before he treats addition and subtraction. In doing this he effectively deals with multiplication which, together with division he is thereby able to dismiss very briefly after he deals with addition and subtraction.

Recorde's treatment of the multiplication of fractions is the usual one of multiplying the numerators together to give the numerator of the product and dealing similarly with the denominators. He shows some unease about the explanation he can offer to the Scholar, who finds it difficult to understand why the product of multiplication is less than either of the multiplicands. Thus, using the given rule, $3/_5$ multiplied by $5/_{12}$ is $15/_{60}$, but using the crown of 60d. as the unit, the scholar finds that three-fifths is 36d. and five twelfths is 25d which multiplied together give 900d. or 15 crowns rather than the 15d which is fifteen sixtieths of a crown. The Master does not attempt to point out the fallacy inherent in multiplying two sums of money together to produce a third sum of money. He says that whilst the product of multiplication of two integers is always greater than either for fractions the product is always less. He does attempt to persuade the Scholar that the same rules of proportion apply in both cases and invokes the aid of another monetary example, again the 'Old Angel', to explain why if three fifths multiplied by five twelfths by the rule amounts to fifteen sixtieths then the latter bears the same proportion to $3/_5$ as $5/_{12}$ does to one. An old Angel contains 180 half pence, $15/_{60}$ of which is 45 ob., $3/_5$ is 108 ob., $5/_{12}$ is 75 ob, 45 is to 108 as 75 is to 180 viz. $5/_{12}$. However he will defer 'exact reasoning' until later. He then attempts another demonstration using a crown worth 5s. as the unit. Three fifths is worth 3s.; five times this gives 15s. or 3 crowns; twice gives 6s. or one crown and one fifth; once gives 3s so less than once must give less than one unit. Therefore one fraction multiplied by a second fraction must be less than the first fraction. At this point the Scholar accepts the contention, but the master says he will return to the topic when he deals with Reduction following the treatment of Division. However he soon runs into a similar problem in explaining the process of division. The general rule for cross multiplication is given, the Scholar shows that he can manipulate numbers accordingly but does not understand the reasoning. He calculates correctly that $^2/_3$ divided by $^3/_4$ gives $^8/_9$ but does not understand why, as the Master says, the quotient $^8/_9$ expresses the proportion of the dividend $^2/_3$ to the divisor $^3/_4$. Once again recourse is had to an example using coinage; a shilling contains 12d., two thirds of which is 8d. and three quarters is 9d. It is conceded that the argument is less easily understood for the case of $^5/_8$ divided by $^2/_6$ which yields the improper fraction $^{30}/_{16}$. Again it is stated that there will be difficulties until proportions are understood and so it is necessary to deal with Reduction. It seems that in making this argument Recorde is implicitly admitting that the order of presentation, Multiplication, Division, Reduction, is not unequivocally one of moving from the easiest to the more difficult. His treatment of the multiplication and division of fractions is probably the least satisfactory of all the instructions offered by Recorde throughout his texts. It is not evident that at this juncture Recorde himself had resolved in his own mind how best to present the arguments for the procedures to be followed.

The Master lists five varieties of Reduction the distinctions of which the Scholar appreciates but does not understand. Reduction is first defined as 'an orderly alteration of numbers out of one forme into another, which is never done ordinarily but for some needful use...'. This is a broader definition than that used for integers 'Reduction is by which all summes of grosse denomination, may be tourned into summes of more subtle denomination: And contrarywise....', but the practical need remains a key element.

The first variety of reduction deals with the conversion of two fractions that are to be united, into ones having a common denominator. This is accomplished in the obvious way by multiplying the two differing denominators to give the common denominator and then multiplying the numerator of the first fraction by the denominator of the second fraction to give the new numerator for the first fraction, and generating the new numerator of the second fraction in a similar way. Working according to this rule $^3/_{16}$ and $^4/_6$ are reduced to $^{18}/_{96}$ and $^{64}/_{96}$ respectively. This is illustrated with reference to the 'new Angell accompted at 8s. which contayneth 96 pence'. [This convenient equivalence might well have been the reason for Recorde using the formal value of the Angel, rather than its cried down value of 9s.8d., which at 116 pence would not have served as an easy example.] The use of the rule is then extended to treat three fractions. A second form of the first rule is then given, to no superficially apparent advantage. Two shortcuts are given for calculations involving fractions have simple numerical relationships between their denominators. Proof of correctness of calculation is by reversing the operations, as will become clear when the fourth variety of reduction is treated.

The second variety deals with the reduction of fractions of fractions to one fraction and so to one denomination. This is carried out by multiplication, which the Scholar declares that he understands and proceeds to demonstrate by multiplying together $^3/_4$, $^2/_3$, $^6/_7$ and $^7/_9$ to give $^{252}/_{756}$. This he declares to be a fraction too difficult for him to understand. The Master says that the fraction is indeed no more than $^1/_3$, which he demonstrates by a 'figure of measure' the whole of which is divided into nine parts as follows

1	2	3	4	5	6	7	8	9
1	2	3	4	5	6	7	$7/_9$	

.1.		.2.		.3.		$6/_7$		
1	2	3	4	$2/_3$				
1	2	3	$3/_4$					

The Scholar sees that the answer to the calculation is three parts of nine or one third. This demonstration is identical in form with that used by Gemma Frisius to demonstrate the multiplication of $6/_7$ by $2/_3$ which he gives as $36/_{84}$ but does not reduce it to the simplest form of $3/_7$ implied by the figure. This suggests strongly that Recorde had Gemma Frisius' book to hand when he was compiling this segment of his own book. The third form of reduction deals straightforwardly with the inter-conversions of mixed numbers and improper fractions.

The fourth kind of reduction is concerned with the simplification of fractions. This may be desirable but sometimes there may be value attached to having a specific denominator that corresponds to a practical unit of measurement. To find the greatest common divisor of numerator and denominator the Euclidean algorism is explained. This is another feature held in conmmon with Gemma Frisius. Recorde points out that inspection, which comes with practise, can sometimes ease this form of reduction.

The last kind of reduction deals with turning any fraction into one that possesses any other denominator e.g. one relevant to a system of measurement. The new numerator is obtained by cross multiplying old numerator and new denominator and dividing the product by the old denominator. The scholar however goes astray when new numerator so calculated is a mixed fraction and he divides only its integral part by the old numerator. This defect of understanding is corrected.

Given the practice that the Scholar has already received, the topics of Addition and Subtraction were quickly disposed of.

The sections on the Golden Rule, Backer Rule and Rules of Fellowship for integers given in the first edition were extensively revised with respect to the examples provided of their applications to sums involving fractional amounts. Before doing so, the Scholar is referred back to the rules of proportions using integers which are then supplemented to embrace fractions. The first numerator of the first number of the question is to be multiplied by the denominators of the second and third numbers, which total is to be divided into the product of the denominator of the first number of the question and the numerators of the second and third numbers, to give the fourth number or answer. Thus if $3/_4$ yards of velvet costs $2/_3$ of a sovereign (esteemed at 20s.), what shall $5/_6$ yards cost? Following the rule, the divisor of the answer is $3 \times 3 \times 6$ and the dividend $4 \times 2 \times 5$, yielding $40/_{54}$ simplified to $20/_{27}$ of a sovereign. Two further examples involving coins are given. Proof by inverting the operation is recommended.

The Master then moves on to the more serious topic of the 'Statutes of Assize of bread and ale, which in all Statute bookes in Frenche, Latine and Englishe is much corrupted for want of knowledge in this arte' viz. application of the Golden Rule. Probably the most lucrative part of the book publishing business of the first half of the sixteenth century lay in the printing of the books of Statutes, reflecting a wide interest in the laws embedded in them. Of these Statutes, few were more important to the common man and hence the concept of 'Commonwealth' than those relating to bread and ale. The importance of bread can be seen from the fact that its weight was measured in Troy ounces, as for gold, silver and other precious materials such as spices.

Legislative control of weight as a function of cost of raw material was in place in thirteenth century England, probably continuing a practice of the previous century.[10] The types of loaves were classified by quality and by selling prices of a farthing, halfpenny and penny. The Assize listed the weights of such loaves as functions of the costs per quarter of wheat. As this cost rose, the weight of loaf obtained for a given price fell, in inverse proportion. Assizes were held locally to check for compliance with the regulations listed. Records exist for three very similar, but not identical regulations for 1,202, 1,266 and 1,290.

The first problem the Master poses to the scholar uses a set of parameters that are common to all these Assizes viz. when the price of a quarter of wheat is 2s. a farthing loaf shall weigh 68s., so what shall such a loaf weigh when a quarter of wheat is sold for 3s.? The Scholar wants to deal with the matter using whole numbers and not involving fractions. However the master reins him in, pointing out that the rate given in the Statute is one of weight measured in parts of pounds Troy and that it is also convenient to measure money in terms of pounds value also, 'seeing shillings do change often, as all other moneys do'. The Scholar then calculates correctly that the revised weight of the farthing loaf will be $68 \times {}^2/_3$ or $45^1/_3$ shillings or 45s.4d.

The Master points out that the Statute books constantly give 48s., 'one great error.' He goes on to wish that 'all Gentlemen and other students of the Laws would not neglect the Arte of Arithmetike as unneedful to their studies.' To encourage them he provides two tables of corrections to two sets of Statutes. These latter correspond approximately to those of 1266 and 1290 A.D., but not exactly so and their accurate provenance has not been found. What the Statute of 1266 and that used by Recorde have in common is coincidence in values of the cost of wheat for which the calculated weights of loaves were correct viz. for wheat costing 1, 1½, 2, 4, 6, 8, 8½, 12, 16 and 17 shillings a quarter. All of these values lead to calculations yielding results that can be expressed accurately in integral values of shillings and/or pence i.e. division was easy. The remainder of the calculated values, given for increments in value of 6d. per quarter of wheat, were in error to a greater or lesser extent, but even in the most extreme case only by 5% and usually a tenth of this. Nonetheless, there were errors in the Statute that had persisted uncorrected for several centuries despite the existence of means for their correction.

---

[10]Connor RD (1987) The assize of bread and ale. In: The weights and measures of England. HMSO, London, Chap. XI, pp 193–207.

The words that Recorde puts into the mouths of both Scholar and Master suggests that he had tried a direct approach to getting the Statute corrected and had been rebuffed: Scholar '...sith never good man that would reform error, could escape ye venomous tongues of envious detractors, which because they either cannot or list not to do any good themselves, do delight to bark at the doings of others.' It was presumably for this reason that the master says about his corrections '...And for annoying of offence I have rather done it in this private book, rather than any book of the Statutes self, trusting that all men will take it in good part.' As has already been noted such books of Statutes were widely published during the first half of the sixteenth century. Sight must not be lost of Recorde's expressed desire to demonstrate the usefulness of numeracy to the makers of Statutes.

The Master goes on to translate the corrected rates for the weights of loaves, from pounds, shillings and pence into pounds ounces and pennyweights. The original Statutes were made at a time when a penny weighed a penny weight. By the beginning of the sixteenth century pennies were coined at the rate of 37 shillings and sixpence to the pound weight i.e 450 rather than 240/lb. The Master chooses to use the contemporary rate of 60 shillings or 720 pennies to the pound weight. The cost of bread would thus be higher by a factor of three than that given in the Statute. An Assize of the beginning of the century did give the weights of bread in Troy units of weight and not in monetary units. From this compilation it is clear that the price has increased compared with that of the original statute by a factor of more than two, as might be expected from the decrease in weight of the penny, amongst other things. Connor has pointed out that there were costs additional to that of the grain that contributed to the cost of the loaves, such as the baker's allowances. At the time of the original Statutes these allowances were about 10 pence per quarter baked plus 2 loaves, which could constitute a sizeable fraction of the of grain costing 3 to 4s. a quarter. By circa 1500 the allowances had risen to 2 shillings for a similar price and by 1592 the figure had reached 6s.10d. when the best wheat cost 21s.4d the quarter.[11] The Statutes, however accurately the relationships they expressed were calculated, were inadequate to give more than temporary guidance to the authorities and then only for a limited range of grain pricings. This seems to have been acknowledged by the authorities, so Recorde's arguments for accuracy were of theoretical rather than practical concern.

He did however continue his campaign to revise the Statutes by examining the errors in the 'Statute of the measuring of ground [which] is common that it it touches all men, and yet no more common than needful, but so much corrupt that it is out of all good rate, not only in the English books of Statutes commonly printed, but also in the latin books, and in the French also, for I have read of each sort, and conferred them diligently....'. The Master define the units of linear measurement, inch, foot, yard, and perch in terms of one another and then defines the acre as containing 160 square perches. The Scholar is asked, if an acre of ground is four perches broad and 40 perches long, how long will it be when it is 13 perches broad? The Scholar uses the golden rule to deduce that the breadth will be $12^4/_{13}$ perches. This is converted into sub-units of length, using the fact that a perch contains $11^1/_2$ ft which multiplied

---

[11] Connor, loc. cit..

by 4 and divided by 13 gives $5^1/_{13}$ ft, or nearly 5 ft 1 in. The printed books in English and Latin give the breadth as 13 perches 5 ft and 1 in., or 13 square perches too many in the acre, an error also found in much older Statutes. One French copy was however found to give xij perches $1^1/_4$ and one foot which 'misseth little of the truth.' The Master then produces a correct table of lengths per acre for breadths starting with 10 perches in steps of one up to 45 perches. On this occasion the master declines to give the erroneous values because he is sure that 'it were too vile to judge of so Noble princes and worthy Counsellors that have set out and authorised this Statute, that they wouldmake one acre in any form greater than an other...'. This is again deemed to show how much need there is for the law of arithmetic. At this time there was a considerable market in land, free and forced, so accurate calculation of areas was very necessary. However this topic was dealt with much more extensively by Thomas Digges in his *Panometria* published at about the same time as the 'Grounde of Artes'.

Having finished with the Golden Rule, the master moves on to deal with other 'severally' named rules, which he describes as branches of the former rule. He dismisses the various rules of fellowship as having been explained adequately using whole numbers. However he gives one example as a refresher. Four men share a prize of £8,190 unequally, receiving respectively $^1/_3+^1/_{10}$; $^1/_4+^1/_{10}$; $^1/_6$ and $^1/_{20}$, fractions which together make up a whole The actual amount each receives is calculated by finding a common denominator of 240, and converting the fractional shares to integers which are then converted to real money using the golden rule. The results are checked by taking the fractional parts of the whole directly, adding them to give the appropriate shares. The indirect route may have been taken just to exercise the reader, but as Easton has opined, it may have been used to set the scene for the next examples to be presented and worked.

The Master prefaces these examples as follows, 'Now will I propond certayn other questions, whiche have been set foorth by certaine learned men, all be it not without some oversyghte, which questions I protest hartely I do not repeate to deprave those good men, whose labors and studies I much praise and greatly delyte in, but onely accordinge to my profession, to secke out truth in all thinges, and to remoove all occasions of errour, as much as in me lieth: and for that cause I will only name the questions without hurtinge the Authors name.' Four examples are given, each of which is identical [apart from the substitution of crowns fror aureos] with those given in the corresponding section of Gemma Frisius' treatise, although presented in a different order. There can be no doubt that Gemma Frisius was one errant author that Recorde had in mind and also therefore confirms that the latter had the former's book to hand when compiling his own work. Importantly, it also establishes that Recorde was not prepared to accept other men's work uncritically.

The point that troubled Recorde is made clear at the outset of the approach to solution of the first question he poses. This is Frisius' third question. A house is built for 3,000 crowns to which the first man contributes $^1/_2$ plus 6 crowns; the second man $^1/_3$ plus 12 crowns; the third man must lay out 2/3 abating 8 crowns and the last man contributes $^1/_4$ plus 20 crowns. What are the individual monetary contributions? The Scholar says that he cannot answer and the Master sympathises, because it is an impossible question, for the parts added together amount to 'the whole sum and $^3/_4$ and 30 more' or alternatively taking the fractions of the whole and summing them to 5,280 crowns.[12] '...and therefore can that question not be accepted as a possible thing, but yet do certain learned men propound such questions, and answer to them. Therefore somewhat to say to their excuse rather of their good meaning rather than for their dooing, I will anon declare for their defense: but in the mean season I will propond he question as it may be wrought by good possibility'. The summation that he demonstrated may have a greater significance than just that of using the = and − signs albeit not explaining them. It can be interpreted as a nascent exercise in cossike arithmetic, showing that Recorde had in mind intentions of dealing with this subject 5 years before he actually pronounced on the topic in the *Whetstone of Witte*.

The interpretation of the question proposed by Recorde is that the fractional amounts are intended to partially define the relationship of the individual shares to one another and not to the whole. Thus for the partition it means that for every 6 crowns the first man pays, the second pays 4, the third 8 and the fourth man 3 crowns. The same set of proportions are arrived at by considering that the first man pays double that of the fourth, the second man $^2/_3$ and the third man double that of the second. Both approaches rely on inspection. To these proportions have to be added or abated, the stated monetary contributions. The question may now be solved using the Golden Rule. The first number in the latter operation is given by the sum of the proportions viz. 21. The second number in the calculation is obtained by subtracting from the 3,000 crowns the overpluses and adding the abatement leaving 2,970 crowns. The third number is the portion of the first number associated with the individual share. Thus the first man would contribute 6 × 2,970/21 or $848^4/_7$ crowns and likewise the second man gives $565^5/_7$, the third man $1,131$ $^3/_7$ and the fourth man $424^2/_7$ crowns. The Scholar omits to add or subtract the monetary adjustments and is prompted by the master to do so, thus giving the first man $854^4/_7$ crowns. The master returns to his original argument, asking whether this was a half and 6 crowns more of 3,000. The Scholar computes this to be 1,506 with sums of 1,012, 1,992 and 770 for the other contributions, none of which agree with the

---

[12]For the first time Recorde uses the + and − signs in his books. Thus

$$\frac{1}{2} + 6$$
$$\frac{1}{8} + 12$$
$$\frac{1}{8} - 8$$
$$\frac{1}{4} + 20$$
$$\overline{1\frac{3}{4} + 30}$$

He says '.... whereof I will be bold to use first the representation of nombres in their aptest forme, (although I have not yet taughte you that manner of woorke) bicause it may appear plainly that the question is not possible. ' Some works of reference credit Recorde with the introduction of the + and − signs into England in his *Whetstone of Witte* of 1558 when he uses it to construct equations, but he was clearly aware of their existence and value before 1552, when the first edition of Edition B was printed.

solutions proposed, and marvels '…that so wise men could be so much overseen.'
The master takes the opportunity to emphasise the need to examine the works of
'elder writers' with a critical eye before accepting their findings, one of his continuing themes. He then excuses the learned men he has taken to task saying, 'But these
learned men did not mean anye other thing by this question, then to fynde suche
numbres as should bear the same proportions togyther, as those numbres in the
question proponed dyd bear one to an other,' This is a statement of the obvious as
far as Gemma Frisius is concerned for he uses the same method as Recorde to obtain
exactly the same results, but without pointing out the potential for error. The same
cannot be said for Tunstall who when dealing with a very similar problem, goes
badly astray in his reasoning and gets the wrong answers.

The remaining three problems of Gemma Frisius deal with questions of inheritance and in Recorde's eyes, suffer from the same defects in presentation as the
problem just discussed and are resolved in a similar manner.

Sections dealing with the topic of alligation are found in many of the texts, manuscript and printed, in Latin or vernacular, starting with Fibonacci and continuing
up to the time in question and beyond. Recorde's description of 'alligation' is clear.
'Now will I go in hande with the rule of alligation which hath his name, for that by
it there are divers parcels of sundry prices, and sundry quantities alligate, bound or
mixed together, whereby it might well be called the rule of mixture, and it hath great
use in the composition of medicines, and also in the mixture of metals, and some use
it hath in the medicine of wines, but I would wish it were less used therein than it is
nowadays.' His description of the method of calculation is equally clear. 'The order
of the rule is this: when any summes are proposed to be mixed, set them in order one
over another, and the common number whereunto you will reduce them, set on the
left hand, then mark what summes be lesser than that common number, and which
be greater, and with a draft of your pen, evermore link two numbers together, so that
one be lesser than the common number, and the other greater than he, for two greater
or two smaller cannot well be linked together, and the reason is this, that one greater
and one smaller may be so mixed that they will make the mean or common number
very well: but ij lesser can never make so many as the common number, being taken
orderly: no more can two summes greater than the mean, never make the mean in
due order, as it shall appear better to you hereafter. And as it is of necessity to link
every smaller (once at the least) with one greater, and every greater with one smaller,
so it is at liberty to link them oftener than once, and so may there be to one question
many solutions. When you have so linked them, then mark how much each of the
lesser numbers is smaller than the mean or common number, and that difference set
against the greater numbers which be linked with those smaller, each with his match
still on the right hand, and likewise the excess ofthe greater numbers above the
mean, you shall set before the lesser numbers which be combined with them. Then
shall you by addition bring all these differences into one sum, which shall be the
first number in the Golden Rule: and the second number shall be the whole masse
that you will have of all those particulars: the third sum shall be each difference by
itself, and then by them shall be found the fourth number, declaring the just portion
of every particular in that mixture. As is now by these examples I will make it plain.

There is four sorts of wine of several prices, one of 6d. a galon, another of 8d. a galon, the third of 11d. and the fourth of 15 pence the galon, of all of these wines would I have a mixture made to the sum of 50 gal, and so that the price of each galon may be ix pence. Now demand I how much must be taken of every sort of wine?' The master then works the first example at the request of the Scholar according to the method described

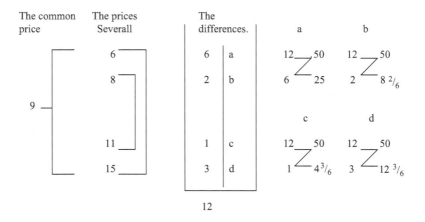

Proof of the correctness of the answer is made by multiplying by the common price by 50 to give the total cost of 450 and comparing this with the costs of the sum of the individual parcels of the various wines. The Scholar then repeats the calculation using a different combination of lesser and greater components to get a different but equally correct answer.

The next question deals with the purchase of a parcel of mixed spices, cloves, nutmeg, saffron, pepper, ginger and almonds, each spice having a different price, to a total weight of 300 lb of value 5s. a pound. First the Scholar solves the problem by combining one lesser with one greater component as in the previous example. Then the Master offers a solution in which he combines two of the lower priced components each with two of the more expensive sorts to give another solution. The Scholar wants to demonstrate further variations, but the Master expresses himself as '... loth to stay to see all the variations, for it may be varied above 300 wayes although manye of them would not serve well to this purpose.' The Scholar is surprised by this assertion, but the master says 'Marvaille not thereat, for some questions of this rule may be varied above 1,000 wayes, but I would have ye forget such fantasies, till a time of more leysure.' The Scholar then provides another six variations, noting on the way that in four of the answers the sum of the differences was the same but the relative proportions of the components in the solutions were different.

The Master finishes his very full exposition of Alligation with two questions dealing with the mixture of metals '... because the use of it serveth often in the charge, not so much for goldsmiths as for coinage in mints.' The Scholar first gives five solutions to a problem concerning the mixing of gold of six different purities and then deals with a similar problem involving silver of six different finenesses, to which the scholar gives five solutions and the Master a further two, using a less confusing way of presenting the data.

The final words of the Master show his concern with the topic of the wide variety of combinations giving correct answers that are possible, when he extends its possible significance to other areas. 'Truth it is, that this consideration may fall in practice as well politic, as philosophical, and sundry ways in them be applied, therefore when time shall fall fit for the discussing of this consideration, you shall not want my helping hand.'

Recorde has thus given a clear structure to methods for the application of the 'Rule of Alligation' and also has shown a full appreciation of the way in which a wide range of combinations of the given components provide correct solutions to problems posed. For problems involving six components, 343 combinations are possible. Recorde, speaking as the Master, says that above 300 ways of resolving the problems were possible, although some of them might be trivial. He must have travelled some distance along the path of understanding combinatorials. Superficially it might seem odd that he did not deal with the case of just two components, but if the method for dealing with six components is understood, its application to just two starting materials is trivial. Equally, given the demonstrations of how one component might be combined with more than two appropriate partners, there would be no difficulty in dealing with problems involving an odd number of components.

Just how innovative was Recorde in his treatment of alligation? The topic was introduced to Western Europe by Fibonacci early in the thirteenth century. His treatment of the topic was both extensive and showed depth of understanding. Thus he was clearly aware of the availability of more than one solution to the mixing of components when more than two were involved. Tracing the progress of the mathematics of the alloying of coinage from Fibonacci up to the close of the seventeenth century, Williams concluded that, following Fibonacci, Recorde provided the first advance in the methodology of alligation. This was a matter of some surprise, since the Italian maestro d'abbaco, knowledge of algebraic-type procedures that was more than adequate to put the subject on a firm basis. Manuscript texts generated by these teachers usually contained a few worked examples concerned with mixing and hence application of the rule of alligation, but were largely limited imitations of Fibonacci's handling of the subject and certainly showed no advance in its treatment.[13] Philippe Calandri's *de Arithmetica opusculum* of 1492 offered solutions to some 26 problems concerned with mixing but he appears to apply mechanically the methods used by Fibonacci, without showing any of the latter's deeper understanding of the arithmetic involved.

---

[13] Bodleian Library, Canon Ital 188, *Zibaldone di aritmetica*.
Bodleian Library, Canon Ital 244, *El Modo de risignare ogni raxone*.
Bodleian Library, Canon Ital 236, ff. 102v–103r, *L'Arte de Abbaco secondo il modo di Luca di Matteo da Firenze*.
British Library, Ms. Add 8784, *Libro di Ragione*.

The first step in taking the treatment of mixing beyond that of Fibonacci is found in the 'Nobel opera de arithmetica' of Pietro Borghi, first printed in 1484, followed by a further 15 editions up to 1577. Borghi starts with a problem involving three components of differing compositions that are to be mixed to give one of intermediate composition. The only correct solution possible using the Rule of Alligation is deduced. He then proceeds to analyse the mixing of five grains of differing prices to produce a mixture having a value laying below the two most valuable and the three of lesser value.[14] The grains are given as having values of 44,48,52,60 and 66 soldi di piccioli per bushel, and the mixture of 50 bushels required is to have a value of 56 piccioli. The manner in which the solution is worked and displayed, if rotated through 90° is very similar to that used by Recorde. If Borghi's book had been available to him, then it could have provided him with the starting point for his more complete and methodical treatment of the subject. No evidence has so far been found of possible access by Recorde to this text.

Clear evidence that Recorde did have Gemma Frisius' text to hand when he was compiling the second edition of *The Grounde of Artes* has already been noted. As Recorde's treatment of the rule of alligation followed immediately after that dealing with problems of fellowship, so did that of Gemma Frisius. The problems he deals with involve the same materials to be mixed and in the same order of presentation, viz. four sorts of wine, six sorts of spices, as did Recorde, but only one example of the mixing of two coinage alloys to form a third. His diagrammatic illustration of the method of solution is the same as that used later by Recorde, but without the use of the Golden Rule to obtain the final proportions. Only one solution is given to the problem of mixing of four wines, whereas Recorde contrives to show two solutions. To the problem of mixing of spices, Gemma provided only five solutions, but concluded, 'Atque huiusmodi infiniti fire extant modi. Interiunti memineris oportereut quidlibet numerusemi minmum alligetur, posse tamen saepius iaque ad varies comparari, atque huius modis ingeniis discentum relinque.' It seems that he was aware that there were potentially a large number of solutions, but was not prepared, or possibly not able to quantify them. It is possible therefore that the originality that Recorde contributed to the topic is to be measured just in terms of his clearer quantitative appreciation of the number of solutions possible as determined by the number of components available for mixing. Neither Recorde nor Gemma Frisius were aware that an infinity of solutions was possible, given the freedom to mix components together generally to form an infinity of component compositions prior to dealing with the question of blending them to produce an alloy of defined intermediate composition. Nearly a 100 years were to elapse before this matter was formally understood.[15]

The final section of the book deals with the Rule of False Position. More accurately it was what is generally called the rule of double false position. This is basically a means of solving linear equations for one unknown. As such it was a precursor of

---

[14] Ibid. fos. 96v–101v.

[15] Williams J (1995) Mathematics and the alloying of coinage 1202–1700: part II. Ann Sci 52:235–263, 256–260.

Algebra and was recognised as such by Gemma Frisius. Recorde only dealt with this evolution in his final mathematical text *The Whetstone of Witte* when he introduced 'Algebra'. Gemma Frisius introduced algebra immediately following his treatment of the rule of false positions, which was appreciably less expansive than that of Recorde.

Both authors take time to explain that the term 'false' relates to the tactics employed in the method of solution and not to the results obtained. Recorde's exposition gives what may have been a glimpse of his social life, 'Mayster, in the Golden rule, the rule of fellowship and the rule of Alligation, although the numbers that you work by, be not the numbers that you seek for, yet are they numbers in just proportions, and are found by orderly work: whereas in this rule, [falsehood] the numbers are not taken in any proportion, nor found by any orderly work, but taken at all adventure.

And therefore I sometimes being merry with my friends, and talking of such questions, have caused them that proponed such questions to call unto them such children or ideotes, as happened to be in the place, and to take their answer, declaring that I would make them solve those questions that seemed so doubtful.

And in deed I did anwer to the question and work the trial thereof also, by those answers which they happened at all adventure to make: which numbers seeing they be taken as manifest false, therefore is this the rule of False positions, and for briefness, the rule of falsehood, which rule for readiness of remembrance, I have comprised in these few werdes following in some form of an obscure riddle [not given]. The sense of these werdes and the sum of this rule is this........'.

The few words turn out to be rather many and are best appreciated by following the first example given, which concerns the payment of a mason for building a wall. A mason covenants to build a wall in 40 days. For every day that he worked he would receive 2s 1d. and for every day that he did not work, he would repay 2s.6d. At the end of the 40 days the mason received only 3s.[5s.]5d. the question is, how many days did he work and how many were not worked? The first guess made, called by Recorde the first position, is that he worked for 28 days and hence was at leisure for 12 days. This yields a payment to the mason of 28 × 25d. i.e. 700d. and a payment by him of 12 × 30d. or 360d. giving a final payment to the mason of 340d. This compares with the actual payment of 65d. and is thus in excess by 275d. This Recorde designates the first error which being too much he qualifies with a + sign. Had it been too little he says that he would have used a − sign. He then sets the first position 12 in a column vertically above the first error 275+. This procedure is repeated for a second position of 16 days play and 24 days work. The resultant error is 24 × 25d. or 600d. less 16 × 30 i.e. 480d giving a wage to the mason of 120d. compared with the actual payment of 65d. This gives a second positive error of 55d. The second position and the second error are arranged vertically above one another placed alongside the first assemblage and opposite corners of the resulting figure joined to form a St. Andrew's cross. The values at the opposite ends of the joins are multiplied to give 660 and 4,400 respectively. According to the rule, as both errors were too much and therefore alike, they had to be subtracted from one another giving 3,740, which Recorde calls the dividend. The errors are also subtracted to give 220, designated the divisor. Dividing then gives the quotient 17, which were the number of leisure days taken by the mason, so the answer to the question is that he worked 23 days and played 17 days. It is pointed out that substituting the numbers of days worked in the calculation for the number of days played would lead to the same answer, the important point being that both numbers were of one type, either working or playing but not to be mixed. Proof of the answer is made by direct calculation.

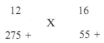

The Scholar then wishes to try his hand and does so using guesses that lead to errors of opposite nature. This allows demonstration of the rule that in such cases the errors have to be added for the purposes of the final calculation. The three following examples dealing with purchase of cloth, calculation of debt and tithes payable following losses of lambs all of which are pertinent to the time but not particularly politically sensitive. The next example trod on rather more sensitive ground as commented on by Easton.[16] 'Scholer. There is supposed a lawe made that (for furthering of tillage) everye man that dothe keepe sheepe, shall for everye 10 sheepe ear and sowe one acre of grounde: and for his allowance in sheepe pasture, there is appoincted for every 4 shepe 1 acre pasture: Nowe is there a riche sheepe-master whyche hathe 7,000 acres of grounde, and would gladly keepe as manye sheepe as he might by that statute, I demaund how many sheep shall he keep?'

Using the rule of falsehood the answer deduced is that he

> ....may keepe 20000 shepe: and therby I conjecture that many menne may keepe so many shepe for many men (as the common talke is) have so many acres of grounde.
>
> Master. That talke is not likelye, for so much ground is in compasse above $48^3/_4$ miles [only if as a very long narrow strip], but leave thys talke, and retourne to your questions, leaste your pointing be scarse well taken.
>
> S. In deed I do remember that the Egyptians did grudg so much against the shepherds, tyl ay length the[y] smarted for it, and yet they were but smal sheepmasters to some menn that be now, and the sheep are waxen so fierce now and so myghty, that none can withstand them but the Lyon.
>
> M. I perceave you talke as you hear somether: but to the work of your question …

However Recorde did not leave the matter rest and chose to rework the question using the Golden rule and the rule of Fellowship. Easton points out that the question of enclosure and the balance between sheep-land and land for tillage was a matter of major political concern at the time she believed this text to have been written – 1549. Protector Somerset had attempted to introduce some measures to control these issues, but whilst welcomed by many in the farming community, failure to enforce them resulted in the uprising in the West of England, brutally suppressed by Pembroke and Russell. Pembroke had a vested interest in this suppression as he was one landowner whose attempts at enclosure had been attacked by the revolt. At this juncture Somerset lost power and his measures went with him. Easton conjectures that subversive opinions of the sort implicit in this example together with other examples of Recorde's

---

[16] Easton JB, loc. cit., 528–529.

commonwealth sentiments might have been enough to delay the publication of this edition until 1552. As will be seen later, Recorde's standing in the eyes of authority in general and that of Pembroke in particular, only went downhill from this point in time. It seems more likely that the publication was delayed until Recorde's book on Geometry had been published, for he clearly felt an understanding of some of the topics therein was required before this edition was published. The final examples are found in other practical mathematics of the time, the last being that of Archimedes distinguishing between pure and alloyed golden crowns.

The Master concludes the book by advising the Scholar that this extension [of Edition A] should not be taken for the second part of the Arithmetic he had promised. This latter would deal with all Roots at one time but he does not mention any other components of the extension.

Chapter 8

The Pathway to Knowledg Containing the First Principles of Geometrie, as They May Moste Aptly Be Applied unto Practise, Both for Use of Instrumentes Geometricall, and Astronomicall and Also for the Projection of Plattes in Everye Kinde, and Therfore Much Necessary for All Sortes of Men

Abstract The Pathway to Knowledg presents the contents of the first four books of Euclid, as adapted to meet the needs for an understanding of the principles of Geometry by mathematical practitioners and also as an essential precursor to Recorde's mathematical treatises yet to come. Thus, in the first definitions, those of point and line he eschews the Euclidean versions as being fit only for theoretical speculation and calls a point 'that small print of pen, pencil or other instrument which is not moved nor drawn from his first touch, and therefore hath no notable length nor breadth'. In a similar vein, whilst not touching the content of Euclid's text he rearranges the order of its presentation in a way that he thinks will better engage the reader. Recorde adds nothing new to the subject, but his comments disclose the wide range of sources he consulted. A number of these have been identified and show how widely Continental scholarship in the form of printed books had spread to England. The large number of errors in the Pathway illustrates how great were the obstacles to be overcome in printing new mathematical works accurately in English.

This book is the second in the sequence of Recorde's mathematical texts. The need for an understanding of Geometry before dealing with matters relating to areas in the extended, second edition of The Grounde of Artes was presaged in its first edition, as already mentioned. Such a need was made evident as a precursor to the later text, The Castle of Knowledge, in that text where both Scholar and Master refer back to topics dealt with in the Pathway, the Master saying at one juncture, 'That Pathwaye wyll lead you rightlye to this woorke, if it be well travayled as it oughte to bee before you come to this woorke.' and later 'It [geometry] is a common interest to many arts, and infinite conclusions: and if you procede to farther knowledge of higher artes, without good exercise in it before, you do as a carpenter that goeth to woorke without his tooles.' The book was dated 'the xxviij daie of Januarie. M.D.LI., which translates into 28 January 1552 in our calendar. Recorde does however say that the first part of the text had been written 5 years past. He had intended it to be made up of four 'books', but it comprised only the first two because of

J. Williams, *Robert Recorde: Tudor Polymath, Expositor and Practitioner of Computation*, History of Computing, DOI 10.1007/978-0-85729-862-1_8, © Springer-Verlag London Limited 2011

misfortune. This presumably refers to Recorde's difficulties of early 1551, chronicled in Chap. 2.

The book differs from the *Grounde of Artes* in three respects. The title shows that it is aimed at application in specific areas of practice viz. the use of geometrical and astronomical instruments and for the preparation of 'plattes', which can be interpreted as an umbrella term for two-dimensional representations of a variety of subjects. Practitioners in these areas would have already been familiar with the instrumental tools of their trade and also be numerate to a limited extent. This concern with practical application does not diminish Recorde's support for the place of Geometry in the pantheon of the seven liberal arts, but this support is confined to polemic in the Preface. Unlike the topic of arithmetic where he had to generate his own text, in the case of geometry the text had to be that of Euclid in one form or another. These two considerations relating to audience and text made a dialogue form of presentation inappropriate.

Recorde's text is derived almost totally from the first four books of Euclid. What needs to be explored are the ways in which Recorde has varied the presentation of material from that of Euclid, where the material differs from that of Euclid and why he has made these changes. This exploration brings in its train the need to examine Recorde's sources and their possible influences upon his choices. Finally there is a requirement to discuss reasons for the relatively large number of errors in the book. The only published study of the Pathway is that of Easton, which avowedly does not deal with the topic of sources, but in other respects remains largely authoritative, needing only to be brought up to date in the light of more recent study.[1]

Book I

Definitions

Dealing first with the definitions, Recorde is clearly sensitive to the fact that he is producing the first English geometrical text, which has to be comprehensible as well as being rigorous. With respect to the latter consideration he shows that he has an understanding of the deeper issues, which are not necessarily relevant to the immediate pedagogical ends of his writings. Thus in the first definitions, those of point and line, he writes, 'A Poynt or a Prycke, is named of Geometricians that small and unsensible shape, which hath in it no partes, that is to say: nother length, breadth nor depth. but as this exactness of definition is more meeter for onlye Theoricke speculacion, then for practice and outwarde worke (considering that myne intente is to applye all these whole principles to woorke) I thynke meeter for this purpose, to call a poynt or prycke, that small printe of penne, pencyle, or other instrumente, which is not moved, nor drawen from his fyrst touche, and therefore hath no notable length

---

[1] Easton JB (1966) A Tudor Euclid. Scripta Math XXVII(4):339–355.

nor bredthe..... Now of a great numbre of these prickes, is made a Lyne, as you may perceive by this form ensuying. .... where as I have set a numbre of prickes, so if you with your pen will set in more prickes between every two of these, then wil it be a lyne, as here you may see and this lyne, is called of Geometricians, Lengthe without breadth........But as they in theyre theorikes (which ar only mind workes) do precisely understand these definitions, so it shal be sufficient for those men, whiche seke the use of the same thinges, as sense may duely judge them, and applye to handye workes if they understand them so to be true, that outward sense canne fynde none erroure therein.' By this avowal Recorde shows that he is fully aware of the distinction between continuous magnitude and discrete numbers. Whilst acknowledging theoretical exactitude of the former, for practical purposes he will work with the latter, hinting that it can be made as exact as practical ends demand. This theme he elaborated later in the augmented second edition of the *Grounde of Artes* when he refers to the infinity of fractions available, as already noted in Chap. 5. The topic recurs in the context of the extraction of square-roots in Chap. 11.

A straight line is defined as the shortest that may be drawn between two points i.e. the Archimedean definition rather than that of Euclid who describes it as 'a line which lies evenly with the points on itself', which is perhaps less easily interpreted practically. Recorde does not follow Proclus' account of Geminus' classification of non-straight lines exactly. Thus in dealing with 'Croked lynes', Geminus calls a broken line forming an angle a composite line, but for Recorde it constitutes two lines. He describes incomposite lines as crooked lines that 'boweth any waye such' rather than subdividing them further as Geminus does.

At this juncture Recorde found it convenient to deal with general case of the meeting of two lines. 'So that whenever suche meetyng of lines doth happen, the place of their metyng is called an Angle or corner'. He provides examples of acute (sharpe) and obtuse (blunt or brode) angles formed by the intersection of curvilinear lines. For the right angle he uses only straight lines, deliberately omitting other lines 'for their true judgement doth appertain to arte perspective ...'. He then returns to the Euclidean sequence to treat surfaces, using language less formal than that of Euclid. His treatment of the definitions of boundary and figure are also quite graphic and perhaps more extensive than found in other geometrical texts. Thus he includes figures made up of assemblages of points as well as those made of lines. His defini- tions of figures made up of lines are as given by Euclid, other than that for parallel lines. Whereas the latter defines them as being straight lines, lying in the same plane, which never meet, Recorde prefers the definition 'Paralleles, or gemowe lynes be suche lines as be drawen foorth still in one distaunce, and are no nerer in one place then in an other', which approach is to be found in Proclus' commentary. Recorde does not confine himself to straight lines but gives examples of parallel tortuous lines, circular lines, concentric circles, a twist line and a spiral line. This definition thus lies out of the sequence found in the Elements as well as being broader in concept. The definition is also of more practical appeal as appears later, when Recorde departs from the process used by Euclid to construct one line parallel to another, by setting off circular arc from the latter and then drawing the common 'touche lyne'[tangent]. He makes the interesting comment, 'here might I note the error of good Albrecht Durer, which affirmeth that that no perpendicular lines can be parallels. which errour doth spring partlie of oversight of the difference of a streight line, and partlie of mistayking certain principles geometrical, which I will pass until an other tyme, and wil not blame him, which hath deserved worthyly infinite praise.' This comment refers to sentences found on the first page of Durer's book Elementa Geometrica.[2] Significantly, Durer defines parallels in terms of equidistance and gives illustrations of tortuous and part-circular parallels, as did Recorde. The latter's definitions of circle, its centre, diameter and semicircle are as in the 'Elements' but they are augmented by those for segment and sector taken from Euclid's book III. He also illustrates egg and barrel forms.

Recorde's definition of the triangle is more general than that of Euclid, 'The nexte kynd of figures are those made of iij lynes other be allright lynes, all crooked lynes, other some right and some crooked .... they are named generally triangles, for a triangle is nothinge else to say, but a figure of three corners. And this is a generall rule, looke how many lynes any figure hath, so many corners it hath also, yf it bee a platte forme, and not a bodye. For a bodye hath dyvers lynes metyng sometim in one corner.' Consequent on this inclusion of both curvilinear as well as rectilinear figures, Recorde increases the number of illustrations given by Euclid, from 6 to 12. He provides Latin, Greek and his own English names for equilateral, isosceles and scalene triangles viz. aequilaterum, isopleuron and threlike; aequicurio, isosceles and tweylike; scalene, scalene and nowelike; respectively.

Recorde's definitions of quadrilaterals also give his English names for the five possible varieties. The square is named as such but the rectangle is called a long square. The term rhombus is not employed and the figure is described as a losenge or diamond; rhomboides are designated losenge- or diamondlike. He further classifies those shapes having all sides equal as 'likesides' and those with opposite pairs of sides equal as 'likejammis'. The fifth sort '...doth containe all other fashions of foure cornered figurs, and ar called of the Grekes trapezia, of Latin mensulae and of Arabitians, hemuariphe, they may be called in englisshe bordeformes, they have nosyde equall to an other as these examples shew, neither keepe they any rate in their corners, and therefore are counted unruled formes .....'.

He concludes his definitions with cinkangle for five-sided figures, sise angles for six-sided figures and septangle for a figure with seven angles. As an afterthought he describes the carpenters square as an example of a siseangle. He metions briefly a number of solid forms but defers their treatment to his book of Perspective.

None of Recorde's proposed English terms survived in neither late sixteenth century English texts, nor into the next century, by which time Geometry as a subject was receiving attention at the hands of the academic community, who could be expected to prefer Greek or Latin terms.

In her analysis, Eaton says 'Although the Pathway is primarily a translation of Euclid, no attempt has been made here to determine Recorde's sources. One difficulty is that if he followed his usual procedure he would not have used a single edition.' Identification of the sources of some of the names used to describe different

---

[2]Durer A, Elementa Geometrica. Christianum Wechelum, Basiliensis MDXXXII.

geometric figures has, perhaps unexpectedly provided some of the sources that Recorde might have had to hand in compiling his text.

Two names are of Latin origin, one of Greek and one of Arabic viz. aequicurio and mensulae; isopleuron; and helmuariphe respectively. The term aequicrurius, meaning of two equal legs, isosceles, is found both in earlier editions of the Oxford English Dictionary and also in Lewis and Short's Latin Dictionary. The latter gives only one source for the word, viz. Martianus Capella's *De nuptis Philologie & Mercurii*, written in the fifth century A.D. This latter is presented as an encyclopedia of the liberal culture of the time dealing with the seven liberal arts. The sixth book embraces Geometry and Geography. Whilst its treatment of geometry obviously owes a good deal to Euclid it is far from being an introduction to the subject. However it does name geometrical figures in both Latin and Greek more comprehensively than most subsequent texts. Of specific interest are the definitions '....;nam trigonus aut isopleuron, quod latine aeqilaterum dicitur, quod tribus paribus lineis lateribusque concurrit; aut isosceles, quod ex tribus lineis duas aequales habet, quibus quasi cruribus insistit, denique aeqicrurium vocitatur; aut scalene, quod omnes tres lineas inter se inaequales habet. ....'[3]. Whilst the term 'ysopleurus' is found in the same context in Version II of the 'Elements' attributed to Adelard of Bath, no other occurrences of 'aequicruria' have been found. There are therefore grounds for believing that Recorde used the Geometria of Martianus Capella as one of his sources. As is obvious from Willis' edition a large number of manuscripts of the work survived. It is also to be noted that it exists as one of the early printed books, 'Cura Fr. Vitalis Bodiani Vincentiae per Henricum de Sancto Urso [Rigo di Ca Zano] 1499'.[4]

Recorde gives mensula as the Latin form of trapezium. Only one occurrence of this equivalence has been found in Geometrical texts viz. that of Boethius. Single names either of Latin or Greek origin are given to the various geometrical figures, with the exception of the trapezium

. ...Praeter haec autem omnes quadrilaterae figurae trapeziae, id est mensula, nommanture. ...[5]

In his 'Preface unto the Theoremes', Recorde refers to 'Boetius that wittye clerke' so it seems safe to conclude that he had a version of Boethius to hand.

'Helmuariphe' which Recorde gives as the Arabic equivalent of trapezium is to be found in variously distorted renderings in a number, but not all, of translations of Euclid into Latin from the Arabic and in their redactions. Busard gives the Arabic version as 'al-munharifa'. In Adelard Versions I and II of the Elements it appears as 'elmunharifa'.[6] The version by Herman of Carinthia is 'almunharifa'.[7] It is not found

[3] Willis J (ed) (1983) Martianus Capella, De nuptis Philologie et Mercurii. Bibliotheca Scriptorum Graecorum et Romanorum, Teubneriana, pp 252–253 (712, 22–26).

[4] Sel.a.253, Main Library. The Queens College, Oxford.

[5] Folkerts M (1970) "Boethius" Geometrie II, Ein Mathematiches Lernbuch des Mittelalters. Franz Steiner Verlag GMBH, Weisbaden, p 116, lines 54–55.

[6] Busard HLL (1983) The first Latin translation of Euclid's elements commonly ascribed to Adelard of Bath. Pontifical Institute of Medieval Studies, Toronto, pp 392–396.

[7] Busard HLL (1967) The translation of the elements of Euclid from the Arabic into Latin by Herman of Carinthia. Janus LIV:1–140, 10.

in the version by Gerard of Cremona but appears as '…idem tamen nec rectis angulis nec equis lateribus continet: Preter has autem omnes quadrilatere figure helmuariphe nominantur….' in Campanus of Novara's edition of the 'Elements', first printed by Erhard Radolt at Venice in 1482.[8] This version is repeated in Pacioli's edition of 1509 edition[9] and also in the compound version of Campanus and Zamberti's translation of the Greek manuscripts of 1516.[10] Zamberti vilified the use of the Arabic terms in his original translation of 1505 and no subsequent translation of the Greek based Euclid used the Arabic terms for trapezium, rhombus and rhomboids. It appears therefore that Recorde had access to the works of Campanus in either manuscript or printed form and recognised elements of its Arabic origins.

Constructions

Recorde rearranges the order of Euclid's constructions, themselves distributed between his first four books although mainly in the last book so that they are all grouped together. It follows that the constructions preceed the theorems on which they are based. This was a deliberate move. In the preface to the Theorems Recorde writes, 'For I dare presuppose of them, that thing which I have sette in myselfe, and have marked in others, that is to saye, that it is not easie for a man that he shall travaile in a straunge arte, to understand at the beginninge both the thing that is taught and also the juste reason whie it is so. And by experience of teachinge I have tried it to bee true, for whenne I have taughte the proposition, as it imported in meaninge, and annexed the demonstration withall, I didde perceave that it was a greate trouble and a painefull vexacion of mynde to the learner, to comprehend both those thinges at ones. And therefore I did prove firste to make them to understande the sence of the propositions, and then afterward did they conceave the demonstrations much soner, when they hadde the sentence of the propositions first ingrafted in their mindes. This thinge caused me in bothe these bookes to omit the demonstrations, and to use only a plaine form of declaration, which might best serve for the firste introduction. Whiche example that beene used by other learned men before nowe, for not only Georgius Ioachamus Rheticus, but also Boetius that wittye clark did set forth some whole books of Euclide, without any demonstration or any other declaration at al. But & if I shall perceave that it maie be a thankefull travaile to sette foorth the propositions of geometrie with demonstrations, I will not refuse to dooe it, and that with sundry varietees of demonstrations, both pleasaunt and profitable also. And then will I in like manner prepare to sette foorth the other bookes, which now are left unprinted, by occasion not so muche of the charges in the cutting of the figures, as for other juste hynderances, whiche I truste hereafter shall bee remedied. In the

---

[8] Bodleian Library, Byw E.i.6., 1.
[9] Christ Church College Library (Oxford), Rare Books, phi.A.3.12.(2).
[10] Merton College Library, MER. 40. J.16.

mean season if any man muse why I have sette the conclusions before the Theoremes, seynge many of the Theoremes seeme to include the cause of some of the conclusions, and therefore oughte to have gone before them, as the cause goeth before the effecte. hereunto I saie, that altough the cause doo go beefore the effecte in order of nature, yet in order of teachyng the effect must be fyrst declared, and than the cause thereof shewed, for so shal men best understand things First to lerne that such thinges are to be wrought and secondly what thei ar, and what thei do import, and thirdly what is the cause thereof. An other cause why that the theoremes be put after the conclusions is this, whan I wrote these first cunclusions (which was .iiij. yeres passed) I thought not then to have added any theoremes, but to next unto the conclusions to have taught the order how to have applied them to work for drawing of plottes, and such like uses. But afterward considering the great commoditie that thei serve for, and the lighte that thei do geve to all sortes of practices geometricall, besyde other more notable benefites, whiche shall be declared more specially in a place convenient, I thoughte beste to geve you some taste of theym, and the pleasaunt contemplation of such geometrical propositions, which might serve diverselye in other bookes for the demonstrations and proofes of all Geometricall woorkes.'

Johnson and Larkey, comment that this idea of splitting Euclid's propositions into problems and theorems may have originated from passages in Proclus' commentary on the first four books of Euclid, printed in Grynaeus' Greek edition of Euclid in Basle 1533. Proclus, in turn, is quoting from Carpus[11] 'Carpus the engineer in his work on astronomy has revived the discussion about problems and theorems – whether opportunely or not may be ignored for the present. In any case he falls upon this distinction and says that problems are prior in rank to theorems because problems discover the subjects whose attribute are under investigation. And the enunciation of a problem, he says, is simple, requiring no additional technical knowledge at all; it only demands that something clearly possible be done, such as constructing an isosceles triangle or, given two straight lines, cutting off from the greater a length equal to the less. What is unclear or difficult about these? But the enunciation of a theorem, he says, is a laborious matter and needs much precision and scientific acumen if it is not to appear redundant or lacking in some element of truth, as is illustrated by the first of the theorems [which appears fourth in order in the Elements]. And for problems one common procedure, the method of analysis has been discovered, and by following it we can reach a solution; for thus it is that even the most obscure problems are pursued. But the handling of theorems is a difficult matter, and no one to this day he declares, has been able to teach a uniform way of approaching them. Consequently the ease with which a problem can be handled would make it the simpler form'. Proclus goes on to say

> Now problems do rightly come before theorems in order of presentation, especially for those who are coming to science from the arts concerned with sensible things; but in worth theorems are superior to problems.

---

[11] Morrow GL (1970) Proclus, a commentary on the first book of Euclid's elements. Princeton University Press, Princeton, pp 188–190.

The similarities between the arguments advanced by Recorde and Carpus are evident and led Johnson and Larkey to speculate that 'Both Recorde and Ramus doubtless derived the idea from certain passages in Proclus' commentary on Euclid, first printed in Grynaeus 'Greek edition of Euclid (Basle 1533).'[12] It is from the same source that that Recorde most likely derived his appreciation of the debate that had taken place over the theoretical meaning of a point, as discussed earlier. If indeed Recorde had access to Proclus' commentary there seems to have been only one document involved but two ways in which it might have been provided for him. Grynaeus says that the document he used was placed at his disposal by John Claymond of Oxford. Claymond had been at both Merton and Balliol colleges during Recorde's time at Oxford. Grynaeus addressed his preface to Cuthbert Tunstall, author of the Latin text on arithmetic 'De arte supputandi', who in 1533 was Bishop of Durham and to whom he presumably sent a copy. It may also be relevant that Recorde's publisher, Wolfe, had acted as a go-between for Cromwell and Grynaeus and was also an importer and seller of books. As will be seen in Chap. 9 dealing with The Castle of Knowledge, Recorde quoted in Greek from Euclids Book XII, which was only available at the time in Grynaeus' Greek text, so he must have had access somehow to this work.

Recorde gave Boethius and Rheticus as examples of authors who had taken the same presentational route as himself. Others were Reisch, specifically in the section of his Margarita Speculativa that deals with practical geometry. The same is also true of that portion of Pacioli's Summa Arithmetica dealing with practical geometry. However Pacioli's presentation of the Campanus' version of the 'Elements' follows the standard presentational sequence.

As pointed out by Easton, apart from the rearrangement of the order of the conclusions, Recorde's selection of constructions was strongly influenced also by practical considerations.[13] Thus three of Euclid's constructions using the Euclidean compass are omitted altogether, whilst some are simplified by the use of a fixed compass, a practical contemporary instrument. A few approximate constructions are added using other practical instruments such as plumb-line, carpenter's square and ruler. Some awkwardnesses arise from the modified sequence of presentation adopted by Recorde, again as documented by Easton. Thus conclusion V, the construction of a perpendicular at a point on a line is dealt with in three ways, the normal one (Euclid I,11) and two others of a strictly practical nature designed to cope with the situation where the point is near to the end of a line which cannot be extended. One method invokes the construction of a right angled triangle at the point using the Pythagorean Theorem to construct a 3:4:5 sided triangle, which precedes the presentation, as conclusion XII, of the more general method for construction of a triangle of three given sides. Easton's comment, 'Recorde's construction of a parallel to a given line (…; XI) is necessarily approximate since it is placed

[12] Johnson FR, Leakey SV (1935, April) Robert Recorde's Mathematical Teaching and the Anti-Aristotelian Movement. Huntingdon Libr Bull 7, footnote p 68.
[13] Easton JB, loc. cit. 344–345.

before the construction of an angle equal to a given angle (..;XIII)'. The construction involves the use of a compass set to the desired distance separating the two lines, the striking of two arcs from points in the given line and the drawing of a tangent common to the two arcs. This choice of methods is entirely consistent with the 'equidistance' definition of parallelism adopted by Recorde.

The emphasis on the use of practical tools in the constructions just discussed may throw some light on a potentially more serious 'aberration' in Recorde's presentation discussed by Easton. It relates to the fact that those conclusions associated with constructions involving the pentagon give no method for construction of a regular pentagon, knowledge of which is assumed to exist. Seemingly this sets up circular arguments. Thus to inscribe a regular pentagon in a circle (XXXVII.), Recorde merely instructs his reader to 'Fyrste make a circle, and devide the circumference of it into five equall partes...', the pentagon then being obtained by joining adjacent intercepts on the circumference. The two different sets of instructions for circumscribing a regular pentagon start from the same initial construct. A similar approach is adopted by both Reisch and Pacioli in their practical geometries. Easton's comment 'Whether Recorde himself did not know how to inscribe a pentagon, or whether, in view of the complexity of Euclid's construction, he considered approximate methods sufficient, is not clear.', needs to be explored further.

Dealing first with the second issue, another class of instruments in use at the time were those based on the divided circle. The backdrop against which this set of constructions has to be viewed has been described by Chapman in *Dividing the Circle*.[14] 'Renaissance circle dividers, it seemed were masters of a highly conservative craft, and capable of producing very uniform results amongst themselves. What also became clear from this study [his examination of the accuracy of division of a range of astrolabes] was the lack of improvement in accuracy over the 200-year range of instruments examined, as the astrolabe of 1450 was not significantly inferior to later pieces. One might interpret this as the growth of a conservative craft to meet the demand of a stable market of satisfied customers.' Clearly practical means were available to subdivide a circle with adequate accuracy and Recorde's instruction, to divide a circle into five parts was credibly based on existing practices used in the fabrication of astronomical instruments.

The regular pentagon also was a shape that had remained one of iconographic significance from the time of the Greek geometers onwards and had been used as such by masons and artists. Practical methods for constructing the shape of both given side or as inscribed in a given circle must have been available. Two such approximate methods are described by Heilbron in his discussion of the pentagon.[15] The formally accurate method was devised by Ptolemy and is found in the Almagest I 10, 'On the size of chords'. He computed the chords of 36° and 72° using a theorem from Euclid's Elements XIII.9, which showed that the sides of a regular pentagon, regular hexagon

---

[14]Chapman A (1995) Dividing the circle. Wiley, Chichester, p 155.
[15]Heilbron JL (1998) Geometry civilized: history, culture and technique. Clarendon Press, London, pp 221–228.

and regular decagon formed the sides of a right-angled triangle. Ptolemy showed how such a triangle could be constructed within a given circle, hence defining the lengths of chords needed to construct these regular figures within that circle. It provided a theoretically accurate means for dividing the circle into five equal parts. Did Recorde know of these constructs, Ptolemy's in particular? This latter was reproduced in Rheticus' extract of the geometry ultimately found in Copernicus' De Revolutionibus, which work Recorde has quoted. In his Geometria, Durer gives both the more accurate of the approximate methods for constructing a regular pentagon of given side and also the accurate method arising from Ptolemy's construct.[16] It has been noted earlier that Recorde has both quoted from this work and borrowed illustrations from it. It seems highly likely therefore that Recorde was aware of the Ptolemaic construction.

His decision to not include it seems most likely to have been a conscious one. Possibly it was because the theorem used by Ptolemy came from Euclid XIII.9, which was far in advance of any knowledge that Recorde was attempting to convey at this juncture. Also it is quite a complex construction. In application it requires some seven operations with ruler and compass and the probability of accumulation of errors is high. The method of construction of an inscribed pentagon using well established practices in dividing the circumference of a circle must have provided a much more appealing approach.[17]

Recorde then dealt with constructions based on the hexagon (siseangle), where he again follows the path of setting up the hexagon within a circle by subdivision of its circumference into six parts and joining the chords so created. This subdivision is of course straightforward, the lengths of the chords being the radius of the circle. Continuing the sequence given in the Elements, he concluded

'To make a figure of fifteene equall sides and angles in any circle appointed. This rule is generall, that how many sides the figure shall have, that shall be drawn in any circle, into so many partes justely muste the circle bee devided. And therefore it is the more easier woorke commonly to drawe a figure in a circle than to make a circle in another figure. Now therefore to end this conclusion, devide the circle firste into five partes, and then eche of them into thre partes againe: Or els first devide it into thre and then each of them into five other partes, as you list, and canne most readily. Then draw lines between every two prickes that be nighest togither, and ther will appear rightly drawen the figure, of fiftene sides, and angles equall. And so do with any other figure of what numbre of sides so ever it bee.'[18] If Recorde's

---

[16]Durer A, 'Geometria,' Wechelman, Basle MDXXXII, pp. 54–55.

[17]Thomas Digges edited his father Leonard's book on surveying, Pantometria [1571] and appended to it, his own text on the five Platonic geometrical solids. As a precursor to the main work he gives constructions for the inscribing of planar figures, including the pentagon, essentially repeating Ptolemy's method. The illustration of the construction is as badly misleading as any of those to found in the Pathway.

[18]In the *Castle of Knowledge* when describing the division of the equinoctial circle into degrees, the Scholar is instructed to divide it first into three parts, and then each part successively into a further three parts which are halved and then further into fifths. Alternatively, the 3,2,3,5 sequence of sectioning can be used. These two sequences are those advised in descriptions of instrument fabrication by other authors.

constructions for the inscription of regular planar figures into a circle were indeed based on the existence of established practical methods for the subdivision of a circle into segments appropriate for astronomical instruments, his final generalisation needs qualification. For a circle graduated in degrees, only regular polygons having numbers of sides that were sub-multiples of 360 could be generated.

The author believes that Recorde was familiar with geometrical methods for the inscription of certain regular polygons into a circle, but that his text was intended to instruct practical users of geometry in how to derive such results using established practical techniques with which they were familiar.

Book II

Postulates

Before he presents the Theorems, Recorde gives 'certaine grauntable requestes which serve for demonstrations Mathematicall'. The first five of the six postulates are as given by Euclid, but written in plain English. The sixth postulate 'Two right lines make no platte form' Heath classifies as post-Euclid and found in two manuscripts as a postulate whilst a third lists it as common notion 9.[19] Purists regard it as unnecessary, but Recorde uses it to distinguish between the properties of 'right' and 'crooked lines'. One of each sort may cross in two places and so define a space, as also may two pf the latter variety. Such distinctions may not be needed by a theoretical geometer, but would be relevant in practical applications.

Common Notions or Axioms

Nine of the ten 'Certayn common sentences manifest to sence, and acknowledged of all men', listed by Recorde are to be found in the MSS and editions, according to Heath.[20] The only minor deviation from the sequence given by Heath is found in that sequence of four axioms, similar to the first three which are accepted as being genuine Euclid. Three of these four were admitted by Hiedberg in his edition, but the fourth was not. In texts where the latter was found it was placed between the first and second in the sequence, whereas Recorde places it first. There would seem to be no reason for him doing so. However Recorde does add a tenth axiom 'Every whole thinge is equall to all his partes taken togither'. Easton notes that this was credited to Clavius by Heath. As she notes, Clavius' geometry post-dated Recorde who probably did not originate it, for he says in his commentary on it 'It shall be mete to

---

[19] Heath, vol I, 232.
[20] Ibid., 223.

express both with one example, for of thys last sentence many men at the first hearing do make a dout.'[21] It was clearly a matter of concern to him for he devotes as about half as much space to its explanation as to the previous nine axioms.

The 'dout' appears to be that the axiom is interpreted by some to mean that ' ...all the partes are of the same kind that the whole thyng is: but that meanyng is false...'. This contention he demonstrates by showing that a diagonal divides a square into two triangles. He then elaborates this example by dividing a rectangle by its diagonal and reassembling the resultant two triangles into five different shapes. This attention to a matter of practical concern mirrors that found in his inclusion and treatment of the tenth postulate. Before he moves on to theorems, he reaffirms '..the certenty of geometry and consequently of the other artes mathematical, which have the grounds (as arithmetike, music and astronomy) above all other artes and sciences, that be used amongst men.'

Theorems

Recorde introduces the theorems by saying '...now will I go on with the theoremes, which I do only by examples declare, minding to reserve the proofes to a peculiar boe which I will then set forth, when I perceave this to be thankfully taken of the readers of it.' There is no original material and all the 77 theorems are taken bodily from Euclid. The presentation is clear and the examples given are biassed in favour of potential practical utility. Easton draws attention to Recorde's free use of arithmetical and pseudo-algebraic approaches in his explanations.[22] Thus for the first seven axioms he resorts to numerical examples, incidentally using Hindu-Arabic numbers for the first time in the book. It may be argued that he presages his algebra in The Whetstone of Witte in his treatment of Theorem xlvi (Euclid II, 13) which reads 'In acute-angled triangles, the square on the side opposite an acute angle is less than the sum of the squares on the other two sides by twice the rectangle contained by one of the sides about the acute angle, namely that on which the perpendicular falls, and the straight line cut off within by the perpendicular towards the acute angle.' By mental transposition of terms Recorde makes a new theorem, saying '... so mighte I out of it, and the other that goeth nexte before, make as manny as woulde suffice for a whole booke.'

Errata

The Pathway to Knowledg provided Easton with a range of errors whose origins required explanations and for which blame might be apportionable.[23] The first class of error she isolated might be described as due to difficult, rather than ragged copy. Recorde's texts generally were introducing not only new words and numbers but new

---

[21] Easton JB, loc. cit., 348.

[22] Ibid., 349.

[23] Easton JB, loc. cit., 349–353.

types of illustrative material with which English printers would have been unfamiliar. This manifested itself in the Pathway in the form of many careless slips but also as a generic fault in which the drawings were presented as a mirror image of what was intended. These printers' errors would have made understanding of the theorems with which the errors were primarily associated quite difficult. Nonetheless, as Easton points out, readers did go to the trouble of inserting manuscript corrections into the text, which was not common practice. The xxv theorem was however omitted altogether.

The second class of errors primarily concern the constructions which Easton feels that have to blamed on Recorde himself, involving two sets of mismatches of references between constructions and the relevant theorems. They constitute errors of collation for which both author and a corrector could have been held responsible.

At the time of printing of the *Pathway*, January 1552, Recorde was still in Ireland, having left England the previous July. As described in Chap. 4, he was embroiled in some very difficult negotiations in Ireland and he may not have returned to England until after March 1552. He was therefore in no position to either modify the contents of the text or correct it. Why then was the decision taken to go ahead with publication. Easton's conjecture that the decision was motivated by personal considerations is credible. At the time Recorde's enemies, Pembroke and Northumberland were in the ascendancy and the former had managed to exact some financial revenge on Recorde for his obstinacy over matters relating to the Bristol Mint during Recorde's absence in Ireland. As Easton half suggests, the sentiments expressed in the contents of the Dedication to Edward were probably of potentially greater practical value to Recorde in obtaining political backing in his struggles than the mathematical contents of the book, a copy of which Sir John Cheke, the King's tutor, presented to the monarch. The Pathway was probably the least successful of Recorde's books, but not because of its errors. In the sequence of his books, the need for an understanding of planar geometry for practical purposes did not rise above the elementary level. The practical needs were for spherical geometry which topic Recorde did not address. The two subsequent editions of the work remained uncorrected.

Eighteen years later, Billingsley was to write in the Preface to his translation of Euclid:- 'Marvaile not (gentle reader) that faultes here following, have escaped in the correction of this booke. For, that the matter in it contayned is strange to our Printers here in England, not having bene accustomed to Print many, or rather any books contayning such matter, which causeth them to be unfurnished of a corrector skilfull in that art: I was forced, to my great travaile and paine, to correcte the whole booke my selfe. And in deede sometime for want of Argus eyes, and just consideration, notwithstanding my diligence in correcting, faultes escaped through me; sometimes also for lacke of diligence in the Printer to amend my corrections, faultes remayned uncorrected by his meanes. So that betwene us both these faultes have escaped uncorrected: which faultes yet, to say the trouth, for the most part are such, as a very young student without noting them unto him, mought easily of him selfe find and correcte. And in this I dare boldly affirme, that not many bookes, if any, concerning this art in other tounges, Greke. Latine, or Italian, are with so fewe faults of importance printed as this booke is. ....'. Recorde was unlikely to have been better blessed with a competent corrector than was Billingsley and he certainly would not have been able to do the correcting himself.

Chapter 9
The Castle of Knowledge

Abstract The book is a sequence of four treatises that approximate in terms of presentation, to that found in the *Pathway*. The first treatise defines the terms used to describe the parts of earth and sky and of the sphere on which they are to be represented. Great care is exercised in choosing between authorities on these topics. The second treatise details the construction of solid and armillary spheres. The manufacture of these instruments is intended not only to consolidate the instruction of the previous treatise, but also to circumvent the need for knowledge of spherical geometry and trigonometry. The third treatise deals essentially with mapping of the celestial spheres onto a terrestrial sphere and is then used to deduce physical consequences such as climate, calendar, time and length of day. The fourth treatise deals primarily but not exclusively with Cosmographical matters, the model employed being that of Ptolemy. Alternative models are demolished by demonstrating their inability to explain experiential data. A similarly critical approach is adopted in assessment of the views of writers on the Ptolemaic model. Recorde's calculations showed that he understood the implication of Copernicus' theory, but was not prepared to commit to it. The major part of the text is used to up-date and correct Sacrobosco's writings and displays many computations in the form of Tables.

This work, completed some time between the end of September and the end of November 1556, is described as 'Containing the explication of the sphere both celestiall and materiall, and diverse other thinges incident therto. With sundry plesaunt proofes and certain new demonstrations not written before in any vulgare woorkes.' The book is divided into four treatises.

'The first treatise is an introduction into the Sphere, declaring the necessary partes of it, as well for the materiall Sphere as for the celestiall: And that no partes of it are admitted without profitable use.

The second treatise doothe teache the making of the sphere, as well in sound and massive forme as also in Ringe forme, with hoopes: and the proportions of eche of them justly described.

J. Williams, *Robert Recorde: Tudor Polymath, Expositor and Practitioner*
of Computation, History of Computing, DOI 10.1007/978-0-85729-862-1_9,
© Springer-Verlag London Limited 2011

The thyrde treatise dooth briefly declare certaine thinges appertaininge to the use of
the Sphere, and other matters thereunto incidente: without proofe or demonstration: and that briefly, for easiness of learninge and remembringe.

The fourthe treatise doth approve many thinges, that were noted in other partes
before: and beside then addeth divers other maters, concerninge the necessarye
use of the sphere, which were not touched before, and doth bring demonstration
or other certaine proofe for the perswadinge of them: wherein are many Tables
set forth very pleasaunte and profitable.'

This presentational sequence follows that used in the *Pathway* presumably for
the reasons already adduced in that book, first the definitions, then the effects they
may describe and finally their causes.

The topics covered are to be found, described vaguely, in the list of contents of
works- in-hand given in the preface to the second book of *The Pathway to Knowleg*
as 'The use both of the Globe and the Sphere, and therein also of the arte of
Navigation, and what instruments serve beste ther unto, and of the trew latitude and
longitude of regions and townes.' What influenced Recorde to shift the balance of
presentation towards astronomy away from navigation is not clear. Perhaps he felt
that the needs of the navigators would be better served by a firm grounding in
astronomy than he had previously imagined.

It was in the period around 1553 that he was known to have been involved most
directly with the activities of merchant adventurers. A certain John Phillips claimed
in 1586 that he had set out on an exploration of the North-West passage with two
ships, 23 years earlier. He claimed that '...upon the encouragement of Mr. Chancellor
that first found for us the Musco, and Dr.Recordes conference in my house, and
specially the noble pilott Pintiago the portugale encouraged me.' He further declared
that a plot of the West India 'doth agree with the opinion of Doctor Recorde,
Mr. Bastian Cabotta, Harry Estrege his soone in lawe, Mr. Chancellor that founde
the Muscovia and noble Pintagio that was with Windam in Guinea.'[1] Pintiago died
on that Guinea voyage which left England in August 1553, so Recorde's involvement
must have predated that event.[2] It also demonstrates that he was to be found in the
company of explorers at that time. The geographical examples of latitude he uses,
are drawn from all parts of the known world of the time and must have come, first
or second-hand, from such venturers, as late as 1556.

The book is not the work on navigation that he intended to write, for in the dedication of his last book *The Whetstone of Witte* to the Muscovy Merchants, he promises
'....I will for your pleasure, to your comforte, and for your commoditie, shortly set
forthe a booke of Navigation, as I dare saie, shall partly satisfie and contente, not only
your expectation, but also the desire of a greate nomber beside.' The book is thus best
regarded as an introductory text in astronomy and cosmography for navigators.

Recorde follows the dialogue form of presentation again.

---

[1] British Library, Harleian MSS 167, fos.106–108

[2] Andrews KR (1984) Trade, plunder and settlement. Cambridge University Press, Cambridge/
New York, pp 106–107

The First Treatise

As was his wont, Recorde opens the book with a brief discussion between the Scholar and the Master on the benefits of knowledge in general and science in particular. The opening letter 'T' is illustrated by a man using a pair of dividers to measure distances on the Zodiacal circle of an armillary sphere against a rural background

The Scholar opens the debate by saying, 'Syth my last talk with you about the knowledge of the worlde and the partes of it, I have readd dyvers bookes that intreate of that matter, as namelye Proclus sphere, Ioannes de Sacrobosco, Orontius cosmographye, and divers others, whose woordes in manye thinges I remembre, but of the matter I have sondry doubtes, and therefore desire much your healpe therein. For although I have consulte with divers men therein..... but lyghte of understandynge, I have gotten lyttle yet.' The reference to a previous talk on the topic is somewhat mystifying as none of Recorde's extant books that predate this present one contain such a conversation. Might it have referred to the missing 'Gate'?

The Scholar then lists some of the confusions generated by the texts he had studied. This allowed the Master to excuse their presenters and to propose his own order. This latter is divided into two parts, the first of which deals with the cosmos and the second with the earth. To ensure that the Scholar has understood properly, the master has him summarise each part calling them the first and second repetitions. The presentations are conveniently discussed using the headings of these repetitions.

The fyrste repetition

'This is the summe of your doctrine hitherto.'

1. *That the worlde is that entiere body, which containeth in it all the heavens and the elements, with all that in them is.*

   Recorde spends a deal of space establishing this point that the 'world' is the cosmos as then known. The Scholar quotes Orontius in support of this contention whilst the Master cites Aristotle, Cleomedes and the Sybil to the same end. The reason that so much effort is expended on this point may be explained by the Master's contention '...: for this name Worlde, hath not the derivation of cleannes in englysh as the Latine and Greeke names have in their tongues: nother can I well tell whereof this englyshe name is derived, although I remember from other significations of this worde, as first it is used in Scripture for a name of long continuance of tyme when we say: Worlde without ende. and, thorough worlde of worldes: which signifieth for ever.' The first quote presumably came from the Book of Common Prayer reprinted in 1549.

He also spends some time in putting the earth in its proper place in the scheme of things

> … I mean the Earthe, which in comparison to the whole worlde is not onlye a parte without notable quantitye, but also leaste adourned with mervailous woorkes, and more subject to all frayle transmutation and chaunge, styll replenished with continuall corruption.

and further

> …whiche earthe in comparison to the whole worlde beareth no greater vewe than a mustarde corne on Malborne hylles, or a droppe of water in the Ocean sea. for of all the partes of the worlde, the earthe is the leaste…

2. *The partes of the worlde are two especial, the heavens which are eight in numbre, and the elements whiche are .iiij. in kinde.*

   The Scholar complains that in some other texts, nine or ten spheres may be mentioned compared with the eight proposed by the Master. The latter refuses to be deflected in his immediate purpose and say that he will deal with this diversity of opinion in the fourth treatise.

3. *The ordre and situation of all these partes, as well elementes as heavenly spheres, beginning at the highest, and proceding to the lowest, is this. the Firmament, Saturne, Jupiter, Mars, the Sonne, Venus, Mercury, and the Moone*
   THE FOURE ELEMENTES.
   Fyer, Ayer, Water, and Earthe. and ever the hygher incloseth all that is under it.

   The Master goes to some trouble to explain the difference between stars and planets. Some say that the difference is because the stars twinkle, but the Master avers, 'But this is their moste certaine difference, that all those starres, which be in the firmament, do stande and continue in one forme of distaunce eche from other, and chaunge not their places in their sphere, and therefore they be called Fixed starres.' Aratus is cited but not quoted in support of this point possibly because his view is more diffuse, 'The numerous stars, scattered in different directions sweep all across the sky every day continuously for ever.'[3] In contradistinction the Planets are similarly carried around the earth every day, but change their places in their own spheres.

   The peculiar status of the elements is also addressed. They lay under the sphere of the moon, first the fire, then the air, next the air and then the water 'whiche with the earth jointlye annexed, maketh as it were, one sphere only'.

4. *The worlde and all his principall partes are round in forme and shape, as a globe or ball.*

   The Scholar has some difficulty with the concept of celestial spheres, which difficulty, according to the Master, he shared with Joachim Ringelberg. Settling of the issue is deferred until later in the text.

---

[3] Kidd D (1994) Aratus: phaenomena. Cambridge University Press, Cambridge, lines 19–20

5. *The earthe is in the middle of the worlde, as the centre of it: & beareth no view of quantitye in comparison to the worlde.*

   The Master reiterated this point in several places, but also makes it clear that the earth is greater than Venus, Mercury, or the Moon.

6. *The earthe hathe no motion of it selfe, no more than a stone, but resteth quietly: and so the other elements do, except they be forcibly moved.*
   The Scholar initially found this idea difficult to accept, but was persuaded otherwise
   'Yet in the water and in the ayer we see everye day notable movynge. and sometimes I have heard of movynge of the earthe, by earthquakes: and as for the fyer that we see, it alwaies moveth and flickereth in burninge.
   Master. And so you have seene a stone move swiftelye, when it fell from anye hyghe place. but the motions have an ende quicklye, except they be continued with violence, as hereafter I will sufficientlye declare. But as the stone although it wyll movei n fallinge, yet in his place lyeth quiet without motion: so the earthe of itselfe, and the other elementes must be accompted quyete by nature, and without motion'. A body at rest will remain at rest!

7. *The heavens do move continually from the easte to the west, and that motion is called, The dayly motion: and is the measure of the Common day.*

8. *The Mone hath a severall motion from the west towards the easte, contrary to that movinge of the day course, and that motion is ye juste measure of a moneth, and every quarter dooth make a weeke.*

9. *The Son also hath a peculier motion from the west toward the easte, which he accomplisheth in a yeare, and of that course the yeare taketh his measure and quantitye.*
   With respect to these last two points, the Master says that they can both be seen daily but for the sun, is best observed over the course of a year, 'for sometimes he is hygher and nearer over our headdes, and sometime farther from our headdes, and lower in the southe: yea sometime he shineth with us almost 18 howers, (as in the middle of the Sommer) and in the middle of Winter hee shineth but 6 houres or lyttle more: this every childe dooth see, althoughe they know not the reason thereof.' The Scholar then suggests that the sun has to shine longer when the day is longest and shorter when the day is shortest, which allows the Master to deliver a homily on what is the difference between cause and effect.

The second repetition

Scholar. This it is as I remmebre,

1. *Fyrst you taught me what a sphere is, and howe it is made, also what is his Centre, his Axetree, his Diameter and his Poles, and what the Poles are named.*

This bald statement covers a very careful approach to the definitions of sphere and globe. The opening presentation is a quotation from the eleventh book of Euclid, given first in Greek then in Latin and finally in English, 'A Sphere is a round figure, made by the tournynge of half a circle, tyll it ende where it began to be moved, the diameter of that halfe circle continuyng steddye all the mean whyle.' Recorde calls this a description. Heath translates this as Definition 14. 'When, the diameter of a semicircle remaining fixed, the semicircle is carried round and restored again to the same position from which it began to be moved, the figure so comprehended is a sphere.'

The Scholar continues by saying that a sphere is thus a round solid body closed by a single surface which, in the *Pathway*, was called a Globe. With this the Master agrees and goes on to say that as the sphere has a centre so the globe also has a centre, from which all lines drawn to the surface of the globe are equal. The definition of Theodosius is then quoted. 'A sphere is a massye bodye, inclosed with one plat forme, and in the middle of it there is a pricke, from which all lynes drawn the sayde plat forme, are equal eche to other, and that pricke is the centre of the globe, and so sayeth Euclid also.' [Definition 16] This care in separating description from definition is significant. Heath's comments on Definition 14 are relevant. 'The scholiast observes that this definition is not properly a definition of a sphere but a description of the mode of generating it. ....' For the real definition the scholiast refers to Theodosius *Sphaerica*. Did Recorde know of these comments by the scholiast, or did he think out the differentiation by himself? Heiberg associates the scholiasts remarks with one particular manuscript, one to which Recorde seems unlikely to have had access.

The Master is equally careful to distinguish between Axetree and Diameter reverting to Euclid's definitions of these two terms, translating first into Latin and then into English. The Poles are defined as the two points about which the sphere or globe turns, which are the two ends of the axetree and do not move. The Scholar is disabused of the existence of a physical form of Axtree. Proclus is quoted in Greek and the Latin translation of Linacre is used to define North and South Poles. But the Master introduces his own description of the North Pole as being the one always seen from where he dwells and the South Pole as never being seen because it is below the *Horizonte*.

2. *Nexte you declared two circles, that is the Horizonte and the Meridiane circle, whiche (I perceave) stand styll, and tourne not with the worlde, but keepe their places.*

The Scholar attempts to define the Horizon in terms of the rising and setting of the sun, which the Master accepts as being not greatly amiss, being that of Hyginius also. However he prefers the definition used by Proclus as being the circle which parts the world we can see from that we cannot and thus divides the world into two equal parts, half thus being always above the ground and half always under the earth. This determines the rising and setting of sun moon and stars and leads to a consideration of noon-time and the equivalent time at night. Proclus is quoted in Greek, Linacre in Latin and, in English, 'The Meridian is a cyrcle drawen by the poles of the world, and the point right over our heads.

in which circle when the Son is, he maketh the middle of our day, & the middle of the nyghte.'

3. *Then did you describe five parallele circles, the Eqinoctiall, the twoo Tropikes: the Sommer tropike, and the winter tropike, and then the other two Paralleles, that is the Northe circle, and the Southe.*

4. *After that you shewed me what the Zodiac was, and the 12 Signes that be in him, and of their division.*

5. *And last of all, you described the twoo Colures, which divide the Zodiake into four equall and principall partes, according to the four times of the year.*

The definitions of these circles offer nothing out of the ordinary other than of a practical help in placing their positions. Thus for the ecliptical circle, at about the eleventh of March or the fourteenth of September, the sun rises precisely in the East and sets in the West. Similarly for the Tropic circles the upper limit of the Summer tropic is reached on the eleventh of June and that of the Winter tropic on the twelfth of December. The description of the Zodiac is of interest for the attention that the Master devotes to explaining that degrees, of which there are 360 in the Zodiac are not measures but represent parts. A degree is a thirtieth part of a sign and a sign is a twelfth part of a circle. The Zodiacal circle is divided into four equal parts, representing the four seasons by the two great circles, called the Colures, passing through the poles and the equinoctial and ecliptic points.

Throughout this presentation, Proclus is quoted as the authority, in Greek and also in Latin and English. At the back of Recorde's mind in reshaping and clarifying Proclus' presentation was the provision of the basis for manufacture of a 'materiall sphere, as may serve both to lerne by, and also to worke by, in practising the observations needefull to this arte.'

The Second Treatise

As promised in the general description, this Treatise quantifies the general description of the 'worlde' given in the first treatise. It does so by detailing the construction of both a solid astronomical sphere and an armillary sphere. The Master lists some 13 astronomical instruments available, but chooses the sphere for his instructional purposes because '...it is the grounde and beginner of all other instruments..... and the rather bycause it doth more aptly represent the form of heaven, than anye other instrument canne doo.'

The text is in essence a constructional manual for the two instruments mentioned and as such, is the first such document written in English on this subject. In terms of detail and also of illustration it is not matched either by publications of Continental authors, which is somewhat surprising given the relatively advanced state of scholarship to be found on the mainland of Europe at the time. Little such detail might not perhaps be expected in more general texts such as the *Cosmographia* of Apian even in its form as edited by Gemma Frisius. But detail is also absent from the compilation *Annuli Astronomici*, published in 1558 and containing contributions dated to about 1536 from Beausardo, Gemma Frisius, Johan Dryanda, Boneta Hebraeo, Orontio Fineo as well as an item dealing with some of the instruments originating with Regimontanus.

The Master starts with the construction of a solid sphere, first made by turning on centres and then refined by the use of a sharp edged former or shaping tool manufactured according to the description given by Euclid for the generation of a sphere i.e. his definition 14 as discussed in the first Treatise. Care is given to the alignment of the centres, but then the Master opts out saying 'What more is to be doone, I leave it too the studiouse devyse of your awne practise for such thynges are better taught by hande, then by mouthe.'

The centres are defined as the poles of the sphere and then used as fiduciary points to set up the equinoctial circle using a form of bowed compass, which is illustrated. One foot of this compass, whose legs have been set to approximately half the circumference of the globe, is placed first in one pole and a light mark is scribed on the globe. Another line is then scribed using the same compass setting, but with the other pole as centre. If the two lines just touch the centre has been found; if they overlap or fall short, a point halfway between the marks is chosen to define the corrected radius and the scribing process repeated. When the correct centre has been found, the compass is set to the distance between pole and centre and a circle lightly scribed around the sphere. This process is repeated with the other pole as centre. If the two circles coincide that is the correct equinoctial, if not the diameter used has to be corrected. When the correct radius has been found, this setting of the compass is used to divide that equinoctial circle into four equal parts by stepping. At this point the Student acknowledges the need for attention to the sharpness of the compass points and to due care during the scribing operation.

Having divided the equinoctial into four parts the Scholar is then told to divide it successively into three parts, a further three parts, then two and finally five parts. The Scholar recognises these subdivisions as degrees. No indication is given of the methods for subdivision. Recorde has this undisclosed approach in common with all contemporary writers on division of the circle. The only point of variation in practice is whether the sequence for division of the quadrant advocated is 3,3,2,5 or 3,2,3,5. Recorde uses both.

The backdrop against which this portion of the Treatise should be viewed has been described by Chapman in *Dividing the Circle.*[4] 'Renaissance circle dividers, it seemed were masters of a highly conservative craft, and capable of producing very uniform results amongst themselves. What also became clear from this study [his examination of the accuracy of division of a range of astrolabes] was the lack of improvement in accuracy over the 200-year range of instruments examined, as the astrolabe of 1450 was not significantly inferior to later pieces. One might interpret this as the growth of a conservative craft to meet the demand of a stable market of satisfied customers.' The only light which Recorde's text has thrown on the topic so

---

[4] Chapman A (1995) Dividing the circle. Wiley, Chichester, p 155

far is the description of the practical method of successive approximation using the compass. No hint is given of the means whereby the first approximation for thirds and fifths are arrived at.

The tropical circles are constructed next by setting the radius of the compass to 66 and a half degrees, as measured on the equinoctial circle, and scribing two circles using the two poles as centres. The tropical or substantial colure is next drawn through the poles and the 270 degree point on the equinoctial circle. At this point, the Master gives a general rule about dealing with errors, '…and if it misse [the 270 point] anye whitte, examine it well, and amende the faulte, before you woorke anye farther, which rule you shall observe styll, for els of one fault neglected, may other may ensue.' One is left with the impression that Recorde must have tested out his own recipes and made his own mistakes. The point about accumulation of errors is well made and seemingly unique in the literature of the time. The equinoctial colure is then drawn through the poles and the 0 and 180 degree points of the equinoctial circle. The admonition to correct any uncovered error immediately is repeated. These last two sets of circles are graduated from 1 to 90, rather that 0–360 for the equinoctial. The accuracy of the construction is again cross-checked; the two tropics should cut the colures at the 23 and a half degree point.

If this criterion is met, then the two polar circles can be drawn through the 66 and a half degree points on the colures. This then allows the Zodiac circle to be constructed. With the compass opened to a quarter of the equinoctial circle, a circle centered on the intersection of the North Polar circle and the tropic colure is drawn. This circle should touch the other tropic coloure and intersect the equinoctial circle at 0 and 180 degrees. To prove the accuracy of the circle the same construction is carried out from the other hemisphere. If the two circles do not coincide 'there is some errour which must be amended'. When the correct circle has been found, two circles parallel to it and separated from it by six degrees on either side. The whole band then represents the full Zodiac in length and breadth.

Subdivision of the surface that delineates the Zodiac causes problems to the Student. It is already divided into four quarters by the two Colures. Further division of the quarters into a further three parts, each of which constitutes a part of the Zodiac whilst straightforward for a line, but for a surface 'Those are not easye to drawe, but errour may quickly be committed, in making them wider in one place than another.' This is effectively acknowledging that the lines of the subdivisions are segments of a great circle of the sphere. The Master deals with the error by setting the foot of the compass, which has been opened to the dimension of a quarter of the Zodiac, in each division in turn and scribing a circle across the full breadth of the Zodiac's surface. By this means the placing of each of the divisions is examined and can be corrected if in error. When the signs have been drawn, each of them is subdivided further into two, three and five parts so that each part may be graduated into 30 degrees. Note that the sequence of subdivisions recommended is slightly different from the previously given, possibly leading to lesser inaccuracy.

After distinguishing between the polar circles and the Arctic and Antarctic circles, the Master deals with the subdivision of the Zodiac into 12 equal parts, one for each sign, but firstly into the four quarters of the year. Then instructions are given for the fabrication and subdivision of the annular plate that forms the *Horizont*. The process of subdivision into degrees is via quadration, into three parts and then into 30. A second class of subvision is described to provide the points of the compass as per the 'shipmans compass'. Finally the mount for the Horizonte is detailed, full instructions to ensure the proper suspension of the globe being given.

Attention is next given to the construction of Armillary or Ring Sphere, which the Master sees as possibly more suitable for elementary teaching purposes. Four hoops of one size are made, with one being three times broader than the others. This latter is to be the Zodiac circle, the others are to be the equinoctial circle and the two Colures. All are divided into 360 degrees as before, with the Zodiac suitably annotated with its signs. They are then assembled and joined in the exact form they occupied on the solid globe. The two equally sized tropic circles are next attached at 23 and a half degrees from the equinoctial, followed by the Arctic and Antarctic circles at 52 degrees and the polar circles at 66 and a half degrees of separation. Finally the Zodiac circle is attached such that its middle line touches the middle of each tropic, its edges being six degrees equidistant from both points of contact. Two small holes are made through the crossing points of the Colures to accomodate a polar axis.

The Scholar professes his understanding of the instructions but doubts his ability to execute them, '....I doubte somewhat of the quantity of the parallel circles. for although I know by triall I may att lengthe make them meete, yet woulde I glalye knowe their measure beforehand, if I myght, for so shall I be sure to woorke most certenly.

Master. Your desire is good. and all be it that the writers of the Sphere have omitted it, as they have doone many thinges els, yet will I give you a rate of proportion drawen out of the tables of Cordes and Arkes, called commonly in latine Tabulae Sinuum.' He says that the breadth of the circles can be determined by practical considerations such as their capability to support stress and may thus vary, the exception being that the Zodiac cicle has to be 12 degrees in breadth. Patterns graduated in degrees are then provided for all the sizes of circles required to construct an armillary sphere.

As well as giving the patterns, the Master gives the proportions of the circles relative to the equinoctial circle both in fractions and degrees. The tropic circles are 330 parts of 360 degrees, so they are in proportion 11–12. The Arctic and Antarctic circles are equal to 222 degrees, hence proportionally as 37–60. The polar circles are 144 degree of the equinoctial and thus in proportion 2–5. Which Tables of chords and sines were used by Recorde is not obvious. Those giving arcs in sexagesimal notation would be easily converted into degrees and thence to fractions. A little more calculation would be needed where arcs were given as parts per 100,000. Whether the Tables of Ptolemy, Copernicus, Fale or modern values are invoked, Recorde's values are approximate, but in error by no more than half a degree in the whole circumference. The fractional values derived make scaling of the subdivided linear representation of the largest circumference down to the smaller circles easy, so perhaps this was the criterion used in setting the approximation. The Scholar sees how increasing the size of the instrument might be easily accomplished. The Master concludes 'Here needeth no repetition, bycause all standeth in woorkynge of the former lessons before repeated.'

The most pertinent, near contemporary, comment on this second treatise is that of Fale in the introduction to his *Horologiographia; The Art of Dialling* published in 1593. 'Many have promised (but none yet performed) to write of the Science in our English tongue, which hath bin published in other languages, as *D Recorde.* long since, *M. Digges, M. Blagrave* with other, who if they would take the paines, I knowe could doe it with great commendation.' Fale adds nothing to knowledge of the methods used for the subdivision of the circle. He just quoted the sequence of division of a quadrant viz. by 3, 3, 2 and 5, the second of the sequences used by Recorde. Fale's familiarity with the use of sine tables for other calculations surely means that he was content with the accuracies of subdivision achieved by the methods already well practised by other instrument makers. Chapman's analysis of the accuracies of a range of astrolabes shows that such practices extended back at least to 1,450. They must have involved successive approximations of the sort indicated by Recorde. By insisting on the need for correction of errors for each step of the process, as determined by trial and error, before moving on to the next stage, Recorde was possibly doing no more than recording his observations of actual practice and added little original to the process, but one cannot be certain that this was his only contribution. As demonstrated by his treatment of the dimensions of the various rings of the armillary, Recorde shows his understanding and ability to handle calculations using arcs and chords, which also suggests too that he was quite content with the efforts of the instrument maker's abilities in dividing the circle by empirical methods, properly practised.

Perhaps the last word on this second treatise, unique in English at the time of publication, should rest with Fale. 'Onely thus much I advertise the unlearned, thet they must acquaint themselves with som few Mathematicall principles, as to know what the Elevation of the Pole meaneth, hoe a squire line is to be drawne, and such like, which (if they want a teacher) they may sufficiently learne by themselves out of *Records Castle*, his pathway and ground of Arts, published in the English tongue: for these tearmes could not be avoided, neither plainly described without much tediousnes.'

The Third Treatise

According to Recorde this is the treatise 'Wherein is briefly taught the use of the Sphere, for certaine conclusions of daily appearances and other lyke matters.' Essentially the celestial spheres are mapped onto a terrestrial sphere which is used to deduce certain physical consequences such as climate, calendar, time and length of day.

As in the previous two Treatises, the Scholar provides a 'repetition' of its contents.

1. *Firste the distinction of the Plages of the worlde, accordingly as they be sette forthe in the Horizont of the Sphere.*

2. *Then the Paralleles on earthe, agreable to the Paralleles in the skye, of like names and distances proportionable.*

3. *Thirdly the distinction of the. v. Zones, by their qualities and limites, and of their inhabitantes.*

   There are four principal plages[segments] to the world [all the heavens], east West, Northand South and also other subdivisions of the Horizonte viz.the parallels. These latter map onto the earth. The equinoctial divides the earth into two equal parts, North and South. The start of the tropic of Cancer is the point where the sun is in his highest point towards the north. The Tropic of Capricorn is the equivalent point south of the equinoctial. The Master prefers not to follow the practice of the Ancients in mapping the Arctic and Antarctic circles onto the earth, for reasons to be given later, but accepts the concept of polar circles. These various circles divide the horizonte into five zones, the Zona Torrida between the two tropics, the Zona Temperata's north and south between the tropics and the polar circles and the Zona Frigida's lying within these polar circles. Turning to the inhabitants of these Zones, the Master, clearly influenced by the results of exploration of the earth, rejects the views of classical scholars such as Ovid and Vergil that the 'Burning Zone' was uninhabitable, calling in evidence the evidence from the Portugese's drug and spice trade with the Molucca's and Sumatra which were virtually under the Equinoctial as well as the fact that Calecut was only five degrees north of that line. The first two latitudes are nearly correct and the latter is adrift by about five degrees.

4. *The diversities of Spheres according to their diverse inclinations, but twoo are the generall distinctions, that is a Ryght Sphere, and a Bowing Sphere.*

5. *Fyrstly, you gave me a brefe ordre to take the height of the Pole, or any other Starre or Planete.*

6. *Then followed the divers alterations of the Horizonte, as wel between East and west, as between Southe and Northe.*

   The Right Sphere is defined as that obtaining when the equinoctial cuts the Horizont at right angles. It is thus appropriate only to the zone lying directly under the equinoctial. From this region both [celestial] poles are visible and the sun is overhead. This is demonstrated by setting the solid and armillary spheres so the equinoctial is 'full upright'. For all other zones one or other pole must fall below their Horizonte and the equinoctial decline from the overhead position in the direction of the hidden pole. The position of the model sphere has to be adjusted to mirror this declination and is therefore called a 'Bowing, Leaning, Oblique or Declining Sphere' according to taste.

   The Master takes this opportunity to preach the value of having an Instrument available to demonstrate the points being made 'For that cause did I teach you the making of it [the sphere] before I instructed you in the use of it, knowing how greate a helpe the sighte of the eye doth minister to the righte and speedye understanding of thet, which the ear doth heare.' The correct declination for any place is obtained by measuring the height of the Pole at that point. The Scholar is instructed in making such a measurement using a '..Quadrante (whose composition I have taught amongst other instruments in the Gate of knowledge, but this which you see here, is the form of the moste playnest sorte)'.. The resultant measurement '..may you call Latitude of that region or the height of the Pole ...' As the elevation of the Pole varies so much between countries so does the position of the sphere. Thus in England alone the latitude, in degrees, varies from about 51 at Southampton, 54 at York, 57 at Edinborough to 62 in the Orkneys.

This illustration causes the Scholar to ponder a previous statement of the Master that the Horizonte did not move as the circles in heaven. This is answered by pointing out that the Horizonte moves only if the observer moves, and that each place thus has its own Horizonte. As the latter changes with movement to North and South, so also it does with movements East and West.

7. *Seventhly there was declared the causes of the diversity of the daies, fyrst in divers regions, and then in one region.*

8. *The difference between a Naturalle daye, and an Artificiall daye.*

9. *The quantity of the longeste daye in certen partes of the worlde, and namely under the Poles of the worlde.*
    The timing of sunrise in London and Calecut is used to demonstrate the diversity of days. Calecut is judged to be about a quarter of the earth's diameter east of London and as the Sun moves around the Earth in 24 hours, they must see the sun 6 hours earlier than in England. For similar reasons the Moluccas would see the Sun earlier still, whilst Jerusalem and Constantinople would have sunrise later than Calecut. If the sunrises differ then so also must their noontides.

Consideration then passes to the variation of days within one region. Demonstrations using the Sphere are again brought into play. The point is made once more that the length of day is determined by the length of time the sun shines and not vice versa, so the question to be answered is why the length of sunshine varies. The Sphere is first set with the poles touching the Horizonte. The latter cuts both the equinoctial and the tropics in into two equal parts. Thus the lengths of days and nights will be equal when the sun lies on any of these circles, but only for points on the Equator. Moving from a Right Sphere to a tilted Sphere alters the situation. The Scholar tilts the Sphere to an elevation of 52 degrees, as for Cambridge. Now only the equinoctial is divided equally by the Horizonte, the Tropic of Cancer having about three-quarters of the whole above the Horizonte. Thus when the sun is in the summer tropic he spends some three-quarters of the Natural day above ground giving about 18 hours of daylight. This also provides an opportunity to explain that a Naturall day has 24 hours but an Artificial day is that defined by sunrise and sunset. The pole is then set even higher to 71 degrees as for Wardhouse, the trading warehouse of the Muscovy

Company near the North Cape of Norway. Now the division of the Tropic of Cancer is even more skewed, to the extent that the sun does not drop below the Horizonte fronm 7 May to 19 July. The Sphere is then set so the both poles are vertically above the Horizonte, for the half year that the sun is in the Tropic of Cancer there is continual day, whilst for the rest of the year there is continual night. The value of the Sphere as a teaching aid is emphasised again.

10. *How by this excellente Arte a man maye measure all the compasse of the earthe, and yet abide styll in one cuntrey.*

11. *A distinction of sondrye inhabitantes, according to the diversities of their shaddowes, which are three principallye.*

12. *Then lastlye folowed an other distinction of inhabitantes, accordinge to the agreeablenes and diversities of tymes of the yeare, and the quarters of the daye, and these you named by three severall names also, which are the names of comparison, bicause they take not those names, but in comparison to other nations.*

In treating these matters the intention of the Master is to demonstrate how, given the appropriate data, the seemingly impossible claims made can be substantiated using a theoretical approach.

It is particularly important to bear this intention in mind in assessing the outcome of the prediction of the circumference of the earth based on measurements made only in England. The student is instructed to measure the latitudes of Southampton, Newcastle, Edinborough and Catnesse point [John o'Groats], measure the distances between them and to calculate the number of miles per degree. It transpires that for the cases of individual segments and for the whole distance, one degree of latitude corresponds to a distance of 60 miles. Using this value, the scholar is led to compute the circumference of the earth as $360 \times 60 = 21,600$ miles. The data used to obtain the mile equivalent of a degree is at least partially contrived so as to give the desired answer. The Scholar presages this objection saying 'This I can do with diligence [assemble the data], although it be as harde to marke the myles truly (the reports of them being so divers) as it is to work truly with the Quadrante, but diligence will avoid error in them bothe.' The latitudes for Southampton and Newcastle given are reasonably accurate, that for Edinborough is too great by one degree, but agree with those found on a pocket instrument of the same period.[5] That for Catness point is too great by three and one-half degrees, but is not too greatly different from that derived from maps of the period. How the distances were derived is not evident, nor whether they were statute miles. Preferential selection of mileage data to illustrate the point would seem not to have been a problem and this was undoubtedly done for the segment from Edinburgh to Catnesse point.

It has to be remembered that at this time, the equivalence of a sea mile to a minute of arc at the earth' surface was accepted by English mariners. Leonard Digges gives the circumference of the earth as 21,600 miles in his *Prognosticon* of 1555. There was no need for Recorde to present a derivation based on local

---

[5] Gunter RT, Early science at Oxford, vol II, pp 279–280

measurements of dubious validity, unless he was making a more general point. This latter would seem to be that given the earth to be spherical, measurements of latitude could provide estimates of distances between points on the earth's surface, or given distances, how latitudes could be estimated, with some reservations. That he was aware of possible pitfalls is clear. When the method is applied to the distances between points in the Northern and Southern hemispheres such as England, The Cape of Good Hope and the Straits of Magellan, latitudes derived from distances are additive, whereas within a Hemisphere such latitudes would have to be subtracted. A more subtle point is made when the Scholar calculates the latitude of Wardhouse as 72 degrees based on it being 1,200 miles from Cambridge. This is greater than the 71 degrees quoted by the Master earlier in the treatise, which concerns the Scholar. The master gives the reason for the discrepancy

'....: that rate of 60 myles to eche degree doth serve in going precisely from southe to northe, but nother is Wardhouse northe from us, but somewhat towarde the easte. nother yet in the other two examples any of both places was directlye south from us, for the Forelande of Affrike beareth toward the easte and the Streight of Magellanus bendeth towarde the weste, yet for this tyme it may serve as well for our purpose, as if it were more precisely doone'. Never one to pass up an opportunity to expound his views on teaching, Recorde continues,

'Scholar. Yet I think in teaching there shoulde bee used nothing but certaine truthe.

Master. What so ever is taught to be retained for a truth, oughte to be a very certaine truth in deede: and they do not well that in suche manner doo teache fyrste untruthes for truthes. but where induction is made by examples, it is often tymes more or at the leaste, no lesse expediente to use examples not exactly true, then to take only precyse true examples, for thereby it appeareth the proofe to be of greater force, if it will procede in an example whiche is not precisely true. And in these examples we have so large a scope of triall, that we need not sticke for two or three degrees....'.

Recorde was wise to adopt the approach to precision given above. The question of the relative values of sea-miles and statutory land-miles was not adequately resolved for another 70 years. The account of its resolution is to found in *The Seaman's Practice* of 1637 written by Richard Norwood, who had, by his own account, had taught himself arithmetic using Recorde's *Grounde of Artes*. As the latter has more to say on the subject in the Fourth Treatise, it will suffice for the time being to note that Recorde was aware of some of the problems associated with this matter.

11. *A distinction of sondrye inhabitantes, according to diversities of their shaddowes, whiche are three principallye.*

A good flavour of the presentation is given by the first example. Venice and the Cape of Good Hope were believed to lie on approximately the same meridian and consequently they would have their noon-times at the same hour. However the shadow at Venice would lie to the north of the object casting it, but at the Cape, the shadow would lie to the South. At places such as Guinea which lay directly under the equinoctial, their shadows would sometimes be to the north and sometimes to the south depending on the time of year. It followed that at sometime their shadow must be at their feet. The third type of behaviour was found within the polar circles, where the shadows circled about their feet with a frequency that varied from once a day where the day was of 24 hours to 187 times when the day was of 6 months duration. The point was made that these polar regions also received indirect sunlight as moonlight for 50 days after they had lost the direct sunlight and for 50 days before it rose again.

The Greeks gave names to the inhabitants of these three regions based on the behaviour of their shadows, but the Master was very diffident about giving English equivalents.

12. *Then lastly followed an other distinction of inhabitantes, according to the agreeablenes of and diversities of the tymes of the yeare, and the quarters of the daye, and these you named by three severall names also, which are names of comparison, bicause they take not those names, but in comparison to other nations.*

The subdivision of the earth into 48 climates along lines of latitude between the two poles is described as defining zones wherein the longest day differs by half an hour from that of its neighbours. Every climate is divided into four quarters of the day. The variation of the seasons for places lying within climates equally spaced on either side of the equinoctial are discussed. Possibly of greatest interest is the specification of the antipodeans as persons who live under the same meridian, but are displaced by half of the compass of that circle. Two other types of dwellers are defined and their names given in Greek. The general point is made that the terms used are relative and not absolute.

The Fourth Treatise

This deals primarily but not exclusively with cosmographical matters. It is the part of *The Castle of Knowledge* that has received most attention because of its reference to Copernicus. The basic theory taught is that of Ptolemy, but not uncritically so. Thus the Scholar says that he has found the best writers to be Proclus, Sacrobosco and Orontius. The Master agrees with his choice, but says that he has to read more widely, but selectively, amongst the plethora of writers on the subject for '..I have found great tedious payne in readinge a greate multitude of them' He recommends Cleomedes, but to be read in the original Greek as the Latin translation is much corrupted, Euclid's Phaenomena and particularly Stoffler's commentaries on Proclus' Sphere. He mentions that a number of English men have written well, but were not available in print viz. Grosseteste, Michael Scotus, Batecombe, Baconthorpe and others. Pliny, Hyginius and Aratus and many others are only for those who have sufficient knowledge to separate the corn from the chaff. Ptolemy is only for those who have understood the principles of the Sphere, are well versed in Euclid and exercised in the 'Theorykes of the Planetes.'

The Scholar then presents a table of 25 topics that he wishes to have explained. These include unfinished discussions from the preceeding Treatises as well as matters arising from the Scholar's readings in Proclus, Sacrobosco and Orontius. True to form, the Master wishes to alter the sequence in which he deals with the matters as well as the depth in which he will deal with them. He imposes what he considers to be the best order for teaching the subjects and therefore alters the order of discussion used by Sacrobosco and starts by discussing the roundness of the world. A list of the topic headings used by Recorde is given at the end of this treatise and will be used for purposes of commentary

1. *What occasions moved men to judge the forme of the worlde to be rounde, and namely three principall reasons thereof.*

    The reasons given are those deployed by Ptolemy in his Almagest and the wording used by the Master follows closely that of Ptolemy.[6] The first 'occasion' derived from the regular movement from East to West of the sun, moon and certain stars, their disappearance in the West to be followed after a time interval by their reappearance in the East for the cycle to be repeated. The second reason is the behaviour of some stars which never fall below the horizon but moved about one point in the sky, the size of their circle of their revolution depending on their distance from the centre. Finally certain stars, which appeared close to some that never set, did dip below the horizon, their period of disappearance being proportional to their distance from a centre. '... whereby they were inforced to thinke, that these varieties and formes of movynge could be in none other then in a rounde forme, and that the same movynge was circular and rounde, as it did manifestly appeare in the north part of the skye, where the starres continually move aboute one pointe, and do never set under the Horizonte. And that point about which they noted this motion to bee, they called (as reason inforced them) the Pole of the worlde.'

2. *That the heavens are rounde in forme contrarye to the errour of Lactantius Firmianus, whiche thought it to be flatte, and his opinion confuted by divers reasons, namely by view of the starres, by aptenes of movynge, by reson of capacytie, and avoiding of emptines.*

3. *That the Firmament doth move, thoughe Lactantius thought the contrarye: and how it may be proved, especially by the Milkye waye. And that the starres do not moove as byrdes in the ayer, or as fyshes in the water.*

    Recorde's interpretations of Lactantius' stance differs from that of other commentators, notwithstanding that they all refer to Chap. 24 of the third book of the *Divine Institutions*. That Recorde was commentating on this passage cannot be doubted for the numbers of chapter and book are inverted in the text but corrected at the end of the '*Castle*'.

    The whole of Lactantius' third book is an attack on the ancient Greek philosophers. Chap. 24 is entitled 'Of the Antipodes, the Heavens, and the Stars.'

---

[6] Toomer GJ (1984) Ptolemy's Almagest. Duckworth, London, pp 38–39

The first paragraph is a polemic against the very concept of the antipodes. 'Or is there anyone so senseless as to believe that there are men whose footsteps are higher than their heads? or the things that are with us in a recumbent position, with them hang in an inverted direction ....'. The second paragraph explores the course of argument that led to the idea of the antipodes. This is attributed in the first place to the development of the concept that the world was round, based on the arguments interpreted as 'But since they did not perceive what contrivance regulated their [sun and moon] courses, nor how they returned from the west to the east, but supposed that the heaven sloped down in every direction, which appearance it must present on account of its immense breadth, they thought that the world is round like a ball, and they fancy that the heaven revolves in accordance with the heavenly bodies; ...'. This then led to the supposition 'But if this were so the earth itself must be like a globe; for that it could not possibly be anything but round, which was held enclosed by that which was round.' This assumption of rotundity leads 'to the invention of those suspended antipodes.' The final paragraph describes the defence of 'these marvellous fictions' using the concept of, what was essentially, gravity. Lactantius closes 'I am at a loss what to say respecting those who once they have erred, consistently persevere in their folly, and defend one vain thing by another; but that I sometimes imagine that they either discuss philosophy for the sake of a jest, or purposely and knowingly undertake to defend falsehoods, as if to display their talents on false subjects.'

Notwithstanding the vigour of this attack, it had little impact on the belief held throughout the medieval period that the earth was round. The idea that the preponderant belief during this time was that of a flat earth has been shown to be largely an error propagated from the beginning of the nineteenth century. Taylor wrote in 1943 'It is difficult to understand why the story has gained such ground that prior to Columbus' voyage it was generally believed that the earth was flat.'[7] This thesis was substantiated in depth by Randles (1980), Russell (1991), Simek (1992) and Grant (1994).[8]. However the debate about the existence of the Antipodes did continue, albeit in a low key, throughout the medieval period. The coexistence of these two apparently irreconcilable concepts came to an end towards the end of the fifteenth century. Randles has advanced the idea that resolution of this conflict became necessary at this juncture because of the voyages of the Portugese explorers to the South and East and of Columbus, Cabot and other to the West. The idea of a flat earth had to be formally dismissed and hence Lactantius' writing had to be re-examined. The

---

[7] Taylor ERG (1943) Ideas on the shape, size and movements of the earth, Pamphlet 126. Historical Association, London, p 9

[8] Randles WGL (1980) De la terre plate au globe terrestre: Une mutation épistémologique rapide (1480–1520). Librairie Armand Colin, Paris
Russell JB (1991) Inventing the flat earth. Praeger, New York
Simek R (1996) Heaven and earth in the middle ages. The Boydell Press, New York
Grant E (1994) Planets, stars, and orbs. Cambridge University Press, Cambridge/New York

process of refutation of Lactantius began with the Italian humanist, Galleoto Marzio da Narni (c.1427–c.1497) in his work *De Incognitis vulgo*, and was continued by Rudolf Agricola and Vadianus.[9] The matter was taken up with vigour by cosmographers such as Apian (1524) and Stoffler (1534) in their books on the subject. Copernicus found it necessary to refer to Lactantius' erroneous stance twice in his *De Revolutionibus* (1542).

Recorde was thus in good company in raising and dealing with the matter. Perhaps he had the added incentive of knowing that the *Divine Institutions* was a text highly commended by John Colet for the teaching of Latin style to the pupils of St. Paul's School, which he had founded. Early pupils included William Paget, John Leland, William Rastell and Stephen Vaughan all likely acquaintances of the adult Recorde . However Recorde differed from his continental contemporaries in concerning himself first with Lactantius' supposed view on the shape of the world.

Whilst Lactantius is always viewed as a 'flat-earther' his writing does not explicitly promote the idea. As has been seen, he was violently opposed to the concept of antipodes, which inevitably brings in its train the idea of a planar earth, but he did not pursue his argument to this logical conclusion. This consequence was established by his opponents. As has been seen, Lactantius blamed the erroneous concept of a round earth on the assumption of a round world in which it had to be accommodated. Would Lactantius' implicit assumption of a flat earth justify Recorde attributing to him sponsorship of flat heavens? This interpretation of Lactantius was repeated by Thomas Hood 40 years later.

Recorde chose his words carefully in presenting Lactanius' supposed viewpoint. He has the Scholar say '.... and yet I have heard of certaine great clerks, that in no case thought it reasonable to affirm such a forme of roundness, or such a round motion in heaven: but most of all I mervaille of that famous man Lactantius Firmianus, which doth affirme (as I have hearde) that the heaven is not rounde, but flat and playne.' This has to be taken as an interpretation, there being no direct quotation available. On two occasions later in the book the Master gives quotations from the reference he gives to Lactantius when he wishes to counter the latter's arguments. Clearly Recorde had Lactantius' text to hand and it was in character for him to quote from the Latin texts of other attributed authors, translate them into English and correct the associated arguments they expressed, if necessary. It seems clear that Recorde was interpreting what he thought was a logical conclusion to Lactantius' statements, one which the latter either did not want to arrive at explicitly, or was incapable of deducing. Given that the Master felt 'that he [Lactantius] had conceaved a deadly hatred against all philosophers and against philosophy itself. ...', Lactantius' use of the deductive procedures practised by philosophers seems unlikely.

Who were the other 'great clerks' referred to by the Scholar is not clear. The strange ideas of Cosmas Indicopleures, which included both a partially flat

---

[9] Randles, loc. cit., pp 87–89

heavens and a rectangular earth, whilst written c.540 AD, are thought to have been unavailable in Europe before the eighteenth century.[10] Ptolemy, in dealing with the sphericity of the heavens says 'For if one were to suppose that the stars' motion takes place in a straight line towards infinity, as some people have thought, what device could one conceive of which would cause each of them to appear to begin their motion from the same starting point each day? Toomer notes that according to Theon's commentary (Rome II 338) this belief was Epicurean, but that he knew of no other evidence.[11] John de Sacrobosco quotes Alfraganus' refutation of a flat sky as a 'further proof' of the sphericity of the heavens, so presumably the topic was active at that time.

Lactantius' statement that the heavens presented the appearance of sloping down in every direction because of its immense breadth is probably that which Recorde was interpreting. In Latin it reads '...., caelum autem ipsum in omnes partes putarent esse deuexum, quod sic videri, propter immensam latitudinem necesse est, existinaverunt rotundum esse mundum, sicut pilam,...' As Recorde quotes the immediately following sentence '... ex motu syderum opinati sunt, caelum volui....' later in the book, it is inconceivable that he was not aware of Lactantius' actual words on the nature of the 'caelum'. It is to be assumed that Recorde interpreted these words to mean that Lactanius' caelum was flat as far a human observation was concerned.

Demonstration of the round form of the heavens is logically a necessary pre-cursor to the demonstration of the round shape of the earth so whether the threat posed by Lactantius was real or imagined, it is not surprising that Recorde chose to deal with it first. The Master says 'Whosoever wyll believe him [Lactantius] in this point, must do so without reason: for he alleageth no reason for his purpose, but taketh it as a certaine truthe, thereby to improve the opinion of the Antipodes, as I will more largely declare in prooving the roundnes of the earthe. But seyinge he could bring no reason for his opinion you shall here some reasons against his phantasye, and the judge as you can.'

From Ptolemy via John de Sacrobosco to the fifteenth century cosmologists and Copernicus, none sought to challenge the concept that the world was spherical.[12] Sacrobosco advanced three main reasons to support this thesis viz. likeness, convenience and necessity, supporting them with an argument against the flat world as already mentioned. The main reasons were debated into the seventeenth century, but chiefly with the objective of supporting the thesis of sphericity.[13] Of these the argument based on likeness was of a theological nature and was unlikely to appeal to Recorde who preferred evidence based arguments. It could well have been this preference that caused him to attend first to the demolition of the flat world concept before moving on to his versions of the *convenience* and *necessity* arguments. The material that Recorde uses in his

---

[10] Simek R, loc. cit., 2

[11] Toomer GJ, loc. cit., pp 38–39

[12] Grant E, loc. cit., pp 113–122

[13] Ibid

arguments can be found distributed amongst preceeding writers, who are not acknowledged. To that extent he is not original, but his assemblage of the data to make a logical sequence of argument and his illustrations are differ from his predecessors.

He starts with a little qualitative geometry, establishing the point, obvious to the Student, that if the heaven is flat, however it is placed with respect to an observer, its nearest point to that observer will be overhead.

The master then moves on to question of the motion of the heavens. There is no doubt that the Sun, Moon and stars all move, but the Master points out '... that the starres and Planetes do move in the skye, as fishes do swimme in the water: and that they go forwarde thoughe the heaven stand still.' The issue being raised is one of a continuing debate viz. whether the heavens are solid forms in which the heavenly bodies are embedded, or whether they are fluid. The matter was not settled until Tycho Brahe's observations of comets were made in the early seventeenth century. It also invokes the concept of *place* as defined by Aristotle. Grant cites the analogy of planetary movement with that of fishes and birds as being used by Buridan and Aegidius Romanus prior to Recorde and by Bellarmine, Clavius, Oviedo and Arriaga subsequently.[14] The Master believes such analogies to be inexact, as close observation of flock and shoal behavior shows that individuals in the assemblages vary in their relative positions, whereas the stars '... in 5000 yeare space [to] keepe their place so precisely, that they varye not one minute of a degree.'

However it is the evidence provided by the observed behaviour of the Milky Way that the Master believes provides 'invincible reason' that the heavens move. The Milky Way is said to be called Watling Street in England, but was called Galaxia by the Greeks. The argument deployed is complicated somewhat by the words used. 'This way is in the skye it selfe, as all men hath confessed, and their eyes do testifye, and the starres that bee in it are alwayes seen to keep their places in it: so that it must needes followe, that the same way doothe moove with the starres, and then consequentlye, the skye must needs move also.' *Sky* is an English word and therefore not found in other than English texts. Recorde does not define it. Oresme and Albertus Magnus located the Milky Way in the celestial region in contradiction to Aristotle who located it much closer to earth.[15] The true nature of the Milky Way as a collection of stars was revealed first by Galileo some 70 years later. Whilst Recorde knew of Roger Bacon's fabled telescope and was reputedly an instrument maker himself, the stars referred to must surely be those visible to the naked eye.

The Scholar objects that even if the stars seen to be in the Milky Way move this does not mean that the latter itself moves. He is then invited to observe the position of stars relative to the bifurcation that came to be known as the Great Rift and has to concede that such relative position was maintained. Finally he was invited to observe directly that a notable point in the Milky Way, lying in

---

[14] Grant E, loc. cit., 274 n.16, 348 n.76

[15] Grant E, loc. cit., 400 n.44

the east at the beginning of the night, moved westwards across a quarter of the sky in 4 hours. This shows that the sky moves as well as the stars.

If then the heaven was flat as Lactantius imagined and moved westwards as observed, then he must postulate an infinite supply of heavenly bodies or demonstrate how they recirculate. This can only arise from some sort of circular motion, which he has rejected; 'Ex motu syderum opinati sunt coelum' which implies that the stars move but not the sky. Given this, the Master returns to the argument that things seem largest when nearest. Applying this criterion to the appearance of stars in a flat sky, irrespective of the relative dimensions of earth and sky, the stars must seem largest when overhead and diminish in size as the move away from this position. The Scholar points out that the contrary is true, the sun, moon and stars seem greatest when setting in the west or rising in the east and smallest when overhead. The Master then crowns his campaign against Lactantius by quoting his own rule 'those must be false sentences that do agree with false matters. and so must they be false premises, that do infer a false conclusion.'

4. *That the heavens are not cornered, nother of manye angles.*

A second argument supporting the sphericity of the heaven is then explored. The choice of any other angular figure that may be assumed to rotate is dismissed for two reasons. The first reason is that some position on the surface of such figures must always be nearer to its centre of rotation, i.e the earth, than all others so that the previous argument relating to nearness and size will apply again. The second reason invokes the thesis that nature abhors a vacuum. Rotation of an angular body within its circumscribing sphere requires an infill of some 'subtile and liquide thing, that might change his place as fast as the heavens do turn; for in turning the corners will come anone where the emptiness is now, and so successively eche change place with the other'. The specifications to be met by such a material are deemed unmeet to be matched. This argument corresponds to that of Sacrobosco classified as 'Necessity'. But an additional reason for rejecting angular forms in favour of a sphere is that the latter moves more swiftly and easily. 'S. Every common turner can skil in that reason, and know that a little altering of the one side, maeth the boul to run biasse waies. Master. If the reason be so plaine that common artificers can skyll of it, it were to great a folly for learned menne to doubte of it.'

The final argument presented parallels Sacrobosco's 'Convenience' criterion, and on grounds of capacity 'the Globe is most largest and therefore the most aptest for the form of the skye, which incloseth all thyngs that manne can see.'

The choice of sequence of arguments presented to support the spherical nature of the world is significant, inasmuch as it indicates that Recorde thought that arguments based on things observed should be given priority.

5. *That all thinges shewe greater then they be, thorough vapours, and therefore the starres with the Sonne and Moone doo appeare greatest nigh unto the Horizont.*

The explanations of many previous authors on this topic are repeated. A little additional information on the practical uses of a vacuum is given.

6. *Dyverse opinons of the forme of the earthe: some thinkinge it to be of Cubike forme, others judging it Rygge formed, other affirmynge it to be plaine, other deeming it hollowe as a dyshe, and other esteemynge it longe and rounde, lyke a piller or roller: all whiche beyng sufficiently confuted, it is full proved, that the earthe is justlye rounde in shape.*

This matter is dealt with at length in some 19 pages of text and illustrations. The Master first puts forward the arguments used by the proponents of cube, rygge [three-sided prism], plane and dish forms for the shape of the earth. These arguments are ridiculed by both Scholar and Master. The latter says 'Who lysteth to see the monstrouse opinions of suche dreaming doters, may reade them often touched in Aristotle his naturall bookes, and aboundantly in his boke De philoso-phorum placitis. and in Galene and Eusebius in bokes of the same matter pecu-liarly, writen.' These suggestions arise from the pre-Pythagorean Greek philosophers Anoximander, Anaximenes, Anaxagorus and Democritus amongst others.[16] He then goes on to say that their confutation is dealt with in the first book of Cleomedes' Sphere, which confutations he will present. However he com-ments on a mistake made in Cleomedes' Greek text that must have inadvertently introduced pyramids instead of prismoids, for the latter not the former fit the subsequent arguments. This will be demonstrated later. Whilst the Master gives Cleomedes' arguments he reserves his own position and right to comment.

He does not follow Cleomedes order of presentation of confutations, starting with the plane earth, the latter's third option. If the earth is plane then Sun, Moon and stars must rise and set at the same time for everyone. The Scholar observes that this is false as the sun rises and sets 3 hours earlier in Jerusalem that in England and 6 hours earlier in Calecut [on the south west coast of the Indian subcontinent, where there was a Portugese trading station]. The Master then raises the matter of eclipses as seen in Calecut and in England. He starts bravely with an eclipse of the Moon predicted to take place the coming 17 November 1556 at 3 a.m. to the south-west in England. In Calecut it should take place at 9 a.m. but will not be visible as the moon there will be below the horizon in the north-west. A similar situation will arise in 1562. These predictions will repeat the happening of 5 June 1555, when there was an eclipse of the moon at 3.a.m. but in Calecut the eclipse was not seen as the moon had already set. However at 9 p.m., 20 February 1551 there was an eclipse of the moon in England that was also seen in Calecut at 3 a.m. the following day, as attested by 'Portingales' who were there. These observations demonstrate that the horizons at these two places differ and the earth cannot be flat. It also demonstrates that in the intervening 18 month from this last eclipse, the Portugese had reached England and spoken directly or indirectly to Recorde. The Master then expounds another argument viz. that it is well known that as one journeys north from London, the observed elevation of the North Pole rises and as one travels south the elevation decreases until the the Pole is

---

[16] Heath T (1966) Aristarchus of Samos. Oxford University Press, Oxford

equal with the horizon. Together these arguments demonstrate that the earth is
flat neither east to west nor north to south 'but both waies hath some certain
rising, which I shall prove to be a juste roundnes.'

Cleomedes third reason for refuting the flat-earth concept viz. the absurdity
of commonality of the times everywhere is implicit in the previous arguments.

His fourth reason is more complex and is dealt with at some length. Recorde
presents the whole of Cleomedes argument accurately, translated, but modified
slightly in sequence of presentation. He also inserts some elementary geometry
to aid understanding.

Cleomedes first states that if the earth were planar the whole diameter of the
world would be 100,000 stades, which Recorde has transcribed as furlongs, a
reasonable approximation. The latter then introduces a supplementary maxim,
'If the earth be plain, then al places in the earth are as far asonder, as their
Zeniths or Vertical pointes be in heaven.', to aid understanding of Cleomedes
derivation. The construction given shows a circular section of the spherical
heaven intersected in two points by verticals from two points on a planar earth
roughly equally disposed about a centre that is not marked. The deduction that
the two zeniths are spaced the same distance apart as the points on earth is not
qualified by any indication of how this distance is measured, such by a vertical
to the two parallel lines. Clearly the general case is not established. The Master
then continues with Cleomedes presentation. Syene [Aswan] and Lysimachia
[Hellespont] are 2,000 furlongs apart. The zenith of Syene [Aswan] lies in the
Tropic of Capricorn whilst that of Lysimachia lies vertically below the head of
the North dragon, which are a fifteenth part of the circumference of the heavens
apart, i.e. a fifth of the diameter of the sky [pi equals 3]. Hence the whole diameter of the meridian passing through the two sites is 100,000 furlongs or stades
and the whole compass of the sky is 300,000 furlongs. The approximations in
this deduction are manifest, but perhaps not of major significance. For Cleomedes
continues '... the earth contayneth in compas 250,000, so is the heaven little
bygger than the earth in compas. whiche absurditie may easily be confuted by
the Sonne, whiche in comparison to the skye is a very little part of it, and yet is
bygger than the earthe manny fold:...'. The size of the earth used in this argument is that quoted later by Cleomedes as having been derived by Erastothenes
for a round earth, not a planar one. Applying corrections to Cleomedes' calculations, Goulet has shown that a value of 240,000 stades emerges, i.e. the sizes of
earth and sky are virtually the same, which is not surprising as the calculational
methods are identical and only the data on which they operate differ.[17] Recorde
does not comment directly on this point but deals with it in an oblique manner
following his demonstration using data based on places in England.

The sites that the Master chooses are Arundel with a zenith of 50 degrees
and 30 minutes, which he had measured himself and Newcastle 55 degrees
from the equinoctial, giving a difference of 4 degrees and 30 minutes. On the

[17] Goulet R (ed) (1980) Cléomède. Théorie Élémentaire. Librairie Philosophique. J. Urin, Paris,
200 note 162

earth's surface this is equal to 270 miles, as demonstrated previously in the third treatise. The Master then labours the point that this is true only for the earth's surface and can only be true for the separation of the zeniths if the earth is flat. Given the latter the compass of heaven is 21,600 miles and that of its diameter 6,872 $^8/_{11}$. Recorde does not make the same mistake as Cleomedes of equating the diameter of the earth calculated as a sphere to that for a plate, but has the scholar say 'That is to greate an inconveniencefor any man to affirm. for thereby I see it would follow that if we go any waye from our own cuntry, 3,436 miles, we shal com hard to the sky, which is to childish a fantasye, sith not only reason but dayly travell declareth the contrary. Againe I remember that in the third treatise you declared that the earthe was so much in compasse, which muste needes bee many fold lesse then the heavens, which are so farre distaunt from the earthe on every side.' The Master concludes that Cleomedes reasons against the flat earth are fully justified, but the question remains of why Recorde went to such lengths to dismiss the case. He was not the last to do so as pointed out by Randles.[18]

Recorde then follows Cleomedes in dismissing the possibilities of shapes for the earth other than round. However in the course of doing so, whilst he gives Cleomides' arguments, he criticises them and inserts, at length, what he considers to be stronger arguments but still arriving at the same conclusions.

The case for a dish shaped earth is readily dismissed by Cleomedes as giving the results that rising and setting of the stars would be diametrically opposed to what is actually observed. This confutation the Master observes is agreed by Ptolemy. At this juncture, the former takes the opportunity to adjure the Scholar to not accept the authority of a writer as final justification for his arguments, even Ptolemy. The Master then proceeds '…: for autoritie often times deceaveth mant menne, as here by and by in Cleomedes it shall appeare, whose arguments in confuting the other two opinions are nothing substantiall: which chanced other bicause he saw the sondenes of these opinions so great, that he saw no great reasond to confute them, other els hastinge in his writinge caused him to use the lesse dilgence in framing his reasons. but now will I repeat them.'

Cleomedes argues that if the earth were of cubic form all nations should have days of 6 hours only as the sun should shine on each side for only a quarter of a day. The Master shows that this can only be true for the case when all the corners of the cube touch the heavenly globe which the Scholar agrees is an unlikely occurrence. Cleomedes uses a similar argument to dismiss the proposal that the earth has a three cornered form which would give rise to days of 8 hours. Again the Master shows that this argument is valid only for the case when th corners of the earth touch the sky. He goes on to reassure the Scholar that '… : for a weak confutation of an untruth doth not make that untruth to become true. And bicause you shall not think that these opinions have anye sure grounde, I will repeat Ptolemye hys confutation of them both, by one unfallible reason.' This argument

[18] Randles WGL, loc. cit., 90

is that in the imagined forms of the earth there can be no more horizons than there are sides in the figure e.g. six for a cube and three for a triangular figure and within these horizons there would be but one rising and setting of the sun. This is clearly not so as every 15 degrees movement westwards causes the day to be an hour later. The Scholar suggests that if the earth had 24 flats then this would resolve the issue, but the Master points out that learned men can measure down to seconds of arc. Taken together with the evidence of eclipses, '…there can bee noother forme of the earthe aptly assigned, but a rounde circular forme.' Further, movement to the poles, there can only be a round form in this dimension. This leads to the dismissal of the proposition, probably first suggested by Anaximander c.550 B.C., that the earth was pillar shaped with its axis lying north-south. The Master follows Ptolemy in dismissing the proposition on the grounds that it would predict visibility of the stars, moving axially that were contrary to observation. For the flat surfaces the argument already levelled to dismiss the flat earth concept would apply.

The treatments afforded the foregoing topics by Recorde seem excessive, being far more extensive than his contemporaries and indeed Ptolemy also. He also introduces Cleomedes only to expose the shallowness of some of his reasoning. Maybe Recorde just took this subject as an opportunity to instruct the reader in the need for rigour of argument.

7. *Then followe diverse reasons, approvyng the water to be round, and a declaration with proofe, why the water dooth not, nother can not over-ronne the whole face of the earthe.*

8. *That the earthe and water togither doo make but one rounde Globe and have therefore one common centre.*

The topic addressed is the existence of the terraquaeous globe. Aristotle postulated the sub lunar region as consisting of concentric spheres of the four elements, the earth being at the centre followed by water, air and fire. As dry land existed, the scheme required some adjustments with respect to the relative positions of the spheres of earth and water. The ensuing debate extended over many centuries.[19] For many years the concept of two spheres having separate centres seemed to have resolved the problem. One consequence of this solution was that a large portion of the globe was covered entirely with water, generally assumed to lie south of the equator. However the Portugese explorations of the Southern hemisphere in the late fifteenth and early sixteenth centuries and the discoveries of the land masses there present caused the abandonment of the two sphere concept in favour of that of a single sphere incorporating both earth and water with a single centre of gravity. As Grant states, it was formulated outside the scholastic tradition and was adopted by Copernicus, entering scholastic cosmology with Clavius in 1592.

Record puts forward the terraquaeous concept with no description of any alternative. He first establishes the sphericity of the sea by showing that it has

---

[19] Grant E, loc. cit., pp 630–637

the same properties as that of earth with respect to the change of times of sunrise and sunset with longitude and latitude. The first example he gives is that of a ship sailing due west from Cornwall at midsummer, starting at sunrise and moving at ten miles per hour for 16 hours. As the length of day at this latitude is 16 hours the sun will be setting at the starting point but as the ship has run with the sun, the sun will at that time still be four degrees above the horizon for the ship. The scholar objects that he had been told previously the one degree corresponded to 60 miles, which gives the Master the opportunity to explain that this latter is true for degrees of latitude but only true for degrees of longitude at or near to the equinoctial. He expands further on this point using an example based on a voyage from Iceland when the day was 20 hours long and because of its latitude there would only be 27 miles per degree of longitude. The Master says that 'This varietie coulde not happen, except the water was also rounde as well as the earthe.' This would be somewhat dubious reasoning were it not for the fact the Recorde asserts that with respect to voyage from Cornwall the parameters quoted were within his own experience. Presumably he was referring to his voyages from Bristol to Wexford in Ireland.

He then introduces the proof based on the disappearance of the hull of a ship first when the latter is moving away from the observer on land and the sighting first of the topmast when the ship is moving towards land. He then repeats the argument for a ship based observer. These examples then allow him to move into a discussion about relative heights and to establish that the only true measure of height is with respect to the centre of the earth.

The next proof adduced is that also used by Ptolemy, Cleomedes and John de Sacrobosco viz water when unconstrained takes up a spherical shape as closely as it is permitted e.g. raindrops or the meniscus on water resting on a surface.

He then moves on to gives two reasons that he attributes to the late Erasmus Reinhold, a man whom he holds in great esteem, but who is still capable of error. Reinhold published his edition of the Almagest in Greek and Latin, together with a Latin commentary, in 1549 at Wittenberg. It is from this book that Recorde takes his examples. Ptolemy said very little about the relative dispositions of land and sea, just quoting the experience of the emergence of the tops of mountains first when approached from the sea, thus showing that the surface of the water was curved. Reinhold's commentary on the topic is however quite extensive containing what Recorde describes as two reasons for the curvature of the seas.[20] The latter gives them in reverse order to Reinhold which necessitates a little further explanation by Recorde, but his translation of Reinhold is virtually literal. Reinhold takes the case of the Danube that from Ulm to its mouth at Euxine measures 312 German miles, which is the eighteenth part of the diameter of the earth. A simple geometrical construction shows that given a spherical earth plus water, at the centre point of the traverse given, the river would be higher then at its start and finish by 13 German or 52 English miles. The issue that Recorde takes with this lies not in the concept, but

---

[20] Reinhold E, Ptolemae Mathematicae constructionis. Bodleian library Byw S.14, 51–54v

in the calculation as Reinhold had failed to take into account that the course of the Danube lies not on a great circle but on a lesser one lying on latitude 46 degrees. This reduces the elevation from 52 to 36 miles. The master will not explain the calculation in detail as it relies on an understanding of sines and cosines which the Scholar has yet to learn.

When the Scholar seeks to dismiss Reinhold's argument because of the error his reasoning is admonished as he should have realised that the intent was to prove that'…water doth not run by a right line and downwarde still, as the vulgare sorte doothe imagine, but that it runneth cicularlye..'. But Reinhold is also mildly reproved because that he has not adequately defined highness and lowness which should be defined relative to the centre of the earth.

The Master then progresses to present Reinholds extensive arguments why there are no hollows in the sea such as are found on land. The arguments are based in essence on gravity coupled with the fluidity of water. These arguments were propounded by Copernicus who followed them with a quantitative argument showing that, contrary to the beliefs of the 'peripatetics' that the body of water was ten times that of the earth, even if the proportionality factor dropped to seven the whole of the earth would be covered by water. Recorde then cites the latest discoveries of lands south of the equator as further evidence to support his conclusion '…that land and water together press upon a single centre of gravity; that thearth has no other centre of magnitude; that since the earth is heavier its gaps are filled with water; and that consequently there is little water in comparison with land, even though more water may perhaps appears on the surface.'[21]

Recorde attempts some extensions and reinforcements of these approaches. The Master points out that even the admixture of relatively small amounts of water to earth rapidly diminishes its stability under loads. He then presents a quantitative argument intended to show how small the ratio of water to earth is based on assumptions of the depth of water that exists on the surface of the combined globe. The student first agrees that maybe half of this surface is covered with water and the depth of this water is then debated. The Master states that mariners have measured depths as great as several hundred fathoms in places but perhaps it would be safer to assume an average depth of 1 mile. The Scholar feels this to be excessive but the Master believes 'The more that supposition exceedeth truth, the stronger shall the proofe be of the smallness of the water in comparison to the earthe.' The surface area of the earth is deduced based on the suppositions proposed the volume of water is shown to be less than a two-thousandth part of that of the earth. This demonstration the Master believes shows how foolish the idea is that the volume of water is ten times that of the earth and refers back to Copernicus' calculation.

Whether the additional 'proofe' given by Recorde adds substantially to the substantiation of the terraqueous concept is debateable, but it was reproduced in essence by Clavius in his Commentary on Sacrobosco's Sphere, as part of

---

[21] Copernicus N (1978) In: Dobrzycki J (ed) Nicholas Copernicus, on the revolutions (trans and commentary: Rosen E). Macmillan, London, Chapter 3

the case he presented which finally introduced the aforesaid concept as academically respectable.[22]

Recorde then moves on to demonstrate that the combined body is a perfect globe based on the observed the shapes of the earth's shadows exhibited during eclipses of the moon.

9. *That the earthe is but as a pricke in comparison to the Skye, which is approved by foure dyvers argumentes.*

Ptolemy deals with this issue after he treats the matter of the centrality of the earth in the heavens. Recorde apparently wishes to continue with a topic to which the shadow of the earth is relevant. He discusses the three cases of the earth being greater in size than the sun, equal to it or smaller than it. His thesis is that the Sun is much smaller that the whole sky so that if it is much larger than the earth then the latter must be very small relative to the heavens. If the sun were smaller than or even equal in size to the earth, it is argued that the shadow of the earth would obscure appreciable areas of the sky for noticeable lengths of time. The Scholar is somewhat dubious about this reasoning because he is not convinced that the stars are illuminated only by the Sun. The Master then applies these arguments to the eclipses of the Moon, showing that if the earth was other than sensibly smaller than the Sun, the duration of eclipses would be far longer than those observed.

Recorde then reverts to the series of reasons for the point size of the earth in respect to the heavens proposed by Ptolemy. The first of these is that the sizes and distance of stars from the earth appear independent of the latitude from which they are observed. The second reason is that if gnomons are set up on any part of the earth, the paths of shadows that they cast behave uniformly as if they were at the centre of the earth or of an armillary sphere. Reinhold illustrates this argument using a geometrical figure which Recorde copies exactly. Ptolemy's third reason is that wherever the observer may be his horizons always bisect the heavenly sphere, which could not happen unless the earth was of imperceptible size relative to the heavens. Again Reinhold illustrates the argument with a figure which is copied by Recorde.

10. *The distaunce of everye sphere from the centre of the earthe, with an ordre to trye the quantities of the Sonne and Moone &c, in comparison to the earthe.*

This topic is introduced by the Master to show the Scholar that even if the earth is so small compared with the whole world, it is not negligible in comparison with the diameters of the spheres of the planets Moon, Mercury and Venus, believed to lie between it and the sun. The ratios of 33½, 64 and 167 times the semidiameter of the earth (e.r.) respectively for these planets are those given by Alfraganus. The values the Master gives for the closest distance of approach to the Earth of Sun, Mars, Jupiter, Saturn and the starry sky are also

[22] Clavius C (1594) In Spheram Ioannis de Sacrobosco Commentarius. Bamberg, p 145, in the context of An ex terra et aqua unis, pp 133–151

those of Afraganus viz 1,120, 1,220, 8,876, 14,405 and 20,110 e.r. respectively. They are also the values given by Bacon and are what Recorde called the inner concavity of each sphere i.e. the lower apsis.[23] The Scholar says that Faber gave these values in miles but the Master demurs. 'In that place Faber foloweth the acompt of Alphraganus the Arabitian, which speaketh of myles much longer than the Italian myles be: for 6 of the Italian miles do make but 5 of Alphraganus miles: of which diversity at another tyme I will instruct you, namely in the treatise of Cosmographye: where I wyll set forthdivers varieties and appearante repugnances of sondry writers, for the measuring of the earthe: and prove it to be a disagrement more in words then in meaning: and to come by reason of their divers miles, or other inconstant measures.' He then defines 60 of his miles as being one degree i.e. the mariner's mile. In terms of this definition, the circumference of the earth is 21,600 miles, giving a radius of earth of 3,436 $^4/_{11}$ miles, which is to be compared with the value of 3,250 miles used by Bacon.

Using the former measure he gives the semi-diameters and circumferences of the eight spheres and hence their distance from Earth at their centre. That of the sun is given as 3,848,367 $^7/_{11}$ miles (slightly in error) by Recorde and 3,640,000 milliards by Alfraganus, both using 1,120 e.r, the lower apsis as the basis of their calculations. The Scholar then calculates the value in miles of a degree for each of the spheres and using the measured angle subtended by the Sun and Moon calculates their diameters. A discrepancy between his result derived for the Moon and the accepted value enables the Master to say that he had instructed the Scholar in the method so that he could check other author's results.

11. *That the earthe is in the myddle of the worlde, and the contrary opinions repeated and refuted by sondry proofes.*

Recorde returns to Ptolemy's text, presenting four options for the position of the earth within the sphere of the heavens viz. on the polar axis but displaced from its centre; displaced from the polar axis but equidistant from the Poles; displaced from the Polar axis but not equidistant from the Poles and lastly at the centre of the sphere. The Master says 'And now for the confuting of the three fyrste options I will use Ptolemyes argumentes, augmenting them with a larger explication.' These arguments show that all of the first three options lead to predictions of the lengths of daylight at different points on the earth, of the seasons and of the dispositions of the Zodiac which are at variance with observed behaviour. Additionally eclipses of the Moon would occur at points where the three bodies involved are not diametrically opposed and hence at intervals other than equivalent to a semicircle. The augmentation provided by Recorde takes the form of diagrams, not as numerous as those used in his commentary by Reinhold and not a direct copy of their equivalents. He also provides first hand evidence of the positions of Sun and Moon at eclipses on 20 February 1551, in 1553, on 5 June 1555 and a prediction of their positions for the eclipse to take place 17 November 1556. This information is

---

[23] Bridges JH The Opus Maior of Roger Bacon'. vol I, pp 219–236

valuable not only for the arguments for earth's centrality, but also showing that Recorde was not yet incarcerated and was actively observing the skies. It also helps to date this portion of the book.

12. *That the earthe dooth not move from the centre of the worlde.*

At this point Recorde's presentation departs radically from that of Ptolemy and also that of Reinhold. It is that portion of the book, a mere three pages out of over 300, which has received virtually all of the interest in the work for the last 150 years, occasioned by his reference to the work of Copernicus. This has led to an indecisive debate about Recorde's candidature for the designation of first Copernican in England.

Interpretation of the apparently ambiguous interplay of conversation between Master and Student has given rise to many shades of opinion. The enthusiasm of the nineteenth century commentators on the book for the extent of Recorde's commitment to Copernicanism has been considerably diluted with the passage of time and with closer examination of the text. Nonetheless, in his study of astronomical thought in Renaissance England, Johnson believes that 'Recorde clearly shows in this passage that he believes the Aristotelian and Ptolemaic arguments against the rotation of the earth to be entirely fallacious. The wording of his promise to expound the Copernican system more fully when the student shall have reached the stage where he will be able properly to understand it, definitely indicates that he was ready to go even farther, and accept the revolution of the earth around the sun as well as its rotation.'

Patterson provides a good summary of the history of the matter prior to her own researches.[24] She summarizes her findings as follows:

1. *The Castle of Knowledge* is a Ptolemaic tract which includes, however brief, favourable mention of Copernicus and Reinhold.
2. Recorde stated that the earth does not rotate, but admitted his assertion to be indemonstrable.
3. He stated that the earth is in the centre and does not move from the centre of the world, giving arguments which he describes as invincible.
4. Recorde's discussion of the supposition of Copernicus, however favourable, is nevertheless tentative and noncomittal by contrast with his categorical assertions concerning the earth's stability and central position.
5. The Copernican passage is not sufficiently detailed to indicate the author's familiarity with the theory.
6. The passage appears to have been inserted after the work was complete or nearly so, and is at variance with the rest of the text.
7. Recorde promised to write further on the Ptolemaic as well as the Copernican theory.'

She is clearly much less sanguine about Recorde's adherence to Copernican ideas than was Johnson.

---

[24] Patterson LD (1951) Recorde's cosmography, 1556. Isis 42: 208–218

Easton in her entry on Robert Recorde in the Encyclopedia of Scientific Biography is more circumspect in her comments on the *Castle of Knowledge* and advises that Patterson's paper should be read with caution. Easton describes the book as dealing with the construction and use of the sphere, as presenting an elementary account of Ptolemaic astronomy but with a brief, favourable reference to the Copernican theory.

Russell interprets the passage as meaning that '...Recorde was quite prepared to accept a diurnal rotation. The arguments of Ptolemy, Theon etc. against it can be fully answered. He was at least sympathetic to the full Copernican system but his description of it was very incomplete. The reader is told that Copernicus accepted a diurnal rotation and put the earth "continually out of the centre of the world" by nearly four million miles, but not that he put the earth in orbit around the sun. Evidently he thought that a full statement of the theory would unduly confuse student at this stage of their studies.'[25]

The debate has been revived lately by Lloyd.[26] He was concerned about the source(s) of Recorde's knowledge of Copernicus and his theory and concluded 'So Recorde's direct source for his list of Greek astronomers and, by extension his awareness of that theory cannot have been the *De Revolutionibus* itself.'[27] He goes on to say 'What he [Recorde] knew of those ideas at the time he was completing his *Castle of Knowledge* was, manifestly, incomplete.[28] Further he says '....Recorde states Copernicus' position in terms not of heliocentricity at all, but only of the latter having measured the eccentricity of the earth, its distance from the universe's central point, with no suggestion that that point was occupied by the sun.'[29] He concludes, ' So, while Recorde was alert to new ideas circulating in his time, he falls short of qualifying for recognition as one of England's first Copernicans.'[30]

Further examination of the few pages of relevant text seems to be warranted and will be undertaken piecemeal. 'Master.......And this may suffice for this time touching the earth and his accidentes, principally appertainynge to Astronomye: for although many other things are to be considered in it, they appertaine rather to philosophers or Cosmographers, then to Astronomers, and namely in the doctrine of the principles. As touching the distinction of the zones, I have sayde somewhat before, & somwhat more wil I say anon. But as for the quietnes of the earth I need not to spend any time in proving of it, syth that opinion is so firmly fixed in most mennes headdes, that they account it mere madness to bring the question in doubt. And therfore it is as much follye to travail to prove that which no man denieth, as it were with great study to disuade that thing, which no man doth covette nother any

---

[25] Russell JL (1973) The Copernican theory in Great Britain. In: Dobryzki J (ed) The reception of Copernicus' heliocentric theory. D. Reidel Publishing Company, Dordrecht, pp 189–240, 190–191

[26] Lloyd HA (2000) 'Famous in the field of number and measure': Robert Recorde, renaissance mathematician. Welsh Hist Rev 20:254–282

[27] Ibid., 267

[28] Ibid., 267

[29] Ibid., 269

[30] Ibid., 270

manne alloweth: or to blame that which no manne praiseth, nother any manne lyketh. This is to be viewed alongside Copernicus' comment,

> To be sure, there is general agreement among the authorities that the earth is at rest in the middle of the universe. They hold the contrary view to be inconceivable or downright silly. Nevertheless, if we examine the matter more carefully, we shall see that this problem has not yet been solved, and is therefore by no means to be disregarded.[31]

'Scholar. Yet sometime it chaunceth, that the opinion most generally received, is not most true.

Master. And so do some men judge of this matter, for not only Eraclides Ponticus, a great Philosopher, and two great clerks of Pythagorus school, Philolaus and Ecphantus, were of the contrary opinion, but also Nicias Syracusius, and Aristarchus Samius, seem with strong arguments to approve it: but the reasons are to difficulte for this first Introduction, & therefore I wil omit them till another time. And so will I do the reasons that Ptolemy, Theon & others do allege, to prove the earth to bee without motion: and the rather, because those reasons doo not proceed so demonstrablye, but they may be answered fully, of him that holdeth the contrarye. I meane, concerning circularre motion … '.

Part of Copernicus' contribution to this debate reads 'Moreover since the heavens, which enclose and provide the setting for everything, constitute the space common to all things, it is not at first blush clear why motion should not be attributed rather to the enclosed than to the enclosing, to the thing located in space rather than to the framework of space. This opinion was indeed maintained by Heraclides and Ecphantus, the Pythagoreans and by [H]Nicetas of Syracuse, according to Cicero. They rotated the earth in the middle of the universe, for they ascribed the setting of the stars to the earth's interposition, and their rising to its withdrawal. ....That the earth rotates, that it also travels with several motions, and that it is one of the heavenly bodies are said to have been the opinions of Philolaus the Pythagorean.'[32]

Recorde and Copernicus quote four common sources. Particular attention needs to be given to the 'Nicias of Syracuse' given by Recorde. No such person has been found. The edited version of *the Revolutions* gives a Hicetas of Syracuse who was quoted by Cicero as holding the views expressed. However Rosen notes that the version of this quotation which would have been available to Copernicus mistakenly used the capital letter 'N' in place of 'H'.[33] Indeed in the autograph version of his work, Copernicus notes in the margin that 'Nicetas of Syracuse' is to be inserted into the text. Reinhold has little to say on this debate and cites only the views of Nicias of Syracuse in passing, preferring to expand on some of Ptolemy's views. The glaring omission by Copernicus of the opinions of Aristarchus of Samos on the motions of the earth has caused surprise. This has been moderated to some extent by

---

[31] Copernicus N (1978) In: Dobrzycki J (ed) Nicholas Copernicus on the revolutions (trans and commentary: Rosen E). Macmillan, London, pp 11–12

[32] Ibid

[33] Ibid., 341

the discovery that such reference was made by Copernicus in the autograph version, but was struck out, and therefore not included in the printed versions until the nineteenth century. Recorde must therefore have had some source for his reference to Nicias Syracusius other than Copernicus or Reinhold. Both Plutarch and Diogenes Laertius could have provided the information and the works of both the authors were known to Recorde. It is clear that neither Copernicus nor Recorde intended to indulge in a debate about the axial rotation of the earth.

Continuing, 'M…I meane, concerning circularre motion: marry direct motion out of the centre of the world, seemeth more easy to be confuted, and that by the same reasons, which were before alleaged for proving the earthe to be in the middle and centre of the worlde.

Scholar. I perceive it well: for as if the earth were always out of the centre of the world, those former absurdities woulde at all tymes appeare: so if at any tyme the earth should move out of his place, those inconveniences would then appeare.

Master. That is truly to be gathered: how be it, Copernicus a man of greate learninge, of muche experience, and of wonderfull diligence in observation, hath renewed the opinion of Aristarchus Samius, and affirmeth that the earthe not only moveth circularly about his own centre, but also may be, yea and is, continually out of the precise centre of the world 38 hundredth thousand miles: but because the understanding of that controversy dependeth upon profounder knowledg then in this Introduction may be uttered conveniently, I will let it pass tyll some other time.

Scholar. Nay syr in good faith, I desire not to hear such vain phantasies, so far against common reason, and repugnant to the consente of all the learned multitude of Writers, and therefore let it passe for ever, and a day longer.

Master. You are too young to be a good judge in so great a matter: it passeth farre your learning, and theirs also who are muche better learned than you, to improve his supposition by good argumentes, and therefore you were best to condemne no thing that you do not well understand: but at another time, as I sayd, I will so declare his supposition, that you shall not only wonder to hear it, but also peradventure be as earnest to then to credite it, as you are now to condemne it. In the mean season let us proceede forward in our former ordre, wherein by ordre of your table I should speak of the circles in heaven.........'[34]

The affirmation that the earth 'is continually out of the precise centre of the world [i.e. universe] 38 hundredth thousand miles:' that Recorde attributes to Copernicus needs further examination for two reasons. Firstly Copernicus gave his distances in terms of units of earth-diameters, not miles. Recorde must have converted these units into miles using an earth radius of $3,436 \, ^{3/11}$ miles as he had done earlier in this text (page 125) for a geocentric system, where a distance from sun to earth of 1,120 units converts into $3,848,367 \, ^{3/11}$ miles. Secondly, in his writings Copernicus gives calculated distances for earth to sun but not specifically of earth to the centre of the world. In the *Commentariolis*, he gives the distance of earth from the sun, at apogee as 1,179 units, with the sun lying near to the centre of the universe. Rheticus in his

---

[34] Castle, 263[163]–165

Narratio Prima says unambiguously that the sun is at the centre of the universe and that the greatest distance of the sun from the earth is 1,179 units. In his *De Revolutionibus*, Copernicus wavers between the two concepts. Book IV Chap. 21 gives the higher apsis again as 1,179 units, with the lower apsis as 1,105 units and the mean as 1,142 units. If Recorde continued his previous practise of calculating only the closest approach of the planets to one another, he would have used the lower apsis to derive a value of 3,797,181 9/11 miles. This latter is very close to the earth to centre distance that Recorde gives and attributes to Copernicus. It is reasonable therefore to assume that Recorde accepted that the sun was at the centre of the world. Further, as a value of the lower apse of 1,105 earth diameters is found only in Book IV Chap. 21, it is implicit that Recorde was familiar with the text up to this point at least. Whether this constitutes evidence strong enough to be able to class Recorde as a Copernican is a matter of personal preference.

Even a cursory examination of the *De Revolutionibus* shows that it would have been well beyond the capabilities of the reader for whom the 'Castle' was intended, given that he had fully mastered the mathematics of Recorde's previous two books. It has also to be remembered that Ptolemy's work was considered to be too advanced for the reader. This book could not thus be expected to be the forum for Recorde to present his credentials as a Copernican. What can be said is that he probably understood enough about the new theory not to dismiss it without further experimental evidence, which evidence was to be a long time coming. Of English writers on the subject, only Thomas Digges came out in favour of the new theory.

Whether Recorde would have added anything to this debate other than clarity of presentation can only be guessed at. The activities of William Herbert, Earl of Pembroke ensured that we will never know. Recorde's treatment of the subject fits well into the background of astronomical texts of the second half of the sixteenth century as described by Johnson.[35] The latter suggests that 'The soundest position, scientifically, for the [European] writers of elementary astronomical text-books to take in the last half of the sixteenth century was one of suspended judgement, presenting the arguments on both sides, analysing their validity, and finally, though perhaps indicating a preference, stating that further evidence was necessary for determining which hypothesis was correct'.[36] Chronologically, Robert Recorde was in the van of such a 'scientific' approach in his treatment of the subject in the *Castle of Knowledge*. All the indications are that he would have proceeded to present the arguments on both sides and to analyse them. Given his general desire for evidential support it seems likely that he would have found it difficult to indicate a preference, although he did appreciate mathematical simplicity as found in Copernican theory.

13. *A briefe rehearsall of the parallele circles, with an instruction howe to fynde the distaunce of the Tropikes, and the greatest declination of the Sonne, and of every degree of the Zodiake from the Equinoctiall circle.*

[35] Johnson FR (1953) Astronomical textbooks in the sixteenth century. In: Underwood EA (ed) Science, medicine and history. Oxford University Press, Oxford, pp 285–302
[36] Ibid., 287

The student is instructed in how to determine the angular positions of the two equinoxes by observations of the maximum elevations of the sun at times around midwinter and midsummer using a quadrant whose design is illustrated. The position of the equinoctial is then defined as the 'juste middle' between the two tropics. The greatest declination of the sun is defined as the angle between a tropic and the equinoctial, at that time 23 degrees and 28 minutes. The quadrant is only graduated in degrees. How the minutes were measured is not stated. The reasoning behind the inclusion at this point, of a table giving the declination of the ecliptic and hence all the signs of the Zodiac and the sun, in degrees and minutes at intervals of a degree, is unclear. The Scholar is told that its use will become evident in the next treatise, presumably that on Cosmology or on Navigation, which of course never appeared. Having defined the positions of the tropics and Equinoctial, the discussion then returns to the positions of the remaining circles.

14. *That the Arctike and Antarctike circles are not permanente, but mutable, accordinge to the chaunge of the regions, and so their quantities varieth, and their distaunce altereth, in respect to thother paralleles: and their ordre chaungeth diversly.*

15. *The Zones, beyng immutable, ought not to be distinct by the Arctike and Antarctike circles which are mutable, but rather by the Polare circles whiche persevere styll, and keepe their quantities, their distaunce and their ordre uniformly.*

16. *That there ar no Zones uninhabitable other for heat or could, but may be and are also inhabited, as it is well knowen.*

    The main purpose behind this section of the work is to dismiss the views of certain of the Greek astronomers regarding the Zones into which the earth had been subdivided. It also affords Recorde the opportunity of asserting the primacy of reason, in the form of Aristotelian logic, over authority in determining truth. Parmenides, Aristotle, Polybius, Cleomedes and Proclus described the Arctic and Antarctic circles as enclosing those stars 'which ever appear above our horizont, or never appear above the same.' The Master points out that using this definition, each Climate i.e. latitude has its own circles, which differ from all others. For certain latitudes such circles would overlap the Tropics e.g. for Wardhouse the Arctic Circle would be greater than the Tropic of Cancer which lay above the Horizonte. Such variable boundaries could not be used to delineate the extent of 'certain places' such as the temperate Zones. Aristotelian logic in the form of Ferio, a mnemonic for the fourth mood of the first figure of the syllogistic argument, is used to demonstrate logically that if the sun keeps its yearly course then the temperate regions must remain the same in extent, so they cannot be bounded by the Arctic and Antarctic circles as defined, the Tropics being also defined by the movements of the sun.

    Posidonius and Strabo are credited with having spotted the error of their countrymen, but transmission of the views of Strabo had been hindered by poor translation from the original Greek into Latin. As was his wont, Recorde remedies this deficiency and also provides English translations. The Student points out that partly the obscurity of the matter arose from hypallage, or changing the speaker's person.

The Master then proposes that the example of John de Sacrobosco is followed, by defining the Arctic and Antarctic boundaries of the Temperate Zones as lying where the pole is 66 and a half degrees above the Horizonte, as Strabo may also be interpreted as having implied. To avoid confusion with the earlier established definition of the circles Recorde proposes to call them the Polar circles, so introducing the word 'polar' into the English vocabulary. The Zones so defined each possesses distinct properties with respect to the behaviour of shadows within their boundaries viz within the polar circle men's shadows run around them; within the Temperate Zone the noon shadows move in only one direction; within the region lying between the two Tropics, the noon shadow is cast in two directions. As such shadows are defined by the behaviour of the sun they cannot err. The modern definition of the Arctic and Antarctic circles is that used by Recorde for his Polar circles, worded a little differently but lying at the same latitudes.

Having resolved the extent of the five zones, the Master then points out another error of the 'elder' Greeks in supposing that no life was possible in the zone lying between the Tropics nor in the polar regions had been discredited by voyages both recent and to a lesser extent by their own writers such as Erastothsenes, Posidonius and Ptolemy.

Consideration of the remaining circles of the Sphere is then resumed.

17. *The Zodiake is named of the 12 Signes, which signes are taken in divers significations. and howe any starre or Planete is named to bee in any signe. also what is the longitude latitude and declination of any starres or Planetes.*

The Master dissociates himself from the idea that the signs of the Zodiac actually take the celestial form of the beasts named but the stars ascribed to the signs can be seen there. The Zodiacal circle is divided equally into 12 parts to each of which a sign is allocated as described in earlier treatises. He then describes how the positions of the planets and stars are signified with respect to the Zodiac. The first signification is for the sun alone, which is said to be in a sign when it lies in a direct line between the centre of the earth and that sign. As the other planets may show a declination from the ecliptic, the Zodiacal circle is accorded a latitude of 12 degrees to accomodate the deviation. This is the second signification. A planet is said to lie in the sign when it lies within the solid angle defined by the area of the sign so described and the centre of the earth. The third signification applies to the stars whose declinations place them outside the second signification, such as the pole star. To accomodate these six great circles are drawn, each passing through the boundaries of the signs and the poles of the Zodiacal circle, subdividing the globe into 12 equal segments. If a star falls between two such circles then it is said to lie in the sign associated with that segment. Possibly the most interesting feature of this presentation is the way in which Recorde moves stepwise from the particular to the more general cases. The Declination of the planets is defined as their distance south or north from the equinoctial line and their motion in Longitude their distance from the beginning of Aries. This matter is returned to in later when considering the topic of Ascensions.

18. *The Colures, what they be, and how many in numbre, and whereof they take their name.*

This is a primarily a reiteration of previous descriptions, but Recorde does chide John de Sacrobosco for giving a false derivation of the name colure. However he excuses him '…: but men must bear with the ignorance of that time, for lack of knowledge in the Greeke tonge.'

19. *The Horizonte celestiall and terestriall, how they be distincte: where Proclus sentence is reprehended, and there severall tables set forth for distinction of howers, according to distaunce of myles from easte to weste, and that for diverse climates.*

The Master applauds Proclus for distinguishing between the two horizons. The Celestial horizon divides the heavens into two equal parts of which only one is ever visible to the terrestrial observer and extends in diameter to the eighth sphere. The 'Earthly horizont', as Recorde prefers to call it, 'because it servyth for sightes on the earthe and the water only.' At this stage praise of Proclus turns into criticism for having attributed a radius to this horizon of 125 miles compared with the 22 $^1/_2$ miles of Macrobius and the 20 miles or thereabouts well-known to mariners. The 47th theorem of the Pathway is then invoked to demonstrate qualitatively that Proclus' estimate cannot be true. Presumably the argument was not made quantitative because the Scholar had not yet been introduced to the necessary calculational methods involved. The Master nods in this direction by pointing out that the distance visible increases with height of observer but that whilst such an elevation might permit a man to see maybe as far as 80 miles, but certainly not as far as 125 miles. In fact to see such a distance, an elevation of over 2 miles would be needed but it is not clear whether the Master was querying the ability to find generally land at a height of 2 miles or whether he was querying the ability to see anything at a distance of 125 miles even if the former elevation was achievable.

Criticism of Proclus is continued in his discussion of the ways in which the hours of the day change with longitude and latitude. Proclus said that the meridians change every 300 leagues in going from east to west but every 400 leagues in going from south to north which is clearly wrong. His second error was that, traversing the Climates from south to north, the length of the longest day increased proportionately with distance. This again was manifestly wrong as noted by observations from England, Scotland and Iceland. The Scholar is prompted to observe 'that who soever will travaile in these sciences with profit, must lean rather to reason, then to authoritye, els he may be deceaved.' two tables are then presented, giving the difference of hours with distance in miles from east to west for the latitudes of the equator and 51 degrees 55 minutes north. The latter is included so as to provide material of relevance to 'our own cuntrie.' Further tables will be constructed 'yf you call on me and put me in mynde thereof, els the neccesitye of provision for my family will make me forget suche promises.' The Scholar is taught how to use the Tables and how to interpolate. He is also given a table in which are listed the miles 'agreeing to i. minute of time' at intervals of one degree of latitude from 0 to 90 degrees that would allow the Scholar to construct his own tables for such intervals.[37] This leads into a discussion of the Climates.

20. *The ordre and numbre of the Climates, with the elevation of the Pole and the quantities of the longest daie in eche of them.*

Orontius is quoted as giving an introduction to the subject, but falling short in his presentation inasmuch as his table of Climates extends only as far as 66 and a half degrees, whilst in their explorations the English Muscovy Company had voyaged as far north as 75 degrees. In anticipation of them going even farther north, the Master gives a Table that shows the extent of the Climates from the equinoctial to the North pole. This table collates the positions of the seven Climates of the Greeks with the parallels from 1 to 33 as enumerated by Ptolemy and with the parallels from 1 to 49 degrees, defined by an interval of a quarter of an hour in the length of the longest day for that parallel. The elevations of the parallels are given and the number of climates extended to include one as far north as England, following Reinhold's example. The names of nine zones are given together with their derivation. These Climates of the northern hemisphere are mirrored south of the equator. Further comment on the Climates is to be deferred to the treatise on Cosmography. The Scholar sees how from this Table he may read across from latitude to length of day and vie-versa but wishes to know how to interpolate from latitudes to derive more accurate values for hours of daylight. However he then agrees that to ask for an interval of half a quarter of an hour is 'more precise then our clocks or dials do serve unto,' and accepts the Table as adequate.

21. *Of ascention Astronomicall and Poeticall, and how every one of them is distincte. with certain rulesof ascention Astronomicall, and tables for the same, both in the Ryghte sphere, and also in divers Oblique spheres. with an examination of the rules of Iohn de Sacro Bosco.*

Recorde had to cope with a major presentational problem in dealing with the topics of the risings and settings of the constellations of the Zodiac. Up to this juncture there had been only minor disparities in sequence of topics for him to resolve, which he usually settled in favour of those of Ptolemy and Copernicus rather than that of Sacrobosco. Presentations had been largely of a qualitative or at best of a simple quantitative nature. On the few occasions where trigonometric functions were used he finessed the issue by deferring it

---

[37] Commenting on this Table, the Scholar says that their clocks or dials have an accuracy of less than half of a quarter of an hour.

or by declaring it too difficult. These were not options in dealing with the topic of zodiacal ascensions at the equatorial circle i.e. for the right sphere, or at other latitudes i.e. for the oblique sphere. Ptolemy and Copernicus introduced and used spherical trigonometry to calculate the appropriate parameters as quantitative presentations in reasonably extensive tabular form. Sacrobosco dealt with matters in a largely qualitative fashion and in a condensed presentation, which the Scholar found difficult to follow.

Recorde took the path of presenting the data in tabular form with no attempt to detail its derivation and then using it to clarify or correct Sacrobosco's assertions. The data he gave for right ascensions were however neither those of Ptolemy nor those of Copernicus, but were derived from the Prutenic tables of Reinhold published in 1553. The values were given in the latter in degrees, minutes and seconds. Recorde approximated these to just degrees and minutes with consistency, other than when he had to deal with values that had 30 s as the last term. Here consistency eluded him, but there were few such occasions. The complete table Recorde gives for oblique ascensions is that calculated for latitude 52 degrees. This table is specific to the 'Castle' and must therefore have been calculated by Recorde himself or calculated for him.

There are a number of other more abbreviated tables that are also specific to the text. These findings make it clear that Recorde was fully up to date in the material he was presenting and also understood spherical trigonometry sufficiently to carry out his own calculations or have been aware and in touch with those who could do it for him. There can have been few of such competence in England at the time.

Recorde followed the presentational sequence of Sacrobosco, using literal quotations from the latter's text on which he commented, not avoiding criticism, clarification or correction in the process. This meant that the commentary was almost an order of magnitude greater in length than the original and probably the longest section in the whole of the book.

The Master first defines an ascension as the rising of any star or sign above the horizon. He then makes a distinction between Astronomers, who are interested at all times in whether the stars rise right or obliquely and others such as mariners, husbandmen and physicians who mark the risings only at sunrise and sunset. The Scholar queries why, if Astronomers consider only the first form, attention should be given to the other forms in a treatise that is specifically concerned with Astronomy. The Master takes a conciliatory line citing their roots in ancient data. These latter risings were those described by poets, as a result of which Sacrobosco and others called them Ascensions Poetical. Sacrobosco used the descriptions of poets extensively in his text and devoted more time to this aspect of ascensions than did Recorde. The latter uses the opportunity to deplore once more, errors committed by Sacrobosco amongst others in translating poetical texts due to their ignorance of the Greek language. Recorde seems to be underlining the need for applications of Astronomy such as in Navigation to be soundly based on an understanding of the astronomical principles involved.

The Master then turns his attention to the main topic viz. the astronomical ascension, defining it severally as '… the certaine limitation of som pointe of the equinoctiall circle, whiche rises justelye with anye starre, and largely taking the use of that name.' Further, 'It betokeneth also the arke of the Equinoctiall circle, which lyeth betweene the beginning of the same Equinoctiall circle at Aries [i.e. the point where Aries rises], and extendeth to the juste degree that rises with any starre or signe.' Thirdly ' …the ascension of a signe or constellation (whiche includeth a certaine measure in lengthe,) is that just arke of the eqinoctiall, which doth passe the Horizont with the whole signe or constellation.' These are not mutually exclusive and represent a broadening of definition that must have taken place with usage.

The Master then defines Right Ascension and Oblique or Crooked Ascension in much the same words as Sacrobosco. 'Ryghte ascension, is defined to bee that, with whiche a greater portion of the Equinoctiall dooth ascende.[38] And that is called Crooked or Oblique ascension, with whiche a lesser portion of the Equinoctiall doth ascend.' The Student does not understand the meanings of 'greater' and 'lesser', but is told to wait until later for clarification.

This passage draws attention to a deficiency in Sacrobosco's presentation and what is seen as a further deficiency is then exposed. If ascensions pass from right to oblique and vice versa, they must pass through points where the ascension must be equal to the portion of the Equinoctial that ascends with it. This the Master calls a 'Mean ascension'. Whilst logically correct, Recorde's suggestion was stillborn. However it leads into Sacrobosco's first general rule relating to ascensions in the Right Sphere viz. that the four quarters of the zodiac starting with the two solstices and the two equinoxes have ascensions that are equal and also equal to that of the associated equinoctial arising. This rule is used to reinforce the Master's call for the third category of Mean ascension.

The second rule of Sacrobosco, that every two arcs of the zodiac opposite and equal, equally distant from the four fiduciary points already mentioned have equal ascensions, is also accepted by the Master. However, at this stage he introduces the table in which he has set out the degrees of the equinoctial corresponding to the degrees of ascension for every sign in the right sphere, starting with the beginning of the equinoctial at Aries. He uses this to illustrate that the sum of ascensions for each of the four quarters starting with Aries, totals 90 degrees and are thus mean ascensions. But he then goes on to point out that within each quarter two of the signs individually have ascensions less than that of the equinoctial and are therefore crooked ascensions whilst the remaining sign compensates for the shortfall by having an ascension greater than the equinoctial.

These observations provide a natural introduction to a discussion of Sacrobosco' third rule, of which the Scholar says '…. But now cometh to my mind the sayings of Joannes de SacroBosco, whiche longe hathe troubled my minde, and I can not learne of anye man howe to understande him well; for in

---

[38] Measurements of the right ascension of a celestial body, together with that of its Latitude provide coordinates which locate the body on the Celestial sphere and can be mapped theoretically onto the terrestrial equivalent.

mine opinion his woordes import an impossibilite. he blameth the argument as evel: These two arkes are equall and they begin to rise togither, and continually ther riseth a greater portion of the one arke then of the other; ergo that arke will be risen soonest, whose greater portion did alwaies rise. This argument seemeth invincible in mine opinion and yet John de Sacrobosco for improving of it alleageth an example, wherby as he semeth to intend, the antecedent maybe true, and the consequente false: and therefore the argumente must needes be naught.' The example used is that for the quarters starting with the first point of Aries to the last of Gemini and for the first point of Libra to the end of Sagittarius the ascension of the portion of the Zodiac is always greater and yet the two quarters ascend fully together. The same is true for the remaining two quarters but here the portion of the equinoctial ascending is always greater than that of the zodiac, but again they ascend fully together. This the Master describes as '.... a great fallation by Amphibologye, as Logicians do call it, so that in one sense it maye be true and in an other it is false. And fyrste for declaration of John his meaning (as I thinke) marke as many partes of those two firste quarters as you lyste, and still by the former table as well as by tournynge the Sphere it selfe, it will appeare manyfestly , that the portion of the Zodiake is ever greater then the matche portion of the Equinoctiall.' Various selections from the table are given to demonstrate this point. This troubles the Scholar who is wedded to the argument that if two things are equal if you subtract unequal amounts from them unequal amounts will remain, the remainder being least for that from which the greater portion was taken. The Scholar then applies this argument to the first quarter where for the sign of Aries its ascension is but 27 degrees 40 minutes, leaving 62 degrees 20 minutes of the equinoctial to ascend with 60 degrees of the Zodiac i.e. unequal portions ascend together of which the Equinoctial is obviously the greater. By selection of interval between points lying in Taurus and Gemini, the Master shows this to be true, but then proceeds to show that there is an interval lying wholly within Taurus for which the ascensions of Zodiac and Equinoctial are equal. The ascensions for the first degree of Aries and the second of Gemini are equal to one another but neither are equal to their equinoctials. The Scholar concludes that Sacrobosco's fault is in his words not in his meaning. The Master avers that 'Such meane matters must be winked at in other, but not folowed.' The same explanation is then applied to the two quarters beginning with Cancer and Capricorn where, as has been mentioned already, the obverse situation arises.

The Master then makes the important observation 'that all Astronomers commonly do call the Right Ascension so largely, that it extendeth to the ascension of all signes in a Right sphere and so they name the Oblique ascension the rising of all the Signes in anye Oblique Sphere, whereby it appeareth that they give the name of Right and Crooked ascensions, according to the Horizontes or positions of the Sphere and not after the quantity of time in their ascension.' In other words the original precise definitions have been debased.

For the right sphere, descensions are the same as the ascensions but the same is not true for the oblique sphere. In the latter the Master states that signs which ascend right descend crooked and vice versa, so that the descension of any sign in an oblique sphere is equal to the ascension of its geometrically opposed sign. Further, the sum of the ascension and descension of any sign in the oblique sphere is the same as the sum of that sign in a right sphere. The Master then introduces a table which he has prepared giving the ascensions of all signs for the latitude of 52 degrees, set out as for the previous Table for right ascensions. Examples taken from the table are used to illustrate quantitatively the statements of the previous paragraph. These latter contentions are not to be found in Sacrobosco as he did not display such numerical data.

The data in the table are then used to show that for latitude 52 degrees, as for the right sphere signs rise both right and crookedly, so the same is true for the oblique sphere where from Cancer to Capricorn in correct order six signs rise right and from Capricorn to Cancer, the six signs rise crookedly. This rule is stated to be true for northern latitudes from 30 degrees, under which lie Mount Sinai, Memphis, Alcayre, Madiera and Terraflorida, to 66 and a half degrees where are found parts of Norway and Iceland. The master is then provoked to discuss the same subject for more southerly latitudes. Tables are given that list the ascensions for each of the 12 signs of the Zodiac for latitudes of 1,10,11,20,29,30,50,60 and 66 and a half degrees, grouping them according to whether they are right or crooked ascensions. The ascensions are given both as degrees and seconds or as hours and minutes. Each set of signs that constitute half of the Zodiac are summed and shown to be either 180 degrees or 12 hours. The Scholar says that he has not seen the like of such tables before. Again Recorde must have either computed them himself, or have commissioned them. They are then used in an examination of the validity of the seven rules of ascension of the oblique sphere given by Sacrobosco.

The first rule, as given by the Scholar is that each half of the equinoctial ascends equally with each half of the Zodiac. The Master uses the tables to demonstrate that this is true only if the starting point is either of the equinoctial points but not at the tropic points or indeed at any other intermediate point. For these latter points the ascensions are either more or less. The qualification given by Sacrobosco that the halves have to begin with the two equinoctial points is thus justified.

In a diversion, the Scholar draws attention to the singularities that are found for ascensions between 66 and a half degrees and the Pole, where the six signs lying between Capricorn and Gemini are all assigned an ascension of zero. This he conjectures as meaning that the signs have no ascension, which he knows to be false. He is told to take his sphere, put it into the oblique position, put the first degree of Aries at the east horizon ready to ascend and then turn the globe half a degree to the west. The six signs mentioned all ascend simultaneously and instantaneously so they can only be assigned a zero ascension.

Returning to Sacrobosco's rules, the Scholar quotes the second as saying that for the half of the Zodiac between the beginning of Aries and the end of Virgo the portion of the Zodiac which rises is greater than that of the equinoctial but both halves rise together. The table for 50 degrees of latitude shows this assertion to be true. However, the tables for one degree and ten degrees show that the equinoctial show the greater proportion, hence the rule is wrong. The Master then asserts that the rule applies only between 13 and 66 and a half degrees of latitude. The third rule, which states that for the other half of the zodiac a greater part of the equinoctial rises than that of the zodiac, is likewise shown to be true only for latitudes greater the 10 degrees.

The fourth rule states that arcs which succeed Aries upto the end of Virgo in the oblique sphere have smaller ascensions than their comparable values for the right sphere. The Master provides a new table of ascensions for 19 latitudes from 0 to 66 and a half degrees, inclusive, for both northerly and southerly signs. Referring to this table the Scholar find the rule appears to apply for the first three signs, but then the opposite is true for the next three signs. The Master shows that if the reference point for the arc of the sign is the first degree of Aries in all cases, for the southerly signs the rule applies. He provides another table calculated on this latter basis for the northerly signs, to prove the point. The Master then takes the opportunity to return to his earlier thesis that there are potentially three types of ascension depending upon the fiduciary points from which the individual arcs are measured. Further, account has to be taken of the latitude of the oblique sphere in question. Given clarification of the ground rules used by Sacroboco, his first four rules apply. Using the data from a fourth table, similar to the third but for the southerly signs, a similar conclusion is reached.

The sixth rule, that in an oblique sphere, two equal and opposite arcs have combined ascensions equal to the sum of their right ascensions follows from considerations of symmetry, is agreed.

The seventh rule, derived also from considerations of symmetry, is that any two arcs which are both equall and equidistant from one of the equinoctial points, have equal ascensions. This rule the Master gives correctly. Some copies of Sacrobosco's text provide incorrect versions.[39] The point is emphasised that this rule applies to any arc, not just a complete sign.

Recorde's treatment of this topic is clearly more searching than that of Sacrobosco and, inasmuch as it includes all latitudes up to the polar circle, both more comprehensive and more correct. The crucial evidence he uses was not available to Sacrobosco as none of the latter's possible sources provided information for the oblique sphere below 12 and a half degrees of latitude, territory which had become of significance to sixteenth century explorers.

The main weakness of Recorde's presentation is the obvious one that the accuracy of his tables has to be taken as read. He provides no description of their derivation for the obvious reason that the reader would not have been able to understand the calculations, which involved the use of trigonometric functions.

---

[39] Thorndike L (1949) The sphere of Sacrobosco and its commentators. University of Chicago Press, Chicago, Corpus of Medieval Texts 2., 132

22. *The distinction of howers into howres equall, and howers unequall be considered*
    in twoo divers sortes, with tables sette forthe for eche sorte, concerninge their
    quantities.

The master first tackles the confusion in the mind of the Scholar over unequal,
natural and equal hours. He first deals with the vulgar form of unequal hours
which result from the division of the natural day and associated night into 12 equal
parts, irrespective of their variable lengths. To illustrate the variability of these
hours, a table is provided which expresses them in terms of equinoctial hours,
defined as the time taken for 15 degrees of the equinoctial to pass the meridian
line. These values are provided for lengths of day from zero up to 24 hours at
intervals of 12 minutes. It points up starkly the variability of this type of hour.

The second type of unequal hour is that which Sacrobosco calls the natural
hour, defined as the time for ascent of half a zodiacal sign. Of this hour the
Master says '.so againe it seemeth to the wisest sort of men, that the motion of
the Zodiake, whose severall forme of ascension for every halfe signe, dooth
make a severall and distinct quantitie of Unequall howers, and have no fewer
sortes of differences, then there be distincte and several degrees or pointes, at
which the arke of 15 degrees maye beginne his ascension, as partly in this table
following it dooth appeare:..', The table given, was calculated for latitude 52
degrees and shows the variability of the 'natural' hour in terms of the equinoctial hour for intervals of six degrees. Further refinement by linear interpolation
is explained. Again, the value of this table is not immediately apparent.

23. *Of daies Artificiall and Naturall, and what are the causes of diversitie in eche*
    of them, with tables for the quantities of the same; and a declaration of the
    Sonne rysinge and settinge.

The treatment of these topics offers the same explanations as given by treatises from Ptolemy onwards. The natural day is defined in terms of the rotation
of the eighth sphere, i.e. sidereal time. The reason for its difference from solar
time i.e. from rising to rising of the sun is explained and quantified, as also is
its variation as can be determined from ascension tables. A second reason for
variation is the eccentricity of the orbit of the sun, discussion of which is
deemed inappropriate for this particular treatise. The third potential cause of
variations arises from the obliquity of the Horizonte which enters into play if
the day is measured from sunrise to sunrise. Astronomers eliminate this effect
by measuring the cycle from noon to succeeding noon. Summing up the Master
says, 'Touching the diversities of the natural days this may suffice: and for a
common and meane quantitieyou may assign 24 howers and 4 minutes, bicause
that is the common nombre: for although many be greater, yet many other be
lesser, and this numbre is moste nyghest the mean.'

Up to this juncture, Recorde largely follows the treatments by other
authors, contemporary and ancient. His explanations for the diversity of
artificial days, defined as hours of daylight, as a function of latitude, given
earlier when dealing earlier with the topic of climates as arising from the obliquity of the Horizonte are repeated. Whilst understanding the general thesis the

Scholar wants to know how to find the values of artificial days and of rising and setting of the sun. The Master proposes to do this in two ways viz. by use of the Sphere and the data found in the Ephemerides or from calculations which he will give in the form of tables.

Use of the sphere was felt to be the most appropriate for this treatise, but was of limited precision and additionally, knowing the date for which the length of day and its latitude required access to the Ephemerides.[40] Given this data the arc of that day in degrees could then be determined by manipulation of the sphere and converted into hours by dividing by 15. The Scholar gives a conversion table, more extensive than that of Orontius, which is dismissed by the Master as teaching negligence, for the calculation is easily carried out.

The tables provided by-pass the need for the Ephemerides, giving the lengths of day and night, in hours and minutes, for each degree of the signs of the Zodiac, which in turn are associated with days of the month. These lengths are calculated for the latitudes of 51,52,53,54 and 55 degrees, which are of primary interest to England, but will of course be applicable to the same latitudes in other countries. A method for linear interpolation between these latitudes is described. The master declines to work out the values for all elevations of the Pole because it would be of only incidental interest to this treatise.

24. *The names of the constellations, with the numbre of their stares.*

This section constitutes a listing of the northerly and southerly constellations together with the signs of the Zodiac and the numbers of stars they contain. These latter total 1,025, which is three more than given by Ptolemy. This is because he omitted a number of stars in the constellation of Berenice. The master agrees that there are many more stars to be commented upon, but will save them for a treatise on Cosmography. He makes only one correction to the published wisdom on this topic. Proclus says that the star Canopus can be seen at Rhodes only from high ground and not at all at Alexandria. The Master sees this as contrary to commonsense as it rises there by about a quarter of a sign above the horizon. This he attributes to a demonstrable error in transcription.

25. *A briefe declaration of the motions of the Planetes, and consequently a reasonable proofe for the numbre of their spheres. And farther what occasion there was, that men should imagine the ninthe and tenth sphere to be, where as there can be none seene above the eight sphere.*

The conventional Ptolomeic view of the world is presented with each planet having a sphere and the stars occupying an eighth sphere. The Master draws attention to the observation of the apparent eastward movement of the stars relative to the sun at the equinoctial point. There was also the apparent north-south movement of the stars called trepidation. Together these apparent motions

---

[40] Ephemerides were tables listing the future positions of heavenly bodies over certain periods of time.

had been interpreted as calling for two more spheres, but the Scholar is advised against following this path until he learns more of the subject.

In the course of this presentation, the Master notes that Saturn was seen in the 26 degrees of Aries on the first of September 1556, which means that the book must have been completed later than this date.

26. *A shorte explication of the eclipses of the Sonne and the Moone*

The Master is importuned by the Scholar to write on the eclipses of the sun and moon. He is persuaded reluctantly to do so because Sacrobosco had written of the progression, retrogradation and station of the planets as well as of the two eclipses mentioned, but feels that 'His woordes are shorte and therefore obscure and so should my wordes be. beside that, it is a disordrely form to put the carte before the horse: I mean to write of the passions of the Planets, before I have sufficiently taught the full ordre of their motion. Therefore I will saye in fewe wordes, that the reasons of the passions cannot be taught aptly before the Theorikes of their motions.' However he does spell out the standard explanation as clearly as most other contemporary authors.

Where he does differ for these latter is the use he makes of the explanation to comment on a matter which must have caused him some concern. He points out that the eclipse of the sun which happened at the death of Christ could not have been a natural eclipse, because it happened at the time of the full moon. It was reported upon not only in the Scriptures but also by Dyonise the Areopagite and Apollophanes his companion who were in Alexandria at the time. The Master does not volunteer any explanation but says '...there was never any good Astronomer, that denied the Maijestie and providence of God, though many others denied both: but now farewell for a time: I am driven to omytte teachinge of Astronomye, and must of force go learne some lawe.' this was mans law not that of God.

Some 60 corrections are appended to the text, preceded by a poem urging the reader to correct the faults before he starts to read the book.

Consequentials

The material presented in the 'Castle' was up to date and had been critically evaluated. It was, as promised in its title, a description of the fabrication of both the solid and armillary spheres and their relevance to matters both celestial and terrestrial. A second edition of the 'Castle was published in 1596 so a continuing demand for it must have existed. No evidence of a third edition has been found, so this date seems to have represented a high-water mark in its perceived usefulness. Comparison of the contents of publications by English authors on topics related to those covered in the 'Castle' with those of the 'Castle' at this time will help to put the latter in perspective both with regard to its influence and its prescience.

Subsequent to the 'Castle', a bifurcation in the interests of English writers took place, with little work being published on matters relating purely to astronomy and its associated mathematics, but a great deal on matters relating to its application to navigation. This was a course that Recorde himself had intended to follow, having promised future texts on Cosmology and on Navigation.

An exception to this partition was William Cunningham's *The Cosmographical Glass*, published in 1559. It was more extensive than the *Castle* inasmuch as it presented four books, the first two of which deal with the same areas as the *Castle*, the last two being concerned specifically with geography. Where the two works overlap, the *Glasse* is but a much reduced version of the *Castle* and uncritical, even including errors of other authors that had been corrected by Recorde. Debts to Recorde's other works were however fully acknowledged by Cunningham. It did not achieve a second edition, but it was written in English and did accompany a copy of the *Castle*, with Frobisher, to the New World.

The *Castle* is an astronomical text but it specifically avoids other than glancing references to the 'Theorickes of Planetes', promising that a treatise on Cosmology would follow. The preliminary notices on eclipses and the reference to Copernicus' theory provides some indication of what might have followed and the necessary astronomical ground work had been laid so that such a work would have been a natural sequential in Recorde's series of texts. Had it been written it is difficult to see that it could have bettered Thomas Digges' description of Copernican theory in his *A Perfit Description of the Celestial Orbes*, published along with his augmented version of his father, Leonard's, *Prognostication* in 1576. The former was essentially a free translation of Chaps. 7, 8 and 10 of Book I of the *de Revolutionibus*, Chap. 9 dealing with the more complex issue of trepidation having been omitted. Digges' advocacy of Copernican theory was far less cautious than that of Recorde. His commitment to heliocentricity was presaged in his *Alae seu Scalae Mathematicae* of 1573. In this work he presents his observations on the new star that had appeared the previous year in the constellation of Cassiopeia. Digges very obviously hoped that his observations on the diurnal parallax of this star would establish the validity of the Copernican theory. He was disappointed, as good though his observations were for the time, the theoretical bases for his hopes were flawed and later dismissed by Tycho Brahe as irrelevant.[41] However he did developed the geometric theorems needed to deal with the problems of stellar parallax. Dee published a similar book on such trigonometric theorems shortly after Digges, but a longer work containing material dealing with the new star does not appear to have been published. What is known of the commitment of Dee to Copernican theory does not extend beyond a statement he made in his preface to a revision of the Prutenic tables of Reinhold calculated for the latitude of London, published in 1556, a little before the 'Castle', by John Field.[42] In this he stated that his preface was not the place to

---

[41] Russell JL (1973) The Copernican system in Great Britain. In: Dobrzycki J (ed) The reception of Copernicus' heliocentric theory. D. Reidel Publishing Co., Dordrecht, pp 189–240, 192–193

[42] Field J Ephemeris anni 1557 currenti iuxta Copernici et Reinholdi Canones Supputanta..

describe the new Copernican hypothesis, which scarcely commits him either way, but he did praise the new calculations which were based on Copernican theory. The single person most responsible for swaying thought in England in favour of Copernican theory was clearly Thomas Digges, not least because he pleaded the case both in English and in Latin.

Following these efforts Digges devoted himself very much to the practical use of mathematics. As already noted, in 1576 he published a corrected and augmented edition of Leonard Digges', his father, almanac of 1555.... *A Prognostication everlasting.* Part of the augmentation consisted of an appendix entitled 'Errors in the Arte of Navigation commonly practiced'. With the exception of an error in taking observations of the sun, the remainder are those arising from lack of understanding of the difference between spherical and planar geometry. The points he raised had all been addressed in the *Castle* but the only possible reference that Digges makes to the latter is where he says that one of the errors would be manifest to anyone who knew but the first principles of cosmography. His father had addressed such matters qualitatively, Recorde did so quantitatively.

The only treatment of astronomy by an Englishman in an academic context was that of Henry Savile in his ordinary lectures on the subject delivered to undergraduates at Oxford University starting in 1570. Goulding has estimated that the course may have extended beyond 4 years in duration.[43] The lectures were never published and exist only as manuscript notes.[44] An extensive survey of the history of its history is used to introduce the subject, but the main concern is with a very lengthy commentary on the Almagest, probably largely derived from the writings of Regiomontanus. His treatment of the geometry of spherical triangles is noteworthy for its clarity and comprehensiveness.[45] What has attracted most attention however is his presentation of Ptolemaic and Copernican constructions alongside one another when discussing planetary theory.[46] Goulding comments 'In his perceptive conceptualization of Copernicus, Savile differs from other more celebrated writers on the new system of the world such as Robert Recorde and Thomas Digges. Although they were far more unequivocal in their support for the heliocentric form of the universe, they did not provide the reader with the mathematical tools necessary to understand Copernicus's innovations.'[47] As far as the provision of mathematical tools is concerned this is certainly true for Recorde. Neither he nor Digges dealt with the topic of trigonometry. Recorde specifically pointed out that he was not going to deal with the topic of trigonometry, either plane or spherical, at this stage of his writings,

---

[43] Goulding R (1999) Testimonia humanitatis: the early lectures of Henry Saville. In: Ames F (ed) Sir ThomasGresham and Gresham College. Ashgate, London, pp 125–145

[44] Bodleian Library, MSS Savile 29,31,32

[45] Ibid. MS Savile 29ff.83 seq

[46] Ibid. MS Savile 32

[47] Goulding R (1995) Henry Savile and the Tychonic world-system. J Warburg Courtauld Inst LVIII: 152–179, 152
Planets and Orbs, pp 113–122

although he was competent to do so. His book on plane geometry had already been published. Digges presented a new set of geometric theorems relevant to stellar parallax, and was again clearly competent in matters trigonometric. In Savile's compendious list of writers on astronomical topics Recorde and Digges are both accorded a place but no recognition of their contributions is made. Neglect of Digges might be forgiven as his work had not been published at the time of Savile' lectures. A copy of the *Castle* found its way into Savile's collection, but only as the 1596 edition. Digges' *Perfit Description* of 1578 was also found there but not his *Alae*. Whether Savile's lectures had the same influence that Recorde and particularly Digges seem to have had is debateable. How well Savile's audience was mathematically prepared and able to profit from his presentations is far from clear. He expected them to have three terms of arithmetic and two of geometry completed and supported by private study. The arguments he used for the establishment of his chairs of Geometry and Astronomy at Oxford some 40 years later suggests that the fields he was attempting to cultivate still contained a lot of stony ground in the form of lack of basic mathematical competences.

Further advances in astronomical thinking in England, in common with Continental Europe, had to await the advent of more accurate observations based on the development of better astronomical instruments. This sequence was in full accord with the philosophy of both Recorde and Digges with respect to the need for experimental verification of theory.

In the period between the two editions of the 'Castle' therefore nothing had been published in English that assaulted seriously its status as a sound introduction to astronomical or even cosmological thought of the time. The same was less true of matters relating to the use of the globe for terrestrial purposes viz. geography and navigation.

A number of practical aids to mariners in the form of almanacs were published in successive, updated editions during the rest of the century. Arguably the most important of these was those of William Bourne, who published first *An Almanacke and Prognostication for three yeares ..... 1571. and 1572. & 1573. now newlye added unto my late rules of Navigation, was printed iiij. yeres past.* This was the second edition, no copy of the edition of 1567 has survived. The fourteenth rule deals with the longitude and declination of 12 notable fixed stars for navigation. Closing this section, Bourne declines to expand on matters astronomical as being unworthy to do so in the face of a number of authors on the subject, both ancient and modern. Amongst the latter he names '...most famous men, in these our daies as Johannes de Sacrobosco, and Horontius, Jewafrisius, and also in the English tong Doctor Recorde, with a great number more, whiche I passe over, notably seene in the Mathematical sciences.' In his next publication *A Regiment for the Sea*, published in 1574, Bourne corrected the error ridden Calendar of the previous work, augmented his Rules and generally rearranged his material. This work proved popular and was republished throughout the rest of the century. In its preface, Bourne again defers to 'a great number of excellent learned men in the Mathematicall Science [who] have written diverse bookes of Cosmographie and Navigation, yet notwithstanding I have written this Regiment for the Sea with a fewe rules of

Navigation, as it were a nosegay whose floures are of mine owne gathering.'
The learned men were not named. The data presented by Bourne in his first almanac
can be compared with that in the *Castle* in two places. Bourne's twelfth rule
'.. sheweth how many myles will answer to one degree of Longitude in every several
Latitude between the Equinoctiall and any of the two Poles.' has its counterpart
in Recorde's table that tells 'How many myles do answer to one minute of tyme in
every several latitude.' The former is less comprehensive than the latter but where
they coincide Bourne's values, although approximated to the nearest degree, are
the same as those of the latter. The thirteenth rule gives both the latitude and
longitude of many towns in England. Where Recorde gives latitudes for a few such
towns, they are the same as those given later by Bourne.

The last decade of the century saw the publication of a flush of books whose
contents overlapped with those of the 'Castle' to differing extents and were essentially concerned with the application of astronomy and its associated techniques to
Navigation. In 1590 *The Use of the celestial Globe in Plano (1590)* by ThomasHood
was printed, followed 2 years later by *The Use of both the Globes* and *The Mariners
Guide*. Robert Hues' *Tractatus de Globis et eorum usu*, written in Latin, was
published in 1594, as was Thomas Blundeville's *His Exercises, containing six
Treatises*.The next year *The Seamans Secrets* by John Davis was printed. *The Arte
of Navigation* by Martin Cortes had first been translated into English from Portugese
by Richard Eden in 1561. In 1596, it was reprinted by John Tap, who brought it
up-to-date and augmented it. Tap's important work, *The Seamans Kalender, or an
Ephemerides of the Sun, Moone and of the most notable fixed Starres* was not published until 1602. Edward Wright's important work *Certaine Errors in Navigation*
was published in 1599, but his subsequent treatise, *The Description and Use of the
Sphere*, although written in 1600 was not published until 1631.

Hood's first book presented the first English planispheres of the Northern and
Southern hemispheres as a more practical replacement for the celestial sphere.
However his next publication was effectively a manual for the use of the celestial
and terrestrial globes constructed by Emery Molyneux. It is written in the form of a
dialogue, as was the *Castle*, but in a much more turgid style. The introduction is of
interest as appreciable space was devoted to a refutation of Lactantius' assertions
that the heavens are flat. These refutations are largely emotional and do not measure
up to those of the earlier antagonists. Somewhat surprisingly, Hood gives only a
limited amount of numerical data and none in tabular form. His treatment of the
celestial sphere adds nothing to that of his predecessors. The availability of new data
arising from voyages of exploration over both hemispheres, that had taken place
during the preceeding half-century, allowed Hood to demonstrate very effectively
how the terrestrial globe could be used for navigational purposes as well as demonstrating its limitations. In this respect it represented an advance over Recorde's
work, but contained neither reference to the *Castle*, nor corrections of it. Hood does
refer to the four mean points of the ecliptic having equal declinations. The use of the
word 'mean' in this context was promulgated by Recorde, as has already been
mentioned, but did not appear to have been followed by any one other than Hood
who did so only in a rather unthinking fashion.

Hues' treatise was a more learned treatment of the same subject as Hood. Although, perhaps or even because it was written in Latin, it went through numerous editions in England and the Continent being finally published in English in 1638. It was divided into five parts. The first dealt with the construction of the Molyneux globes and their addenda; the second with the celestial bodies not delineated on the Globes; and the third with the geography given on the terrestrial globe, including a debate on the choice of the value in length of a degree. The fourth part, which Hues himself regarded as the most valuable, concerned itself with the use of the globes virtually as instruments to solve some of the navigational problems encountered by navigators. Parameters such as latitude, longitude, the place of the sun, the lengths of days and hours as a function of latitude, declination of the sun and the right and oblique ascensions and others are dealt with. Hues admitted that their determination by mathematical methods was much more accurate, but the difficulties and lengths of the calculations involved made the instrumental methods using the globe the attractive practical proposition.

Up to this point, Hues had largely covered the same ground as Hood, albeit in a more scholarly manner. His fifth part however covers fresh ground, dealing with the relationships between rhumb lines, latitude, longitude and distance on the surface of a globe and how given any two of these factors the other two may be derived. It is basically an exposition of the way in which the solutions of spherical triangles relevant to navigation can be obtained from measurements made on a globe. He admitted that determinations of longitude per se were extremely difficult and unreliable. The treatise proved extremely popular for much of the following century both in Latin, English and Dutch versions. The relation of this work to that of Recorde's is much as has been commented on for that of Hood's work with the added clarification of the fifth part. Hues did also spend some time in his introduction refuting the arguments supporting the flat earth theory by one Franciscus Patricius. The relevant work was not identified so its contentions have not been able to be checked.

The practical counterpart of Hues' book was *The Seamans Secrets* by John Davis, published the following year. It was written in English, clearly and in a manner that would appeal to the mariner. The first half dealt with 'Horizontal Navigation' i.e. using plane charts, but with great attention being paid to accuracy of observations. The second half was given over to the use of the Globe and thus to 'paradoxal' and 'Great circle' sailing, covering similar ground to Hues' fifth part, but in a clear and simple manner. He laid particular emphasis on the procedure for rectification of the globe so as to define the position of a vessel as a precursor to selection of its next course. This rectification required knowledge of the ship's latitude which could be obtained from a variety of astronomical observations. Determination of longitude was another matter altogether and somewhat evaded by Davies. He showed thereby that plane sailing was accurate for only short distances and that the requisite continual correction of the globe would whilst maintaining a fixed course would generate a spiral course or paradoxal sailing- which would only approximately approach the desired end position. The shortest distance of sailing between two points was via the great circle that passed through them. Davies showed how this course could be approximated by successive corrections following each rectification.

The process of rectification described by Davies has strong overtones of the process described by Recorde when using the sphere for the determination of the length of day for any given latitude and for any time of year. Davies' comments on the types of navigation he described are relevant to Recorde's advocacy of the value of Geometry to such ends. 'All these practices of sayling before mentioned, may in a general name be aptly called Navigation Geometrical, because it wholly consisteth of geometrical demonstrative conclusions.'

Possibly the last and arguably the most important adaptation of chart sailing is to be found in Edward Wright's 'Certain Errors in Navigation'. Much of this work was taken up with instructions for the avoidance of instrumental errors and with the correction of Tables of the declinations of Sun and Fixed stars following more accurate astronomical observations of the relevant parameters. Important though these contributions were, Wright's introduction of a chart projection, prompted by Mercator's world map, was of the greatest practical significance. The avowed intention was to produce a plane map by 'increasing the distance of the parallels, from the equator to the Poles, so that at every point of latitude in the chart a part of the meridian had the same proportion to the same part of the parallel as in the globe.' Mercator had of course done this, but had not described his method of calculation. The numerical data needed to carry out such calculations had themselves been calculated by a number of earlier writers who quantified the way in which meridional lines converged as latitude increased. Recorde was amongst these as shown by his table entitled '… how many myles do answere to 1 min of tyme, in every severall latitude.' This was a version possibly perceived as more likely to appeal the reader than the more common version which tabulated '….the minutes which every degree contained in every one of the parallels.' to be found in Cortes and Hues amongst others. Wright provided tables giving the required adjustments at intervals of 10 min of latitude. The practical advantage of this type of chart for the relatively unlettered mariner was that rhumb line traces were straight lines, replacing the spiral lines they produced on the surface of a globe and distances could be read off simply even though a correction for latitude was required. Wright provided the necessary corrections in tabular form. The disadvantages of inability to plot a great circle course on this projection and the fact that distortions at latitudes above about 80 degrees rendered it virtually useless, were insufficient to outweigh its value to navigators for most voyages of the time. The accurate charting of latitude and longitude of places on such charts was a prerequisite the Wright emphasised but, unsurprisingly, without any special enlightenment on methods for determining longitude.

Wright's Tables had of course to be calculated using trigonometric functions. The first Englishman to publish tables of such functions was Thomas Blundeville in his 'Exercises', published 4 years before 'Certain Errors'. These tables lie at the end of the first of the six treatises contained in the substantial book, which treatise deals in a somewhat condensed format with arithmetic. There is no section on geometry. The tables are those as calculated by Clausius. Six examples are given of astronomical calculations using sines and a further 14 using tangents and cosines. Most of these calculations relate to individual cases of relationships that are to be found in more general form in Tables of related values to be found in the *Castle*. By detailing the calculations, Blundeville does what Recorde did not do, viz. expose their laborious nature. Other treatises dealt competently with the use of the sphere and Mercators two spheres. In the treatise that dealt with maps he discloses, with Wright's consent, the latter's treatment of Mercator's projection as just described. The final two treatises describe Blagrave's Astrolabe and a comprehensive description of the '…chiefest principles of Navigation.' Blundeville's work lies on the cusp of the transition from what John Davies a year later called 'Navigation Geometrical' to what he called '… another Knowledge of Navigation which sweet skill of sayling may well be called Navigation Arithmetical, because it wholly consisteth of calculations……….. to a most wonderful degree of certainty.'

Measuring the contents of the 'Castle' against these contemporary publications of the 1590s, apart from marginal corrections to calculations for declinations of the sun, resulting from more accurate astronomical measurements, there is no fault to be found. Recorde was as competent a mathematician as those who followed him. However his work fell short of what terrestrial navigators now required, not least of all as a result of their own activities in discovering and mapping new territories in the 40 years since Recorde had died. The data base for the 'Castle' was out of date. It was to become even more so as instrumental means for acquiring data became more accurate and, most importantly, as instrumental methods were devised for simulating the arithmetical calculations that Blundeville had shown to be impractical for most navigational purposes. Given the facility with the necessary mathematics that Recorde had shown, it is possible that Recorde's promise to the Muscovy Company would have provided a substantial contribution to the subject. '…I will for your pleasure, to your comforte, and for your commoditie, shortly set forthe a booke of Navigation, as I dare saie, shall partly satisfie and contente, not only your expectation, but also the desire of a greate nomber beside. Where in I willnot forgett specially to touche bothe the old attempte for the Northerlie Navigations, and the later good adventure, with the fortunate successe in discovering that voyage, whiche many men before you durste attempte since the tyme of King Alurede his reigne. ……. In that Boke also I will shewe certaine meanes, how without greate difficultie, you maie saile to the Northe Easte Indies. And so to Camul, Chinchital, and Balor, which be countries of greate commodities. As for Chantai lieth so farre within the lande, towards the Southe Indian seas, that the jorneie is not to be attempted, untill you be better acquainted with these countries, that you must arrive at.'

The prospect of what he might have, or in fact had already contributed to advances in instrumentation that propelled English navigation to the forefront of Western European practice during the late sixteenth and seventeenth centuries is tantalising. The description of the construction of the Sphere and Armillary in the Second treatise of the 'Castle' is exemplary, but not original. Similar remarks apply to his illustration and description of the use of a Quadrant. However he had listed the dozen or so astronomical instruments available at the time of writing the *Castle*, from which he chose to use the spheres. In the preface to 'The Pathway to Knowleg' he lists some books as are appointed to be set forth, two of which did materialise. Amongst them are,

'The arte of Measuryng by the quadrate geometricall, and the disorders committed in usyng the same, not only reveled but reformed also (as muche as the instrument pertayneth) by the devise of a newe quadrate newley invented by the author hereof.

The arte of measuryng by the astronomers staffe, and by the astronomers rynge, and the forme of makyng them both.

The arte of makyng Dials, both for the daie and the nyght, with certayn new formes of fixed dialles for the moon and other for the starres, which may be set in glasse windowes, to serve by daie and by nyght. And how you may by those dialles knowe in what degree of the Zodiake not only the sonne but also the moon is. And how many howers old she is. And also by the same dial to know whether any eclipse shall be that moneth, of the sonne or of the moone.

The makyng and use of an instrument, whereby you maye not onely measure the distance at ones of all places that you can see togyther, how much eche one is from you, and every one from to other, but also thereby to drawe the plotte of any countreie that you shall come in, as justely as maie be, by mannes diligence and labour.'

In the admonition to the reader in the *Castle*, Recorde makes it clear that this work was preceded by one called 'The Gateway to Knowledge'

....

2. *Then if he trade that* Pathwaye *pure*
   That unto knowledgeleadeth sure:
   He may be bolde tapproche The Gate

3. Of Knowledge *and pass in thereat.*
   Where if with Measure *he do well treate:*

4. *To* Knowledges Castle *he maye soone get.*

......

On page 68 of the same work there is a comment that the composition of a Quadrante has been taught amongst other instruments in the *Gate of Knowledge*. It seems likely therefore that some work on mensuration certainly concerned with astronomy and possibly with terrestrial observations had been printed but is now lost.

It is unlikely that the work on dialling was ever published, for Thomas Fale, in the Readers Preface to his *Horologia* of 1593, says that 'Many have promised (but none as yet performed) to write of this Science [horology] in our English tongue, which hath been published in other languages, as *D.Recorde* long since, *M.Digges, M Blagrave* with other, who if they would take the paines, I know would do it with great commendation.' Later Fale goes on to say that they must understand some mathematical principles which they may learn by themselves out of Recorde's three published books, of which the *Gateway* is not one. Thus if it had been published it was not available or was irrelevant to Fale's ends inasmuch it contained little appropriate mathematics.

Chapter 10
The Whetstone of Witte*

Abstract Recorde's, second book of arithmetic, the *Whetstone of Witte*, contains expositions on number theory and the roots of abstract and denominate numbers, the rule of Algeber and on surds The number theory described is that of Pythagoras as presented by Euclid, modified by Theon and Nichomacus, copied by Boethius, and modified further to deal with problems raised by the introduction of vulgar fractions. In defining the types of absolute numbers Recorde follows Euclid, whose method of finding perfect numbers he uses uncritically to make one of his few mistakes. The standard methods for derivation of square and cubic roots of numbers are given with special attention to improvement of accuracy of approximation. The greater part of the book is devoted to the arithmetic of denominate numbers of cossike and irrational forms, in which he draws on the works of Stifel, Scheubel and Cardano, but uses his own clearer order of presentation. He fell one step short of dispensing with cossike symbols in favour of a notation that used powers of numbers, but did introduce the sign for equality as part of his description and use of the Rule of Algeber. Whilst there has been wide acknowledgement of the value of the sign, it has to be be recognised that it was introduced specifically to ease the manipulation of equations, a means to an end and a basic feature of algebra rather than of arithmetic. The treatment of surds is both standard and incomplete. Recorde expected that this book would have limited appeal and he was right.

Recorde describes this as the second part of his arithmetic, containing the extraction of roots in diverse kinds, with the Art of Cossike numbers, and of Surde numbers also, in sundry sorts. This structure was determined after he had written the *Pathway*, but before he had completed the augmented version of the *Grounde of Artes* in 1552. Amongst his potential publications listed in the former, his second part of Arithmetic is described as containing working with fractions, the extraction of

---

*The text used is S.T.C. No. 20820 in the Da Capo Press facsimile version (New York 1969). Quotations are referenced using the printer's signatures.

J. Williams, *Robert Recorde: Tudor Polymath, Expositor and Practitioner of Computation*, History of Computing, DOI 10.1007/978-0-85729-862-1_10, © Springer-Verlag London Limited 2011

square and cubic roots and the rules of alligation. The rule of false position is also included along with 'divers examples not only vulgar, but some appertaining to the rule of Algeber, applied unto quantities partly rational and partly surde'. Separately, he expressed the intent to set out in four parts Euclid's works with arithmetical, geometric and linear demonstrations viz. platforms, surd and irrational quantities, bodies and solid forms, and finally perspective. In the event, at the end of his augmented version of the *Grounde of Artes* is an apology for not having included the promised material on the extraction of roots. Recorde says that he had originally intended to demonstrate their utility with reference to the Statute of Assize of Wood, but as he had to write of other roots he would defer their consideration to the 'Second Part' [of his Arithmetic]. He makes no mention of his omission of material on the rule of Algebra, surds and irrationals. These latter topics, together with expositions on Euclidean number theory and roots, make up his *Whetstone*. What occasioned this re-think is not immediately obvious.

In the *Grounde*, Recorde had not defined the types of numbers and had introduced the concept of denominations only in a strictly practical context. In the opening section of the *Whetstone* he presents his version of Number Theory as a necessary precursor to dealing separately with advanced concepts in the manipulation of abstract numbers, and with the arithmetic of denominate numbers, which includes cossike numbers and the rule of Algeber together with irrational numbers or surds.

Number Theory

The basic theory described is that of Pythagoras as presented by Euclid, modified by Theon and Nichomacus and copied by Boethius. However Recorde modifies it further to deal with problems raised by vulgar fractions, a consequence of the use of Hindu-Arabic numbers and further, to define the status of non-abstract numbers. Recorde follows the sequence of the first five definitions given by Euclid in book VII of the 'Elements', rather than those adopted by the latter's later commentators. Unity, Recorde defines as being indivisible but making up everything.

At the Scholars behest, the Master is then diverted into the categorisation of numbers. '…for some are *whole nombers*, and thei onely of *Euclid, Boetius, and other good writers are called nombers*. Other are *broken nombers*, and are commonly called *fractions*. Of these bothe have I written in the firste and seconde parts of *Arithmetike*:…'. [Aii.r] Recorde seems to have forgotten that he has designated the 'Whetstone' as the seconde part of his Arithmetic. He then distinguishes between *Abstract* and *Contracte* numbers, the latter having a denomination e.g. groat attached to it whereas the former is free of such qualification. This distinction is vital for the development of Recorde's approach to the topic of equations but it is not obvious how this categorisation had evolved.

The first chapter of the second treatise of fourth book of Reisch's *Margarita Philosophica* is entitled *De divisione numeri contracti*, but all that it does is to comment on the division of numbers into the familiar classes of digits and articulate numbers etc. It then goes on to deal with the arithmetical manipulation of

Hindu-Arabic numbers. No mention of either contracte or denominate numbers is made in Pacioli's 'Summa Arithmetica', or in the arithmetical texts of Oronce Fine and Gemma Frisius. However both are found in the 'Arithmetica Integra' of Michaele Stifelius, under the heading 'De natura & speciebus numerorum abstractorum'.[1] It is clear from the text of the second edition of the 'Grounde of Artes' that Recorde had Stifel's work to hand when he was writing that book. He may therefore be assumed to have had it available when writing the 'Whetstone of Witte' and hence probably adopted the terms from Stifel.

Recorde then proceeds to discuss whether or not broken numbers can be described as 'Contract'. The Master agrees with the Scholar that some writers believe that the fact that such broken numbers consist of numerators which multiply denominators qualifies them as contract numbers. However he points out that broken numbers can have a denomination such as groats attached to them in the same way as whole numbers, so other arguments have to be advanced to settle the issue, '...therefore must wee either make a more curious distinction of that name of denomination : or els we must seclude fractions, from the necessitie of that name: or els, thirdly to avoid contention, cal them numbers contracte improperly.'[Aii.r&v] Recorde expresses no preference, but goes on to give the reason why Euclid amongst others refused to account fractions amongst numbers, viz. all numbers consist of a multitude of unities so, as every proper fraction is less than unity, fractions can properly only be called fractions of numbers. He seems to be more positively inclined to Euclid's view than does Stifel, but basically leaves the issue unresolved

Of Abstract Numbers

Although he does not make a clear division in his text, from this point forward Recorde confines his attention to abstract numbers. These he subdivides into numbers absolute, relative and figural, reverting to the presentation used by Euclid. He largely follows Euclid's views but does insert his own comments. For instance, after defining even and odd numbers, he points out that half of seven is three and a half which is no number, thus reinforcing the Student's belief that no fraction can be a number. Both even and odd numbers can be compound and simple. The unique status of *one* is established, as if it is a number then all greater numbers as multiples of it are compound, but using it as a multiplier, no number becomes compound. Two is an even number and uncompound. By definition all other even numbers are compound and can be further subdivided according to the product of halving. Recorde follows the classification and definitions attributed to Nichomacus, Theon of Smyrna and Iamblichus by Heath,[2] viz. even numbers evenly, even numbers unevenly and even numbers, even and oddly. Euclid had seemed to admit only the first two classes,

---

[1] Bodleian Library, Savile MS13, Lib.1, Cap II, 7v–8r
[2] Heath TL (1956) Euclid the thirteen books of the elements, 2nd edn, vol 2. Dover, New York, pp 281–284

but the Master points out that the former seems to have approved of the third class by the 34th proposition of his ninth book. This debate presages a similar one introduced by Heath some 300 years later.[3]

Recorde then turns to the topic of perfect numbers, which he defines as those whose parts added together made the whole number exactly, the definition used by Euclid. However he goes on further to deal with abundant and deficient numbers. In doing so he follows the example of Nichomacus. Euclid dealt only with perfect numbers and left his definition of them to the very end of the sequence of definitions of parts, *one* being counted as a part. Euclid had proved the method of formation of such numbers in proposition 36, Book 9 of the *Elements* and all numbers generated by this process are even. A deficient numbers is one whose parts add up to less than the number itself. The examples of such numbers that Recorde gives are of nearly perfect numbers which are deficient by only one and these form the series of powers of two, which are obviously even. More greatly deficient numbers may be both odd and even but dominantly odd in the lower range of numbers. Abundant numbers i.e. those whose parts sum to more than the number, are also both odd and even but the first odd abundant number is 945. Recorde does not return to this classification when he deals with odd numbers but, in the light of present understanding of the topic, can perhaps be excused the omission.

What is possibly less excusable is the incorrect list of perfect numbers, 6, 28, 496, 8,128, 130,816, 2,096,128, 33,550,336, and 536,854,528 that he gives. The method for generating perfect numbers described by Euclid, had been arithmetised by Nichomacus as reported by Boethius. Two terms have to be multiplied together, the first is a power of two $[2^p]$ and the second term is the next higher power of two less one $[2^{p+1} - 1]$ and this latter has to be a prime number. The values for p of 1, 2, 4 and 6 generate the first four numbers in the above sequence and these are those commonly mentioned in mathematical texts. Stifel derived the fifth number using p = 8 and Recorde added the remaining numbers using p = 10, 12 and 14. Neither Stifel nor Recorde noticed that the second term generated using p = 8, viz. 511, is not prime (7×53). Recorde also failed to see that the same term obtained from p = 10, viz. 2,047, also is not prime (23×89). The correct fifth perfect number 33,550,336 is derived from p = 12. $(4,096 \times 8,191)$. Both this number and its correct derivation are to be found in the *Utriusque Arithmetices Epitome* of Hidalrichum Regius of 1536.[4] The basic problem faced by Regius Stifel and Recorde was how to prove that the numbers 511, 2,047 and 8,191 were or were not prime. Regius obviously did this but did not say how. It could conceivably have been done using the sieve of Erastothenes, a labour that Regius may have been prepared to undertake, but neither Stifel nor Recorde. The second term resulting from the use of p = 13 is 16,383 which is divisible by 3, but the equivalent for p = 14 is 32,767 was accepted by Recorde as prime, for he used it to calculate his last perfect number 536,854,528.

---

[3] Ibid
[4] Christchurch College. Oxford, OP. 6. 31(2)., 16v–17v

Recorde then reverts to the sequence of Euclidian definitions and deals with odd numbers, almost as briefly as did Euclid. He defines two categories, compound and uncompound and leaves it at that, not even using the term 'prime'. Stifel was more expansive dealing at some length with the topic of factorisation of odd numbers. The reasons for Recorde's lack of interest in this topic are not obvious. He is only a little less terse in the matter of relative numbers, commensurable and incommensurable, providing some numerical examples of the two categories.

Recorde is however expansive to the point of exhaustion in treating the topic of relative numbers, but does summarise the categories in the form of family trees. He does not introduce any more categories of numerical proportions than his predecessors, but does provide an extensive compilation of numerical examples. In these he follows Stifel's practice of converting these number ratios into mixed numbers. A few asides leave the impression that he was somewhat wearied by the subject. Thus on the topic of nomenclature he has the Scholar question, 'Why do you not name them all by Englishe names?

Master. Because there are no soche names in the Englishe tongue. And if I shoulde give them newe names, many woulde make a quarrelle againste me, for obscuring the olde Arte with newe names: as some in other cases all redy have doen.'

When asked 'What more is there to be learned of these proportions. For by these formes I may easely gather the value or rate of any proportion.

Master. This may stand for their numeration: save that moste aptly thei ought to be sette as fractions, in their least terms: as you have diverse examples.' [Biiij.v].

The Master prefers numbering rather than wordy descriptions. He goes on to point out that proportions can be added, subtracted multiplied and divided but additionally on the subtleties of Geometry, he points out which topics he does not intend to deal with at this time 'considaryng not only the troublesome condition, of my unquiet estate: but also the convenient order of teachynge, whereby it is required that the extraction of rootes, should go orderly before the arte of Proportions: whiche without those other, can not be wrought' [Cii.r]. These were the tactics employed also by Stifel who however followed the treatment of roots with a condensed taxonomy of proportions coupled with a comprehensive algorithm of such numbers. But this he followed by examinations of progressions harmonic, contraharmonic, astronomic and musical, none of which topics Recorde included in his book.

Stifel's and Recorde's interests overlap again in the subject of figural numbers. Their name arises 'because thei doe, or maie represente some figure: And are ever considered in relacion to those formes.' [Ciij.r] Recorde then deals briefly with linear numbers; superficiall or flatte numbers of which there as many varieties as there are of diversity of figures in Geometry; and then there are the sounde or bodily numbers which are formed by further multiplications of sides, the least of which is the cube. Of the multitude of flat numbers he selects only four for discussion, square numbers, long squares [rectangles], diametral numbers and likeflatts. He touches only briefly on squares and square roots, cubes and cube roots before moving on to deal at length with diametral numbers and likeflatt numbers, promising to return later to the topic of bodily numbers of higher dimensions. The reason given for this concentration is that '...to the intent you [the Scholar] maie the better observe and regarde these twoo laste kindes of nombers: which are commonly neglected of artes men,....'.[Dij.r]

Diametral numbers were defined as those whose two factors were such that when squared and added together yielded a square number whose root was the diameter to the diametral number i.e. formed a Pythagorean triplet. In this treatment he again followed Stifel closely, but stopped short of an error the latter made. After a trawl through a range of examples of the properties of these triplets the Master eventually extracts five general properties which may be used to determine whether or not a number is a diametral number. Following a rather exhaustingly exploration of a number of examples illustrating these rules, the Master eventually reaches the point '....And yet some common rule must bee given, that may extende as well to them[diametral numbers], as to other'.

Wherefore let this be it.

That the twoo sides of all *diametralle nombers*, have such a proportion together, as here you see expressed in some one of these formes: if thei bee continued as here thei be begon.

The firste order

$$\tfrac{3}{4} : \tfrac{5}{12} : \tfrac{7}{24} : \tfrac{9}{40} : \tfrac{11}{60} : \tfrac{13}{84} : \tfrac{15}{112} : \tfrac{17}{144} : \tfrac{19}{180} : \tfrac{21}{220} :$$

$$\tfrac{23}{264} : \tfrac{25}{312} : \tfrac{27}{360} : \tfrac{29}{420} : \tfrac{31}{480} : \tfrac{33}{544} : \tfrac{35}{612} : \tfrac{37}{684} : \tfrac{39}{760} \ \text{etc.}$$

The seconde order.

$$\tfrac{8}{25} \ \tfrac{12}{35} : \tfrac{16}{63} : \tfrac{20}{99} : \tfrac{24}{143} : \tfrac{28}{195} : \tfrac{32}{255} : \tfrac{36}{323} : \tfrac{40}{399}$$

$$\tfrac{44}{483} : \tfrac{48}{575} : \tfrac{52}{2704} : \text{etc.} \qquad [\text{Fiij.r \& v}]$$

The first order is when the lesser side is an odd number, the second when the lesser number is even. In both cases the lesser number is the numerator and the greater side is the denominator. Here Recorde is copying from Stifel.

Recorde says that Stifel inverts this presentation of the two sides '... for the more delectable contemplation, to behold their forme of progression, he setteth down as many whole numbers as the fraction will give.' viz.

The first order

$$1\tfrac{1}{3} : 2\tfrac{2}{5} : 3\tfrac{3}{7} : 4\tfrac{4}{9} : 5\tfrac{5}{11} : 6\tfrac{6}{13} : 7\tfrac{7}{15} : \text{etc.}$$

The seconde order

$$1\tfrac{7}{8} : 2\tfrac{11}{12} : 3\tfrac{15}{16} : 4\tfrac{19}{20} : 5\tfrac{23}{24} : 6\tfrac{27}{28} : 7\tfrac{31}{32} \ \text{etc.} \quad [\text{Fiiij.v}]$$

The Master comments that in the first order the whole numbers and numerators ascend in the natural order but the denominators ascend as odd numbers. In the second order however, whilst the whole numbers ascend as before, the numerators and denominators ascend as an arithmetical progression based on four, the former being odd and the latter even numbers. Stifel drew attention to this symmetry but devoted much less time to the overall topic of diametral numbers than Recorde and gave no indication of how he had derived the numerical values displayed in the orders.

Neither of them recognised overtly that the formula attributed to Plato did not give all the possible triplets for lesser even number sides, nor that the formula of Pythagoras gave all possible triplets for lesser odd number sides. However, whilst the Master gives the general rule that if any two parts of any number in one of the proportions found in these tables it will be sufficient to show that it was a diametral number, Stifel was far less cautious. He introduced the tables saying 'Si fuerit numerus diametralis, **necesse est,** ut latera eius diametralis sint sub proportione aliqua earum quae his duobus ordinibus signantur.' i.e. averring that the tables did yield all possible diametral numbers. This error on Stifel's part was recognised by Harriot nearly half a century later and corrected in one of his many unpublished papers.[5]

Whether Recorde departed from Stifel's line because of a deeper under-standing can only be conjectured. It was not the first time he had not followed him uncritically.

The Master then moves on to deal briefly with *like flattes*. Like flatts are taken to embrace right-angled triangles as well as rectangular figures whose sides [subtended by the right-angle] are in like proportion to one another, hence including square figures The treatment afforded is in essence the arithmetisation of the properties of similar right-angled triangles found in the *Elements* in the sequence, translated ver-batim by Recorde, of Book VIII propositions 18 and 20, Book IX propositions 1 and 2 and Book VIII proposition 26. Their relationships to diametral numbers are obvi-ous. Care is taken to point out that whilst square numbers when compared with one another are like flattes, but they also can stand alone inasmuch as they also refer to their roots. As with diametral numbers, this section is most easily comprehended in algebraic format providing a number of illustrations of the basic mathematical oper-ations of algebra. It is a halfway house between Greek geometry and algebra.

The dialogue then returns to the subject of rooted numbers. 'Wherefore to begin, where we lefte a little before, the explication of rootes: I saie, that the roote of nomber, is a nomber also: and is of soche sorte, that by sondrie multiplications of it, by it self, or by the number resulting thereof, it doeth produce that nomber, whose roote it is. And accordyng to the number of times that it is multiplied, the number that resulteth thereof, taketh his name.'[Giij.v] Thus one multiplication gives a square number, two multiplications give a cubic number, three a square of squares and so on. Correspondingly, the root of a square number is a square root, that of a cubic number a cubic root and that of a squared square a squared squared root.

The Master then presents a table of rooted numbers with the admonition, 'But first mark the table well: and it will give you greate lighte, and aptness to understand all that foloweth, moch the better. For examples are the lighte of teachyng.' [Giiij.v]

[5]Tanner RCH (1977) Nathaniel Torporley's 'Congestor Analyticus' and ThomasHarriot's 'De triangulis laterum rationalium'. Ann Sci 34:392–428

With that the Scholar enters the complicated world of multiple dimensions, which is not made easier by the multiple nomenclatures used. The table lists the powers of the numbers 2–10, from unity up to the power of ten. He then gives the vulgar names for these powers, rootes (1), squares(2), cubikes(3), squares of squares(4), sursolides(5), squares of cubes(6), second sursolids(7), squares of squared squares(8), cubes of cubes(9), squares of sursolides(10). This is a simplification of practise. Cossists also used the term *zensis* in place of square, thus Stifel listed the sequence above as zensis (2), cubis, zenzizensis, surdesolidis, zensicubis, ßsurdsolidis, zensizenzensis and cubicubicis(9). The first three terms present no problems to the Master neither in terms of nomenclature nor in terms of visualisation as geometrical figures, i.e. squares and cubes. The Master is concerned about naming these higher powers in such a way as to aid understanding of their content. Thus he takes issue with the use by some writers, Stifel obviously included if not named, of the term surde solide 'seeing thei are not in any waies *Surde nombers*, but have their rootes.'

In typical fashion, Recorde attempts to rectify the situation and make the topic more digestible by constructing his own nomenclature. Also typically he presents his offering in graphical form before he presents his reasoning.

> For all those nombers are considered, in one of 2 formes firste. That is to saie, other thei bee taken as nombers absolute, without any consideration of multiplication. And so maie be named nombers only, without name of relation. Or els thei be considered as nombers multiplied, and thet can be but in 3 varieties.
>
> If thei be multiplied but ones, then thei do make a line of nombers, or a lineric nomber. And that nomber hath onely lenghte, without bredthe, or depthe: And thereforemaie be the roote to a *Square* or a *Cube*. But it is of it self, in that consideration, neither *Square* nor *Cube*.
>
> Secondarily, it maie be multiplied twise, the one number standing for the lengthe, and the other for the bredthe: and so it is a *Square nomber*, and therfore a *flat nomber*.
>
> Thirdly it maie bee multiplied thrice, and thereby get *lengthe, bredthe, and depthe: whereby it is made a* sonde nomber. And because the sides bee equalle, it is specially a *cube* or *cubike nomber*.
>
> Now can there be no fowerth waie, that any multiplication maie increase: for there are no more dimentions in nature.
>
> But if any manne doe multiply the fourthe tyme then must he accoumpte thath he makes a *line of Cubes* and the fifth multiplication maketh a *Square* in whiche every unitie is a *Cube*: So the sixte multiplication maketh a *Cube* of *Cubes*, accoumpting every lesser cube for an unitie. [Jj.r]

This argument is extended to still higher powers and the Master thereby generates his own nomenclature of rootes(1), squares, cubes, long cubes(4), squares of cubes(5), cubike cubes(6), long cubike cubes(7), squares of cubike cubes(8), cubes of cubike cubes(9), long cubes of cubike cubes(10). He argues the case for this nomenclature being more accurately descriptive of the subject, with some justification. Time has dispensed with the need for all of the nomenclature discussed, but not apparently the concept, which re-emerged in the form of Dienes Blocks in the 1960s.[6] The exercise serves to show the persistent need that Recorde felt for logical argument as a pathway to clearer presentation of what were becoming increasingly abstract concepts.

---

[6]Baron ME (1966) A note on Robert Recorde and the Dienes blocks. Math Gaz L(374):363–369

The groundwork having been laid, the Master proceeds to deal with the extraction of roots from such numbers. In the visual aspects of this approach, Recorde follows Stifel. The latter however makes a bald presentation of the representation with no attempt to justify it philosophically. Recorde continues with his concept of square and cubic numbers as building blocks in his exposition of the extraction of roots, which he dignifies with a new Chapter heading. Thus he insists again that the Scholar remembers that a square number can have only a square root and a cubic number a cube root. Only with a compound number does the possibility of more than one root arise. Dealing firstly with square roots, before expounding the conventional method for their extraction the Master tabulates the square numbers below 100 together with their roots and instructs that these 'you have by harte, in readie memorieall soche numbers, whose roots are digits. For as it is superfluous to seke rules for them, so must thei helpe in all greater numbers, whose rootes are above 9.' [Kj.r] The method is then described and illustrated by reference to four large square numbers. These increase successively in size from seven to nine digits and were carefully chosen to draw attention to potential pitfalls in application of the method. Three of these were associated with the presence of *cyphers*, the simplest of which occurred when the number ended with two zeroes. Proof of correctness was obtained by back multiplying the result of the calculation.

Having dealt with square numbers, the Master moves on to deal with *The nigheste root of unsquare nombers*. Here again the two well known methods of approximations are briefly described and illustrated. More attention is given to the unlimited accuracy that can be attained by setting cyphers in pairs before the number whose root is being sought. Using this subterfuge, it is shown how the accuracy is increased by a factor of ten for each pair of zeroes added. The worked example given uses six added zeroes and yields a fractional part of the answer that has nine numbers in its denominator. Karpinski notes that this method is to be found in Cardano's *Practica Arithmetice*, to which Recorde had access, but that the method was in fact much older in origin.[7]

Of the worked examples relating to the practical use of the calculations, the Master gives three concerned with the deployment of troops, one on the scaling of walls, one on encampment and one on plane geography. Similar examples can be found in contemporary arithmetics by Continental authors. Why examples relating to surveying were not included is not obvious. Leonard Digges *Planimetria* which dealt with such matters had not yet been published so there was no question of duplication.

The sequence of presentations used for square roots is also used for cube roots, starting with the need for commitment to memory of the cubes of the first nine digits. The classical method for extraction of cube roots is first given in words. The Scholar complains 'This rule is very obscure in woordes' and so two worked examples are given using eight digit numbers, both perfect cubes, employing two variations of method. The Scholar is then instructed on how to find '*The nigheste roote in a nomber not Cubike*' using a nine digit number [694,582,951]. The standard method is used to generate a root [885] and a residue [1,428,826]. Whilst admitting to a

---

[7] Karpinski LC The whetstone of witte (1557), Bibliotheca Mathematica 1912–1913 (Series 3), 13, pp 223–228, 225–226

variety of methods being available, the Master confines himself to describing two methods of using this residue to approximate an answer more closely.

...Cardano his rule is this.

Multiplie the roote squarely, and again by 3 and that number shall be the divisor unto the remainer,

Where he might have used more plainesse in wordes, if he had saied: and that nomber shal be the denominator to the remainder. Wherefore as here your roote is 885 so is the square of it 783,225 and the triple of that is 2,349,675. So would that fraction bee 1,428,826/2,349,675.

But how nigh this doeth go to the truthe, I leave it till another tyme.

Scheubelius doeth allege an other reason, and infereth an other order, diverse from this, and such as impugneth this saiying:

Triple the roote, and the square of it also, and adde both those nombers together, and one more: And so you have a denominator for your numeratour. [Oiiij.v]

Discussion of the merits of these two methods is deferred until 'a third waie, more certaine than either of these bothe' has been described. This way uses ciphers added to the original number in multiples of a thousand instead of multiples of a hundred, as described for square roots. The example chosen adds twelve zeros to the same number as before. Application of the standard method to this augmented number yields 885 as before but using the augmented method the fraction $^{6,076}/_{10,000}$ has to be added, leaving a remainder of only $^1/_8$ of $^1/_{1,000}$.

To demonstrate the difference arising from application of the three methods, the Master uses a simpler example. It was required to double the size of an altar that was in the form of a three foot cube, What was the size of the side of the new altar, i.e. what was the cube root of 54? Application of Cardano's rule gives 3 $^{27}/_{27}$ or 4 which cubed gives 64 and so is.much too great. On the other hand Scheubel's rule gives 3 $^{27}/_{37}$ or nearly 3 $^3/_4$, which when cubed gives 52 $^{47}/_{64}$, too little by 1 $^{17}/_{64}$. Use of the third way, with six zeros added, gives 3 $^{77}/_{100}$ which leaves a remainder of only about $^1/_{700}$.

A number of examples are then given requiring determination of the cube roots for problems involving spheres and cubes. For the former Euclid's eighteenth proposition of his twelfth book is quoted, the fifth time reference has been made to the *Elements*.

The Master then deals with the roots of numbers compounded of square or cubic numbers treating separately those made of square numbers and of cubic numbers only and finally those containing both types of numbers. 'But as I saied before, that I might not staie long at this present so the use of these greate nombers is rare in practise: ..'.[Qiiij.r&v] A large table of a range of various powers of numbers is appended with encouragement to the Scholar to extend it if he so wishes but with the advice 'And so I counsell you, well to examine this table, and trust not to my castynge. For haste and other troubles, maie often times cause errors in supputation.'[Qiiij.v]

Recorde must have had access to the practical arithmetics of Stifel, Scheubel and Cardan dated respectively 1543, 1549 and 1537 in compiling his treatment of the extraction of roots. To Stifel he owed little new as the latter dealt solely with the extraction of the roots of solid numbers. In the 1549 edition of his *Compendium Arithmeticae Artis*, Scheubel deals with the extraction of square roots both of square and non-square numbers using the first of the two standard methods described later by Recorde. This particular edition does not contain methods for the extraction of cube roots. The worked examples it gives are presented as diagrams with little instruction in the way the workings are arrived at. Cardano's presentation is as complete as that of Recorde in terms of coverage of topics. Thus for square roots both of the two standard methods are illustrated and results compared. Also the most accurate method utilising the addition of zeros is described. For the extraction of cube roots Cardano gives the standard method together with the first method of approximation for non-cubic numbers ascribed to him by Recorde, but he also describes what the latter calls the 'third waie', viz. that employing the added zeros. For this section of his work, Recorde seems to have relied primarily on Cardano for his methods. However in terms of clarity of presentation of methods and proof of results Recorde owes nothing to his fellow authors.

The next section of *The Whetstone of Witte* that deals with cossike numbers has received most of the attention directed to this work.

Of Denominate Numbers

'Thus have I lightly over run the most common kind of nombers *Abstracte*. And now resteth the treatise of nombers *Contracte*, or *Denominate*. Of whiche kinde there bee some called *nombers denominate vulgarely*: and other bee called nombers *denominate Cossikely*. And a thirde sorte there is of nombers *radicalle* whiche commonly bee called *nombers irrationall*: because many of them are soche, as can not bee expressed, by nombers *Abstracte*, nother by any certain rationalle nomber. Other men call them more aptly *Surde nombers*. And although many menne would not accoumpte them, with nombers *denominate*, yet I maie justly doe it, for that thei require a reduction to one denomination, if thei have severalle signes of quantities as you shall heare hereafter. And those nombers never goe alone, without some other signe, and name of rooted quantitie annexed to theim.' [Si.r] Feeling that he had already dealt adequately with 'numbers denominate vulgarely' in the *Grounde of Artes*, he moves on to deal with cossike numbers. This is the point at which Recorde introduces the word 'sign' as a mathematical term.

This general introduction is peculiar to Recorde, not to be found in works of Stifel[8] and Scheubel[9] on the Cossist's art, the works of both of whom Recorde is commonly believed to have drawn upon. It is typical of his meticulous approach to presentations. He believed in explaining himself carefully and this approach he continued into the following chapter. This is not surprising for he regarded the introduction of the 'Rule of Cose,' as the most important feature of this second book of arithmetic singling it out as the subject of an introductory verse to the whole book,

> One thyng is nothing, the proverbe is,
> whiche in some cases doth not misse.
> Yet hereby workynge with one thing,
> Soch knowledge doth from one roote spryng,
> That one thing maie with right good skille

---

[8] Stifel M (1543) Arithmetica Integra. Nuremburg, Petreium, Bodleian Savile R13
[9] Scheubel J (1551) Algebra Compendoisa facilisque de Descripta qua depromuntur magna Arithmetices miracula. Tubingen. Bodleian S 40 Art Selden

Compare with all thing. And you will
The practice learne, you shall sone see,
what thynges by one thing knowen maie bee.

The Arte of Cossike Nombers [Si.v]

Whilst Karpinski thought that Recorde was primarily indebted to Stifel for the remaining part of his book, Hughes' view that Scheubel was the more likely exemplar is better supported by the evidence.[10] Stifel introduces the cossist symbols in Book I of his *Arithmetica integra* when dealing with roots and with irrational numbers, offering scant explanation of their significance. His cossike arithmetic falls in Book II. Recorde's presentation follows that of Scheubel closely, but not exactly. He begins, as did Scheubel, by defining the symbols he is to use.

'Nombers cossike, are soche as bee contracte unto a denomination of some *Cossike* signe as 1.nomber, 1.roote, 1.square, 1.cube etc. But as for compendiousnesse in the use of them, there bee certain figures set for to signifie them: so I think it good to expresse unto you those figures, before we enter any farther, to thintente we maie procede always in certentie, and knowe the thynges that we intermedle withall: for thei are the signes of the arte, that followeth here to be taught.

And although there may be many kindes of irrationall nombers, yet those figures that serve in *Cossike nombers* bee the figures also of irrational nombers, and therefore being ones well knowen, thei serve in bothe places commodiously. These therefore be their signes, and significations briefly touched: for their nature is partly declared before.' [Si.v] The relation between number and sign is clearly defined. Recorde never uses the term 'symbol', which seems not to enter the English vocabulary in a mathematical context until late in the sixteenth century. In the following excerpt he uses the words betoken, signify, represent, and express as well, to describe the relationship between the sign and the cossike number it is used to define.

ꝙ.	Betokeneth nomber abſolute : as if it had no ſigne.
ʒℯ.	Signifieth the roote of any nomber.
ʒ.	Repeſenteth a ſquare nomber.
cℯ.	Expreſſeth a Cubike nomber.
ʒʒ.	Is the ſigne of a ſquare of ſquares, oꝛ Zenʒizenʒike.
ſʒ.	Standeth foꝛ a Surſolide.
ʒcℯ.	Doeth ſignifie a Zenʒicubike, oꝛ a ſquare of Cubes.
ſſʒ.	Doeth betoken a ſeconde Surſolide.
ʒʒʒ.	Doeth repreſent a ſquare of ſquares ſquaredly,

[10] Karpinski LC, loc. cit., p 225
Hughes B, loc. cit., p 70

This presentation may seem prolix, but that of Stifel is virtually non-existent and that of Scheubel intended for the mathematician. Recorde makes the relationship between Cossike numbers and cossike signs very clear. The signs that Recorde uses up to zenzizenzike are also those used by Stifel, Scheubel and Rudolff. For the names of signs of greater value, they go their separate ways. Recorde keeps faith with the arguments he used for his choice of names when discussing figural numbers in his choice of signs.

He goes on to discuss a further classification of the cossike numbers. 'And farther you shal understande, that many men doe ever morecall square nombers *zenzikes* as a shorter and apter name, other men call those squares the *firste quantities*, and the cubes thei call *seconde quantities*: squares of squares thei call *thirde quantities*, and sursolides *fourthe quantities*. And so naming them all quantities (excepte nombers and rootes) thei do add to them for a difference, an ordinall name of nomber, as thei doe goe in order successively' [Sij.v]. Scheubel has to be numbered amongst these 'other men', abandoning his listed symbols for the collateral terminology described by Recorde.

Initially Recorde defines the nature and functions of the signs + and – as follows. 'There be other 2 signes in often use, of whiche the first is made thus + and betokeneth more: the other is thus made – and betokeneth less. And where thei come in any nomber *Cossike*, or other, that nomber is called a compounde nomber bicause it consisteth of 2. nombers. And where neither of theim is, the nomber is then called uncompounde. For the compounde signe, maketh not a compounde nomber.' [Sij.v]. A little later [Siij.v] he adds to these definitions. '…. The signe of addition, whiche is this + betokeneth more'; and 'the signe of abatemente which is this – and signifieth les, or abating.'

Scheubel defines their use but goes no deeper. Also to be noted is that the word 'symbol' does not exist in Recorde's vocabulary.

Of Numeration in Nombers Cossike, Uncompounde [Siij.r]

The Master begins, 'Nombers *Cossike* uncompounde, have no dificulties in their numeration: for ever more the number representeth, so many of that *Cossike* denomination (be thei nombers, rootes, squares, Cubes, squares of squares, or any otherlike) as there be unitie in that nomber.' [Sij.v] He then deals in a straightforward manner with the basic operations of addition, subtraction, multiplication and division. The operations of addition and subtraction of numbers of unlike cossike numbers, using the appropriate signs, introduce the concept of compound numbers without attention being drawn to it. In the section on multiplication an aid is introduced which takes the form of a table in which the cossike signs are numbered, starting with zero for the sign for the absolute number, extending in ordered form up to 13 for the fourth sursolid. When cossike numbers are multiplied or divided, these ordinal numbers are added or subtracted, as appropriate, to give a product that is converted back to the associated cossike sign. In this exposition Recorde follows both Stifel and Scheubel, though as ever he is more explicit in his presentation. These ordinal numbers are what would now be recognisable as powers of numbers or of unknowns. None of the three authors give any indication of appreciation of the possibility of the further simplification of procedures that was within their grasp.

A natural consequence of some divisions was the appearance of fractional quantities. The basic arithmetical operations of fractions is quickly dealt with 'And as for fractions, the workyng is like in every poincte, unto the the worke of nombers *Abstracte*: remembring only that as those broken nombers have a *Cossike* denomination annexed with them, so must that denomination followe the rules, now last declared'. The matter of reduction is dealt with even more briefly.

The title of the last section dealing with uncompound numbers, '*Of Progressions in Cossike figures*,' can be a little misleading. Hughes says of cossike nomenclature 'Apart from the Arabic numerals, plus and minus signs, and the equal sign, the remaining symbols in his [Recorde] equations are strange to modern readers.'[11] Later, in the context of a discussion on solutions of equations Hughes says 'First, for the extraction of the roots of monomials [uncompound numbers], Recorde expects the reader to memorise the powers of 2, 3 and 4 up to the tenth power and to understand the tables of powers of the unknowns.'[12] Both of these statements are true, but this section simplifies things so that even the modern reader need not be mystified by the nomenclature nor the contemporary student too disheartened. As explained by Recorde, to derive the cossike number corresponding to each ordinal number in the natural progression of numbers, only four basic signs are needed. The first three are those for root, square, and cube corresponding to ordinals 1.2.and 3. Each of the remaining primes 5,7,11 etc corresponds to a sursolid sign with a prefix starting with b for 7, c for 11 and so on. All ordinal numbers are compounded of these primes so the corresponding assemblages of signs can be derived for any number given knowledge of its factors and of course vice versa. Recorde gives a table of cossike signs and ordinal numbers [powers] from zero up to 80, '*...which maie increase it self infinitely without any difficultie*' [T iiij(v)]. By way of illustration he factorises 84 as 2.2.3.7 and gives the associated cossike sign. Perhaps this presentation provides a clue as to why Recorde did not recognise the redundancy of the cossike signs. If he had not been preoccupied with the need to provide some physical meaning to powers of numbers in the form of the figural numbers for lines, squares and cubes, he might have seen that use of the ordinal numbers as prefixes to a single symbol would have sufficed.

Stifel fell at the same hurdle. He shows how to derive the symbols for any ordinal number by factorisation, as did Recorde.[13] He then lists the signs for all the prime numbers, viewing the generation of the signs for primes as of 5,7,11 and above as ß, bß, cß and so on, as self evident. He also provides a table for the symbols of the first

[11] Hughes B, loc. cit., p 164

[12] Ibid., p 166

[13] Stifel M, loc. cit., pp 235v–236r

seven composite even numbers, again without explanation. Recorde seems likely to have based his approach on that of Stifel rather than that of Scheubel, but as in a number of other cases already mentioned, preferred to clarify the treatment, leaving less for the reader to worry out for himself.

Scheubel also failed to take the final step to use of the ordinal numbers as powers, perhaps less forgivably. He abandoned the use of cossike symbols in favour of naming the ordinal numbers and thus was only a whisker away from modern practice.

The Arte of Cossike Numbers Compounde

Cossike numbers compound are defined as being made up of two or more cossike numbers assembled by addition and/or subtraction. There is a preference to assemble the terms in order of descending cossike value from the left, though this is not deemed essential, with the proviso '… but not so in other *Surde nombers*, where the order follows of necessitie....'. [Uij.r]

The rules for addition of compound cossike numbers are given and thoroughly illustrated by taking four simple cossike numbers and assembling them in eight different ways to demonstrate these rules. After repeating the procedure for more complex compound numbers, proof of result is demonstrated by substituting numerical values for the cossike signs. A similar procedure is adopted to explain the operation of subtraction. One of the examples gives the Scholar the opportunity to point out that in ordering a compound number '...no number maie begin, with signe of lesse: ...', [Xiij.v] this taking priority over the value of cossike sign.

For both multiplication and division, the relevant rules for manipulation of the signs are given and well illustrated by examples, proof in all cases being by substitution of numbers for cossike signs. The examples chosen for purposes of division are such that there are no remainders to be dealt with. This point is acknowledged by the Master who accepts that there are more numbers that do not fall into this category. However he points out that all types of compound numbers can be divided by abstract numbers. If an abstract number has a cossike sign associated with it, then it may also be divided into a compound cossike number provided the cossike terms in that number are greater than that of the divisor. In other words, inverse cossike terms cannot be accommodated in Recorde's scheme of things. Stifel imposed no such limitation, permitting simplification that resulted in an inverse cossike number.[14] Finally the case where there is a term missing in the hierarchical sequence of the dividend is dealt with and the importance of repairing that omission by the use of a cipher to quantify that sign before proceeding with the division.

The problems of manipulation of fractions are then tackled.

---

[14] Stifel M, loc. cit., pp 237r–237v

Of Fractions, and Their Numeration [Zii.v]

This section treats both the way in which cossike fractions are represented and also the basic arithmetical operations upon them.

Recorde takes a great deal of care in describing how fractions containing cossike numbers are to be written if possible pitfalls are to be avoided. He quotes the case of a 'greate clerke' who makes successive errors in carrying out two divisions involving cossike numbers, ' … (except I shall for his excuse, impute the faulte to the printer)' [Siij.v]. The 'great clerke' was neither Stifel, nor Stifel editing Christof, nor was it Scheubel.

The treatment of the reduction i.e. simplification of fractions is straightforward, the worked examples given having been chosen so as to leave no awkward remainders. One example is however peculiar to Recorde, the resolution of which he justifies by reference to the nineteenth proposition of the fifth book of Euclid. *'If the proportion of the abatemente unto abatemente be, as the whole is in proportion to the whole. Then shall the residue be in like proportion to the residue, as the whole is to the whole'*. [Aaj.v] This is his second reference to the geometry of Euclid in a text concerned basically with algebra.

The addition and subtraction of fractional compound cossike numbers are dealt with straightforwardly with plenty of examples, proof again being by substitution of numerical values for cossike signs. Multiplication is dismissed as involving no more than had been taught before. Nonetheless, the first example is used to make a statement about negative numbers. The numerator in one of the fractions to be used by the Scholar was 32 cubes less 28 sursolids, which the Master calls an *Absurde* number which expresses less than nothing. [Bbiij.r] This finding is demonstrated for whole numbers but the Master says that it may still be used for practising multiplication and goes on to show that if fractional numbers are employed rather than whole numbers, negative numbers need not result 'for as whole numbers by multiplication, maie increase infinitely: so fractions by multiplication, maie decrease infinitely.' [Bbiiij.r&v] He then gives an example which demonstrates how multiplication of two fractional compound numbers can give rise to a negative product and confirms the result by substituting whole numbers to give in this particular case, 6,204 less than nothing. Of division the Master says that no special rule is to be given and provides only two worked examples. This summary treatment is carried over into the application of the Golden rule.

Of Extraction of Rootes [Cciiij.r]

This topic receives greater attention than its predecessors. The Master deals first with the roots of uncompound cossike numbers, which is straightforward. He then moves on to the roots of numbers having compound cossike denominations which numbers are shown to have more than one root. In both cases simple rules are given for determining whether or not a number qualifies for having a root or roots. In this approach he follows that of Stifel.

Turning to compound cossike numbers he distinguishes two cases. 'Firstly, and generally, all rooted numbers, other are specially framed, by orderly multiplication, or els are nombers equalle to some one rooted nomber *Abstracte*.' [Ddj.r]

To determine whether or not a compound cossike number comprising a sequence of added and/or subtracted cossike numbers, i.e. a polynomial, is a square number, a set of five rules are given. Given such a square number, a method for extracting its square root is explained. The final example uses a number containing 11 terms having a root with six terms. Proof of accuracy is by the inverse operation of multiplication. Stifel does not treat this topic but moves directly to Recorde's second case of what in essence is the numerical solution of quadratic type equations.

Recorde's presentation of this topic is more akin to that of Scheubel, who however does not classify it as the determination of roots of cossike numbers. Recorde was clearly uneasy with his own classification, for he closes his discussion by saying '...But this by the waie, I must admonishe you, that I doe follow here, the common forme of writers, in callyng these rootes, that rise in equation, where as thei are not the rootes of those nombers, but are the value of a roote. For of a *Cossike* nomber, the roote must neades bee a *Cossike* number also. And soche as by multiplication will make the rooted nomber: But so can not those nombers doe.' [Eeiij.r]. Stifel did not have such reservations.

Basically Recorde solves the examples of quadratic equations that follow using the quadratic formula $x^2 = bx + c$; $x^2 = c - bx$; $x^2 = bx - c$. The instructions given for extraction of the roots of the right-hand side of these equations, using the usual plethora of examples, give single positive roots for the first two and two roots for the third, negative number solutions not being considered. Hughes believes that this approach suggests a geometric mindset on the part of Recorde, albeit an unconscious one. In the two of these cases considered by Stifel, the latter presents the solutions openly in geometric form.[15] Recorde was not one to avoid the opportunity of a geometrical illustration other than by intent and Stifel's book was clearly to hand. Abstinence from this approach, together with his reservation about the classification of the topic already mentioned, suggests that he was moving towards an algebraic stance, but had not yet resolved that issue in his own mind at least to the extent that would allow him to present a strong case to a reader. Recorde's motive in including this topic is made evident by his concluding remark, 'And now will I applie them to practice in the rule of *equation*, that is commonly called *Algebers* rule.'

The Rule of Equation, Commonly Called Algebers Rule [Eeiiij.v]

This portion of the *Whetstone of Witte*, has received the most publicity but only limited analysis. 'Hetherto have I taughte you, the common formes of worke, in nombers *Denominate*.. Whiche rules are used also in numbers *Abstracte*, and

---

[15] Stifel M, loc. cit., pp 240v–242v

likewaies in *Surde* nombers. Although the formes of these workes be severalle, in eche kinde of nomber. But now will I teach you that rule, that is the principall in cossike workes: and for which others do serve.' [Eeiiij.v] By this stage in his presentation, Stifel had dealt with surds, but Scheubel had not and Recorde was following Scheubel's example. It also appears that Recorde is accepting that the rule of algebra subsumes all previous rules other than those of the four basic arithmetical operations. He continues 'This Rule is called the Rule of Algeber, after the name of the inventoure, as some men think: or by a name of singular excellence, as others judge. But of his use it is rightly called, the rule of *equation*: bicause that by *equation* of nombers, it doth dissolve doubtful questions: And unfolde intricate riddles. And this is the order of it.' [Eeiiij.v] Recorde is happier with an operational rather than an emotional reason for its excellence. He continues with a description of the rule.

   '*The somme of the rule of equation*

When any question is propounded, appertayning to this rule, you shal imagin a name for the number, that is to be sought, as you remember that you learned in the rule of false position. And with that nomber shall you proceed according to the question, until you find a Cossike nomber, equalle to that nomber, that the question expresseth, which you shall reduce ever more to the leaste nombers [i.e. simplify] And then divide the the nomber of the lesser denomination, by the nomber of the greateste denomination, and the quotient doth answer the question.[for monomials] Except the greater denomination, doe beare the sign of some rooted nomber. [for polynomials] For then must you extract the roote of that quotiente, accordyng to that signe of denomination.' [which is where the works of the previous section become relevant] [Eeiiij.v].

The Master then enlarges on this summary through the medium of questions from the Scholar. The latter thinks that rule is the same as that of false position, which allows the master to explain that the number taken under this new rule is a true one, not false and might be more appropriately called 'the rule of dark position'. 'And for the more easie and apt worke in this arte we do commonly name that darke position 1 zensus[sign]. And with it doe we worke, as the question intendeth, till we come to the equation.' [Ffi.r]. For 'dark position' read 'unknown' and for '1 zensus' read 'x'and we have the nomenclature of the present day.

Recorde then divides the rule of equation into two parts, explaining that Scheubel used three sub-divisions and other men even more. This description would certainly apply to Cardano's categorisation in his *Ars Magna*. 'But I intende (as I thinke beste for this treatise, which maie serve as farre as their workes do extend) to divide it onely into twoo partes. Wherof the firste is, *when one nomber is equalle unto one other.* [monomial]. And the seconde is, *when one nomber is compared as equalle unto 2 other nombers'*. [Ffi.v]

Having derived the equation the Scholar is adjured to reduce the terms to their least forms and to rearrange the resultants so that the number with the greatest sign stands alone, equated to the rest. Before he sets out examples of how these ends are accomplished, the Master introduces the most quoted sentences from the book 'Howbeit, for ease of alteration of equations I will propounde a fewe examples, because the extraction of their rootes, maie the more aptly bee wroughte. And to avoide the tedious repetition of these wordes: is equalle to: I will sette as I do often in worke use, a pair of parallels. of Gemowe lines of one length thus : =, bicause no 2 thynges can be more equalle.' [Ffi.v] He then presents six equations using the signs +, − and =, using no terms greater than square illustrating how monomials and binomials can be reduced to the form he prefers. Based on these examples, Hughes has credited Recorde with being the first of all writers on mathematical topics to publish a symbolic statement that can be properly named an equation.[16] Heeffer argues that the sentence that opens the paragraph gives the motivation for introducing the sign, viz. the manipulation of equations. In addition to to the direct reference it makes to arithmetical equivalence, it represents the combinatorial operations to which an equation may be subjected. Some of these are demonstrated subsequently. Heeffer continues, 'The introduction of the equation symbol completes the basic stage of development towards symbolic algebra, as initiated in Germany by the end of the fifteenth century. .... As the minus sign facilitated the the acceptance of negative numbers so did the equation sign contribute to the further development of algebra towards the structure of equations'.[17]

The Master then explains why he thinks there are only two forms of these equations that have to be considered instead of Scheubel's three. En passant he says 'As for the manyfolde varieties, that some other do teache, I accompte it but an idle bablyng, or (to speake more favourably of them) an unnecessary distinction.' [F(f) iij.r&v] Recorde must have felt somewhat irritated by this unnecessary distinction having mentioned it once already. Cardano's lists of the varieties of first and second degrees equations at the start of his *Ars Magna* is both numerous and repetitive. The repeated comment is therefore of interest inasmuch as it suggests that Recorde had had sight of Cardano's important treatise.

The first kind of equation, with whose meaning Recorde says he agrees with Scheubel, has the form $ax^n = b\ x^m$. The Master distinguishes two cases. When $n = m + 1$, the answer is an absolute number. But if $n > 1$, then the root will be a cossike number. They also agree on the second kind of equation, defined as when one simple cossike number is found equal to two other simple cossike numbers '... of severalle denominations and like distaunce'. [Ff.iiij.v] i.e. of the forms $x^2 = bx + c$, $x^2 = -bx + c$ and $x^2 = bx - c$, or reducible to those forms. As with the first kind of equation there is the variation resulting from the denominations not forming an immediate sequence i.e. $x^n = bx^m + c$ etc., where $n > m + 1$. The roots of both kinds of equations are to be extracted using the procedures established in the immediately previous section. Scheubel gives three sets of rules for dealing with each of these two main kinds of equations. Recorde shows this to be an unnecessary complication, avoided if the equations are translated into the standard form he recommends viz. with the greatest cossike term equated to the rest. This argument for the adoption

[16] Hughes B, loc. cit., p 163

[17] Heeffer A (2009) On the nature and origins of algebraic symbolism. In: Van Kerhove B (ed) New perspectives in the philosophy and history of mathematics. World Scientific Publishing, Singapore, pp 1–27

Heeffer A (2008)The emergence of symbolic algebra as a shift in predominant models. Found Sci 13(2):149–161

of this standard form dilutes Hughes contention that it stemmed from a geometrical mindset on Recorde's part, but does not eliminate it.

As was his wont, the Master then works through a plethora of examples of problems that can be solved using the methods for solution of quadratic type equations that have been described. They are all strongly reminiscent of the problem types dealt with in the *Grounde of Artes* and grounded in the real world. That Recorde does not recognise negative roots or double identical roots provides no impediment in these latter circumstances. In all cases the correctness of the result is checked by back-calculation.

The final problem is interesting inasmuch as the Master provides a solution without giving the method, proving no more than a solution must exist. A number is sought whose square less sixteen, when multiplied by the number added to eight, is equal to 2,560. The equation produced correctly by the Scholar is $x^3 + 8x^2\ 16x - 128 = 2,560$ who then reduces it by moving all the negative terms to the right-hand side, but then does not know how to proceed further. The Master gives the answer as 12 and shows by substitution that it is correct. He then goes on to further manipulate the equation by isolating the cubic term on the left-hand side of the equation, demonstrating that the other side must have a cubic root. This argument he then extends to higher powers of the unknown. He concludes, 'But of these and many other verie excellente and wonderfulle woorkes of equation, at an other tyme I will instructe you farther, if I see your diligence applied well in this, that I have taught you. And therefore here will I make an eande of *Cossike* nombers for this tyme.' [Ll iij.r] As Recorde had dealt with all the algebra to be found in the works of Scheubel and Stifel at that time, it may perhaps be presumed that he was referring to the works of Cardano.

The Arte of Surde Nombers in Diverse Sortes [Lliij.r]

'Now that you have somewhat learned the arte of *Cossike* nombers, with the rule of equation, it seemeth goode time and apte place, to teach you the arte of surde *nombers*, which are diverse in name accordyng as there are diverse nature of rootes, which may give them their name.

For generally, a *Surde* number is nothing els, but soche a nomber set for a roote, as cannot be expressed by any other nomber absolute.' [Lliij.r]. This is the description given by Scheubel. The sequence of development of the topic used by Recorde is also that of Scheubel.

However, starting with Numeration the symbols used for depicting various roots are common to Scheubel and Stifel but the nomenclature Recorde uses is that of Stifel. From that point onwards there is little in common between Recorde and Stifel in presentational practices.

Uncompound Surds [Lliiij.r]

Recorde chooses to deal first with the multiplication and division of surds because they are easier to understand and also because they are needed for the manipulations involved in the operations of addition and subtraction. Examples are given of multiplications of square, cubic and quadratic roots that may or may not make absolute numbers. The simplification of surds is dealt with expeditiously, as also that of their division. It is to be noted that where a surd could be presented as $a\sqrt{b}$, Recorde always uses $(\sqrt{a^2.b})$, or its equivalent for other roots.

Addition is said to be not so easy and has a number of methods of working. The simplest of these just needs the use of the sign of addition and is employed chiefly when expressing the sum of surds of differing roots. The second form of addition is used to add surds having the same roots but of different values e.g. $\sqrt{26} + \sqrt{12}$. The method given stems from the relationship $(a+b)^2 = a^2 + b^2 + 2ab$, the algebraic version of Euclid's proposition 3, Book II of the *Elements*, as Recorde acknowledges later, but which Scheubel acknowledges before he gives the method. The instructions are to add the terms of the two surds together $[26 + 12 = 38]$ and add the sign of the root $[\sqrt{38}]$. Multiply the first two numbers together and then again by four $[312 \times 4 = 1,248]$ to which the root sign is added again $[\sqrt{1,248}]$. These two segments of calculation are then added together to give the final sum. Recorde presents this as $\sqrt{38} + \sqrt{1,248}$, which to the modern eye it clearly is not and he does not clarify this matter until very much later. The confusion may have been because the necessary symbolism was available to neither author nor printer, but nowadays it would be written $\sqrt{(38 + \sqrt{1,248})}$. Proof of accuracy cannot be obtained by back calculation but the validity of the method is demonstrated using square numbers such as 49 and 36 rather than surds.

A method that can be used only for the addition of the roots of commensurable numbers, i.e. those that have a common divisor is next described. The common factor is separated out and then the first arithmetical procedure is repeated on the residual components. The easiest of algebraic analysis would avoid much of the calculation, but whether Recorde's understanding was deficient or whether he did not wish to expose his reader to such processes at this juncture is not obvious. The algebraic approach would lead to an answer in the form $m\sqrt{n}$ whereas the procedure given by Recorde yields $\sqrt{(m^2.n)}$, which is the form he always used.

Changes in the procedure when cubic roots are to be added is considered next, but only for commensurable numbers, i.e. such as can be reduced to what is effectively a form amenable to the method used in the previous examples. '......or nombers incommensurable are added with the signe + without moare worke. I call soche *Cubike* rootes *commensurable*, which beyng divided by any common number, will make cubic nombers in their *quotient*.' [Ooj.r] In the examples given the common cube root is extracted, the cube roots from the separate terms added together, cubed and taken inside the cube root. This procedure seems long-winded compared with its algebraic counterpart. Recorde then compounds the complexity unnecessarily by applying the cubic counterpart of the binomial expansion applied to the addition of square roots viz. $(a+b)^3 = a^3 + 3a^2b + 3ab^2 + b^3$. This is the equivalent of Euclid's proposition 12 Book VIII, but this time is not acknowledged by Recorde as his source.

'Subtraction doeth differ from addition, in little moare then the signe -. Which signe serveth generally, for all nombers *incommensurable*. And consideryng there is litle difficultie in Subtraction: If you remember well the arte of Addition, I wil lightly passe it over in the same examples, that I have wrought in Addition, bicause it maie bee a proofe of that woorke: and that woorke also a confirmation of this.' [Do.iij.(v)] This statement illustrates very well Recorde's belief in continuity of approach to the furthering of understanding. He then does what he said and passes on to treat compound surd numbers.

Of Surde Nombers Compounde [Ppiij.v]

These are described as being made up of two, three or more uncompound surd numbers as well as absolute numbers and employing the signs for addition and subtraction. Where only + is used the number is properly called a bimedial if all its components are of one denomination, but if different denominations are present the the compound number is called a binomial. This distinction is acknowledged to be breached commonly and Recorde does not use it. . If − is used then the number is called a residual. At this point, Euclid is mentioned for the last time. '*Euclids* definitions do not very aptly agree to this place, as at another tyme I will shewe you, and therefore I do omitte them for this tyme.' [Ppiiij.r] Attention is drawn to one further definition. *Universal roots* are not counted amongst compound surds and are to be treated separately as the roots of compound surds.

The basic arithmetical operations are considered to pose little difficulty provided the methods of working already described are properly applied. Worked examples of additions using all combinations of binomials and residuals for compound surds having two components are given. The same approach is used to demonstrate subtraction, but with one warning, raised as usual in the form of a question by the Scholar. Two of the examples used are derived from examples used previously to demonstrate addition. Thus

Addition	Subtraction
$^3\sqrt{320}\ \sqrt{56}$	$^3\sqrt{1,080} - \sqrt{80} - \sqrt{5,376}$
$^3\sqrt{40} + \sqrt{24}$	$^3\sqrt{40} + \sqrt{24}$
$^3\sqrt{1,080} - \sqrt{80} - \sqrt{5,376}$	$^3\sqrt{320} - \sqrt{56}$

The Scholar complains that whilst he can solve the first example and consequently see how the second result must be correct, he does not know how the subtraction of the √ terms is to be carried out. The Master says that he would not be expected to do so, for two of the three terms are not compound surds but are to be grouped together as a universal root of a surd, the manipulation of which had not yet been taught. Interestingly, several of the results of subtractions are negative numbers, but no comments on this feature are made, whether out of ignorance or out of acceptance cannot be judged. They are no less 'absurde' than the negative product of the subtraction of 28^5 from 32^3 which he described as such earlier.

Multiplication of binomials and residuals is dealt with expeditiously, but division when the divisor is a compound number is a little more complex. In this case the divisor needs to be multiplied by its residual to give a whole number which is the new divisor. The new dividend is formed by multiplying the first dividend by the residual. Thus to divide $\sqrt{68} + \sqrt{54}$ by $\sqrt{6} + \sqrt{3}$, both divisor and dividend are multiplied by $\sqrt{6} - \sqrt{3}$. The new divisor is 3 and the new dividend $\sqrt{408} + \sqrt{342} - \sqrt{204} - \sqrt{162}$, each term of which has to be divided by $\sqrt{9}$ to give $\sqrt{45\,^1/_3} + 6 - \sqrt{22\,^2/_3} - \sqrt{18}$.

Proof of the result is by reversing the procedure.

Of Extraction of Rootes [Rriij.r]

Consideration of this subject leads almost immediately to a clarification of the nature of universal roots. The extraction of the root of a number involves only the plac-ing of the sign of the root before the whole number. The significance of the applica-tion to the whole number is demonstrated by adding the square roots of 24 and 144. Taken separately, addition of the roots gives approximately 17, but taken together the result is $\sqrt{(24 + 12)}$ i.e. 6. Understanding then dawns on the Scholar regarding the difficulty he encountered in the immediately previous section. This difficulty could have been obviated much earlier had the brackets () been used to group the components of the compound surds. Recorde had used them in this way for uncom-pound surds, so his omission of their use in this context is somewhat mysterious.

At this point the messenger arrives at Recorde's door, summoning him to deal with the matter of his imprisonment, as has been described elsewhere. Clearly this had been concerning him for some time, judging by a number of comments he made deferring more detailed treatment of topics such as derivation of binomial expan-sion and clarification of the position of Euclid on irrationals, until he had more lei-sure. Whilst he must have had a good idea of what awaited him he still promises to augment his teachings '...when you see me better laiter: That I will teache you the whole arte of *universal rootes*. And the extraction of rootes in all *Square surdes*: with the demonstration of them, and all the former woorkes.' [Rriij.r]

There has been spasmodic debate about Recorde's position with respect to the introduction of algebra into England. He was certainly the first to present the 'cossike arte' in English and by no means the last, despite the fact that the topic did not seem to prove of much interest.[18] He had at hand when compiling his text the works of Stifel and Scheubel, which he used critically and selectively, taking those fea-tures that best fitted in with his overall presentational plans and expressing and explaining them in his own way. The methodical approach of Scheubel appears to have held greater appeal for him than the rather less structured approach of Stifel. The latter was also more committed to a geometrical approach to the topic of surds for instance than Recorde and Scheubel, who show signs of distancing themselves further from such an overtly geometrical mind set.

---

[18] Moore J (1650) Arithmetike

More problematic is the extent to which Recorde might have been influenced by, or had access to, Cardano's *Ars Magna*. Mention has already been made of a comment that Recorde made probably criticising Cardano's profligate enumeration of first and second degree equations. He would however have approved of a statement in Chap. 1 of the *Ars Magna*. 'For as posito [the first power] refers to a line, quadratum [the square] refers to a surface, and cubum [the cube] to a solid body, it would be very foolish for me to go beyond this point. Nature does not permit it. Thus it will be seen, all those matters up to and including the cubic are fully demonstrated, but the others which we will add, either by necessity or out of curiosity, we do not go beyond barely setting out.'[19]

If one accepts Heeffers definition of symbolic algebra viz. 'algebra using a symbolic mode, which allows for manipulations on the level of symbols only', then Recorde was practising it. Recorde might not have realised fully what he was practising but some clarification of the situation might have emerged if the further publications he had in mind had materialised. At the end of his treatment of the rule of equation he says that '... of these and many other excellent and wonderfulle woorkes of equation at another tyme I will instructe you farther ... '. He concludes his section on the art of surds in a similar vein. Given more leisure, he promises to teach the art of universal roots and the extraction of roots in all square surds, 'with the demonstration of them and all the former woorkes.' The events that followed the symbolic knocking at Recorde's door ensured that the *Whetstone of Witte* was to be his last work.

[19]Cardano G (1968) Ars Magna or the rules of algebra, (trans: Richard Witmer T). Dover, New York, p 9

Chapter 11
Antiquary and Linguist

Abstract Recorde's collection of books by British authors was one of the three largest collections listed by Bale in his Index Scriptorum. Their titles, mainly of manuscripts, covered a wide range of interests, religion, medicine, alchemy, prophesy, heraldry, geography, history, arithmetic and astronomy and law. The largest group concerned astronomy, a roll call of past English authors writing on this subject. The greatest number were from the collection of Lewis of Caerleon, an archive of documents of the important English astronomers of the fourteenth century still extant, Recorde having provided a safe haven during a perilous period in their history. The only example of Recorde's writing is as comments in Anglo-Saxon on an early medieval manuscript. This together with other examples of his scholarship in this tongue suggests strongly that Recorde was part of a group of the earliest Tudor antiquarians. His linguistic interests did not end there. One consequence of his decision to write his books in the vernacular was the practice he eventually adopted with respect to the etymology of English scientific terms. Having failed to introduce terms having Old English roots into the geometrical vocabulary, using Latin roots he introduced a wide vocabulary of terms necessary for number theory, cossike arithmetic and algebra, which are still in use today.

In her survey of medieval history in the Tudor age, McKisack writes 'Though primarily a mathematician, Recorde had a large collection of historical manuscripts, including Howden, some of the works of Giraldus Cambrensis, and Fortescue's 'De Laudibus Legum Angliae'[1][Bale's Index]. More remarkable still is the fact that he knew Anglo-Saxon, as is shown by his notes in the margin of a volume of Chronica Varia, now at Corpus Christi College, Cambridge (MS.138).' Exploration of both of these aspects of Recorde's interests' throws light on the scholarly company which was available to him and which respected his capabilities as an antiquarian. Attention has commonly been drawn to his failed attempts to enlarge the English vocabulary

---

[1] McKisack M (1971) Medieval history in the Tudor age. Clarendon Press, Oxford, p 25

J. Williams, *Robert Recorde: Tudor Polymath, Expositor and Practitioner of Computation*, History of Computing, DOI 10.1007/978-0-85729-862-1_11, © Springer-Verlag London Limited 2011

to accommodate mathematical interests, but he did have more successes than failures in this matter which will therefore be re-examined more thoroughly.

The Antiquarian

'Ex Museo Robertus Recorde'

What is known of Robert Recorde's own collection of books derives from Bale's compilation of Famous British writers, in which he lists books 'ex museo Robertus Recorde'.[2] It is one of the three largest collections listed. The titles cover a wide range of interests, religion, medicine, alchemy, prophesy, heraldry, geography, history, arithmetic and astronomy and law. Excluding Recorde's own books, complete or projected, there were upwards of 90 items which were primarily in the form of manuscripts.

The religious texts include a small offering on the Eucharist by Albericus Anglus and books of sermons by Richard of Maydestone and Robert of Bridlington, both of Merton College. There are just two books on medical topics, the standard text on surgery by John of Ardern dating from 1370 and a book on medicines by Richard of Maydestone. There was one book on the transmutation of metals by Thomas Norton and one on heraldry, in four volumes by Nicholas Upton, a canon of Salisbury and Wells.

The subject of prophesy was well served. There were versified sets of prophesies by Donkamen and Scriba. Other texts were provided by John de Muris and Merlinus Sylvestris. A book on prophesy by Galfridus Eglyne had the incipit 'Asinus coronatus turbabit regnum'.[3] There were also a set of nine books attributed to Merlin for one of which 'Merlinus scripsit de honoribus' the incipit reads 'Orietur draco de asino qui in brachio suo...'. Another begins 'Mortuo leone iustice surget albus...'. These incipits taken together with the interest in heraldry, generate some sympathy for Pembroke's contention that Recorde had libelled him by subtle prophesy.[4] These texts were commonly found in monasteries and may therefore have been rescued during dissolution processes. Recorde's first dedicatee, Richard Whalley would have been well placed to expedite such an activity.

Recorde's interest in other countries seems to have been confined to Ireland, in the form of a set of seven books by Girardus Sylvester on both its geography and history. He had two works by Roger Hoveden on the history of England, one being that dealing in two volumes with periods 732–1181 and 1181–1203 and the other a book of his lectures. Gildas was represented by two 'Carmina'. Finally there was one printed book by George Lilius and his son William, written under the aegis of Cardinal Pole, on the lives of the Roman Emperors and the Popes. Interestingly it had a map of Britain prepared in 1546 attached to it.

---

[2] Bale J (1990) Index Brittaniae Scriptorum in the edition by Poole, pp 32, 61, 72, 94, 96, 116, 145, 177, 200, 218, 223, 233, 234, 240, 257, 262, 284, 293 294, 310, 335, 355, 358, 362, 386–388
[3] Approximately 'Fools surround and disturb the crown'.
[4] See Chap. 2.

Recorde seems to have fortified himself well with multiple volumes on the laws of England, presumably in preparation for his trial. They included Glanville (c.1188), Bracton (thirteenth century), Scrope (fourteenth century), Fortescue (fifteenth century), Littleton (fifteenth century) and Rastell (sixteenth century). These were presumably the books he stated in his will which were to be sold to his fellow prisoner Nicolas Adames for four pounds. The law book associated with Rastell requires special attention. Bale's reference reads 'Johannes Rastell, Indices librorum Anthonii Fitzherbert, Ex museo Roberti Recorde.' Rastell published two books containing material attributable to Fitzherbert. Their relationship to one another has been described by Read.[5] Rastell compiled first the Book of the Assizes followed by 'A grete boke of the abridgementes of cases'. The *Tabula libri Assisarum* has been tentatively dated to 1513 and contains references to the cases abridged by Fitzherbert as well as their fuller versions, and therefore was the book held by Recorde. What might have interested Recorde in other than its legal contents, is a page of text at the very beginning of the book in which Rastell describes briefly 'the nombers of algorisme' and the nature of the place value system, together with a table of equivalent Roman numerals. The algorithmic numbers are then used exclusively throughout the rest of the book. Rastell's interest in this format did not end with this slender contribution. Fragments of documents from the Rastell workshop deposited in the Dartmouth College library have been shown by Nash to contain pages from *The boke of the new cardys*.[6] This book intended to demonstrate, by play with cards, how to learn to read and write in English as well as to read both Roman and algorithmic numbers and to cast accounts using both types of arithmetic. Nash points out that Rastell's eldest son William was a contemporary of Recorde at Oxford and wonders whether if any connection to the *Grounde* may have resulted. One may also wonder whether there was a more direct connection between Rastell and Tunstall who dedicated his book on algorithmic arithmetic to Thomas More, Rastell's brother-in-law. In the absence of an exact date for Rastell's work it is not possible to speculate on who might have influenced whom.

The largest collection of titles concerned astronomy and in content provides a roll call of past English authors writing on this subject. The oldest of these documents nominally is the *Theorica planetarum* ascribed to Roger of Hereford by Bale. If correct, this would date the manuscript to the end of the twelfth century, but this ascription may be wrong.[7] The book explains the use of astronomical tables in some 32 chapters. The works of the Merton school of astronomers in the fourteenth

---

[5]Read AW, Early Tudor drama, Appendix III. p 209

[6]Nash R (1943) Rastell fragments at Dartmouth. Library 54 xxiv(1–2):66–73

[7]The incipit reads 'Diversi astrologi secundum diversum....'. Versions of the text are found in Bodley MS.300 ff.1–19v, Digby MS 168, ff. 69v–83v. and in Saville MS. 21 f.42. Criticism of the attribution was first based on the citation of the London star tables for 1232, but tables for London based on the Toledan tables predate Roger of Hereford. A more serious obstacle is that raised by North, who points out that MS. Bodley 300 was probably copied in the mid-fifteenth century at the time it was given to Clare College Cambridge, at which time there was an interpolation in its chapter (17) that mentions the tables of John Maudith.[North JD (1986) Horoscopes and history.Oxford University Press, London, p 119? North notes that in Digby 168, this note was copied out at the foot of the page in a seventeenth century hand.

century in compilation of such tables and almanacs are represented by works of John Mauduith, Simon Bredon, John Ashenden, John Somers and William Rede. Texts on the subject other than from the Merton scholars were by John Walters and John of Northampton. The limited activity by English astronomers in the following century are represented by the tables of John Holbrook at Cambridge and, much later, by the eclipse calculations of Lewis of Caerleon, described by Bale as fragments plus a book. Two members of the Merton school, John Kyllingworth and Simon Bredon also provided texts on arithmetic. Dissertations on astronomical instruments were provided by two books on the concave sphere by William Batecumbe and by Richard of Wallingford on the Albion and the Rectangulus. The latter was also represented by his Extrafrenon, an introduction to judicial astrology, which was the more acceptable face of that discipline at the time.

The work of Kibre on Lewis of Caerleon and of North on Richard of Wallingford , coupled with the data provided by Bale for these two authors provide evidence of both the origin and the fate of all the astronomical texts held by Recorde, other than those ascribed to Richard of Wallingford.[8] Specifically they were originally those held by Lewis of Caerleon, now to be found in Cambridge University MS Ee. 3. 61. In this latter manuscript, Lewis of Caerleon's contributions fall into two groups, the first lying between ff.3–7 and 12–25 and the second between ff.125 and 153. The first group are categorised by Kibre as works attributable to him and the second as manuscripts which he transcribed or annotated. The former Bale described as 'reliquit qaedam fragmenta astronomica: de eclipsiun calculatione li. i ', which could accommodate the two sets of documents attributed to Lewis by Kibre and North. Of greater significance in assigning Recorde's holdings of astronomical manuscripts to E.e.3.61 are other manuscripts held by Lewis lying between ff.31v and 105v of Ee. 3. 61. Correspondence between between their incipits and the works of the Merton school, listed by Bale as present in Recorde's library, is complete.

CUL. Ee. 3. 61.	Ex Museo Robert Recorde.
ff.31v–42v. 'Algorismus Magistri Johannis Kyllyngworth' [inc.] Oblivioni traduntur que certo convertuntur ordine...' [9]	Ioannis Kyllyngworth qoundam socius Mertonis scripsit. Arithmeticum opus li. i.'Oblivione raro traduntur.
ff.43r–47r. Expositio Magistri Symonis super Capitula Almagesti Ptholomei [inc.] 'Nunc superest ostendere quanta sit superest maxima declinacio Ecliptice....'	p.411. Simon Bredon scripsit. Expositionemin quadam capita Almagesti et Ptolomei li. i. 'Nunc ostendere etc.'

[8]Kibre P (1984) Studies in medieval science. Hambledon Press, London, XV, pp 100–108

North JD (1976) Richard of Wallingford. Clarendon, Oxford, vol II, pp 20,33,138,381; vol III, pp 130,136, but particularly Appendix 34, pp 217–220

[9]This text, whilst using Hindu-Arabic numbers operates on them using the abacus method of calculating using doubling and halving procedures. It only starts to deal with vulgar fractions and tails off, but it deal fairly fully with astronomical fractions. See Karpinski LC (1914) The algorism of John Killingworth. Engl Hist Rev 29:707–717

CUL. Ee. 3. 61.	Ex Museo Robert Recorde.
It concludes with two Tables, the first of which was taken by Lewis of Caerleon from one by Magister John Walteri entitled 'Tabula diversitas ascensionis signorum pro omni terra habitabili.'; the second 'Declinationem Almansoris.. .	p.262 . John Walters 'Tabula ascensionium universalem.'
ff. 56r–73v. Two sets of Astronomical Tables by Magister Holbroke.Their incipits read respectively 'Quoniam celestium motuum calculus..'. and 'Gloriosus atque sublimis..'.	p.218. Ioannis Holbroke in Sothraia (ut fertur) natus scripsit.Tabulae astronomicis cum canonibus Prior incipit 'Quoniam celestrium etc. 'Alter incipit 'Gloriosus atque sublimis'
ff. 96v–105v. Ars Metrica M. Symonis Bredon. Boecii, [inc.] 'Quontitatem alia continua que magnitude ..'.	p. 411. Simon Bredon scripsit i. 'Arithmeticam theoricam l.i [inc] 'Quantitatem aliacontinua... '.

There are items in Ee.3.61 not listed in Bale's Index for Recorde. The largest number of these is attributable to John de Lineriis, a Picardian scholar of the fourteenth / fifteenth centuries. They comprise a large set of canons for eclipse calculations [ff.82r–96r], commented on by one of his disciples John of Saxonia [ff.178r–181r.] and his Algorismus de Minuciis [25v–29r.] with the associated Tractatus de probis [30r–31r.]. The treatise on fractions was unique at that time in dealing comprehensively with the arithmetic of both vulgar and sexagesimal fractions using Hindu-Arabic numbers. Bale would not be expected generally to have included works by foreign authors in his compilation although there a few such items listed. The next largest group of manuscripts absent from Bale's inventory of Recorde's library, but present in E.e. 3.61, are those dealing essentially with predictive astronomy or horoscopes, examples of which topics Bale included in his compilation only very rarely.[10] The residual item is a brief treatise on the squaring of the circle, a topic on which Recorde's views are not known.

Recorde held four astronomical manuscripts which are not found in E.e. 3.61.[11] No obvious whereabouts for these texts can be offered.

---

[10] ff.48r–55v., 'A Treatise on Planetary Influences'

ff.74r–v., 'Trutina Hermetis'

ff.108r–120v., 'Tables of the Aspects of Signs for Humphrey Duke of Gloucester.'

ff.159r–175v., 'A Treatise on the Horoscope of the Nativity of King Henry VI'

[11] Bale p 145, 'Guilhelmus Rede episcopus Cisestrensis scripsit

Tabulae astronomicas et canones in easdem' note see Bodleian Library MS 432 f.28

P 200, 'Ioannes Ashendon, quondam socius aule de Merton Oxonij scripsit. De significatione conniunctiorum quaerundum li. i.

Sicut dicit Ptolemeus etc. Composuit alia plura. Claruit ad 1350

P 232.'Ioannes Mauduith scripsit Oxonij ad 1350'

P 257.'Ioannes Somer, Minorita, opus Calendri tertium fecit, ad instantem Ionne matri Ricardi secundi regis li.i. claruit a.d. 1386

Bale lists three books by Richard of Wallingford in Recorde's library :-

Ricardus Wallyngforth, abbas Sancti Albani, composuit instrumentum Albion, et a monasteriosic
vocavit, Albionem id est omnia per unum ut Brittanice[12]
sonat. li. i.
Citantur alie eius questiones frequentur a Ludovico Kaerlion.
Extrafrenon de judicis astronomicis eiusdem Ricardi putatur li. i.
Ad perfectum noticiam.....
Rectangulam (inquit) concepsismus eodem tempore quo composimus instrumentum, quod
nominavimus Albion, hoc est etc.. A.D. 1326.

John North in his biography of Richard of Wallingford has commented on the whereabouts of three of the works by Richard of Wallingford, copies of two of which, the Albion and the Rectangulus were held by held by Recorde. He pointed out that the *Rectangulus* treatise is also found in Laud Misc 657 immediately following Tunsteed's version of the *Albion* treatise. The suggestion that Laud Misc 657 contains Recorde's copies of the *Albion* and the *Rectangulus* is attractive, but not wholly satisfying. The description by Bale of the Wallyngford manuscripts held by Recorde provides the incipit for the *Extrafrenon*, but not for the Albion, nor yet for the *Rectangulus*. Whilst the remarks Bale makes about the *Rectangulus* paraphrase the incipit of all the copies of this work, those for the Albion reflect neither the introductory phrases of Laud MS Misc 657, nor indeed those of the more orthodox introductory phrases of other copies of the work. Tunsteede's name is not mentioned in Bale's presentation. Recorde's copy of the description of the *Rectangulus* might be that found in E.e. 3.61., in which case his copy of the Albion might have to be sought other than in MS. Laud 657. Leland noted that there were a number of astronomical and scientific texts at Clare College Cambridge c. 1535, including an item which might have been E.e. 3.61, the Theorica Planetarum by Roger of Hereford and texts relating to the Albion.[13] There is therefore a possibility that the astronomical manuscripts held by Recorde, as listed by Bale are those identified by Leland at Clare College.

An interesting side issue arises if all the contents of E.e. 3.61 had been available to Recorde. John de Lineriis' text on fractions would have been a very useful source of information on fractions, vulgar and astronomical, as the treatment therein was as sound and comprehensive as that found in later texts on the subject.

Returning then to the question of the fate of the majority of Recorde's holdings of English astronomical manuscripts, it is conceivable that they now reside in CU E.E. 3.61. What happened to them after Recorde died and if indeed they are those in the Cambridge manuscript, how they came into the Library of John Moore, Bishop of Ely is a matter for further investigation. Their final journey to King George I, who gifted this library to the University of Cambridge, is documented.[14]

---

[12] The universal nature of the instrument is echoed by the English pronunciation of its name – all by one.

[13] Clarke PD (2002) The university and college libraries of Cambridge. Corpus of British medieval library catalogues 10. British Library in association with the British Academy, London, pp 151–155, items 6, 22, 20 and 21 respectively.

[14] Kibre P, loc. cit., XV101 note 8

The last words on Recorde's library as well as the first words belong to Bale. The latter had long been concerned about the long term disastrous effects that the dispersal of manuscripts following the dissolution of the monasteries might have. He returned to the theme in a letter to the newly appointed archbishop of Canterbury, Matthew Parker, dated 30 July 1560. In this letter he also named individuals who held collections of 'bookes of antiquite not printed'. The list included Sir John Cheke, Lord Paget, Sir John Mason and Lord Arundel who have already been mentioned in other contexts; John Perkins, William Carye, Peter Osburne, the widow of John Ducket, John Twyne and the antiquary Nicholas Brigham and Robert Talbot a fellow of New College Oxford whom we shall meet later. In this company was included the executors of Robert Recorde. Whatever happened to the manuscripts subsequentially, Recorde had provided a safe haven during a perilous portion of their history.

Other Antiquarian Activities

There may have been a direct link between Recorde and Leland arising from their interests in antiquarian pursuits. Mention has already been made of Recorde's Anglo-Saxon scholarship demonstrated by his notes in the margin of a volume of Chronica Varia, now at Corpus Christi College, Cambridge (MS.138). On the flyleaf of this volume, at the end of a Table made for Parker, is inscribed 'Robertus Recorde erat qui notauit hunc librum Characteribus Saxonicis.' According to Page, attention had been first drawn to these annotations by Dickins.[15] From further studies Page concluded that Recorde had taken some of his material from Cotton Tiberius B I., but the Genealogical Preface which Recorde had copied into MS 138 had probably originated elsewhere. Page identified MS Kk 3.18 (Cambridge University Library) as Recorde's likely exemplar, supporting his argument with evidence of close correspondence between the two sets of texts. Graham has taken the story a stage further and showed that both of the manuscripts had been owned by Robert Talbot.[16] Leland was a friend of Talbot and had had borrowed Cotton Tiberius MSS B I from him. Talbot had entered New College, Oxford, in 1521 and supplicated for his M.A. in 1529. He was therefore a contemporary of Recorde. It has been suggested that he was expelled from college for his vigorous, reforming religious opinions, which he presumably still held in 1541 when he was chaplain to Cranmer. Dickins suggested that Recorde may have used Henry of Huntingdon's Historia Anglorum, Florence of Worcester's Chronicon and his own copy of Roger of Hovedon's Chronicon as sources written in Latin for some of his annotations to MS 138.

---

[15] Page RI (1972) Anglo-Saxon texts in early modern transcripts. Trans Camb Bibliogr Soc 6/2:69–85, 75–79

[16] Graham T (1997) Robert Talbot's 'Old Saxonica Bede'. In: Cowley JP, Tite CGC (eds) Books and collectors 1200–1700, British Library Studies in the History of the Book, London, pp 295–316

MS Tiberius I is likely to have been the source of historical material described by Recorde in the Dedicatory Epistle to The Whetstone of Witte. Speaking to the dedicatees, the Muscovy Company, about his projected book on navigation he says 'Wherein I will not forget specially to touche, both the olde attempte for the Northelie Navigations, and the later good adventure with the fortunate successe in discoveryng that voiage, which noe men before you durste attempte, sith the tyme of Kyng Alurede his reigne. I mean by the space of .700. yere. Nother any before that tyme, had passed that voiage, excepte onely Ohthere, that dwelte in Halgolande: whoe reported that iorney to the noble kyng Alured: as it doeth yet remaine in aunciente recorde of the olde Saxon tongue.' Brewer cites this as having the distinction of being the first piece of Anglo-Saxon to be mentioned in print.[17] Graham was able to identify the version of the Old English Orosius as being that found in the MS Tiberius I, which Leland had borrowed from Talbot. Interactions between Recorde, Talbot, Leland and Wolfe must have been much more than of just a casual nature.

The connection between Leland and Wolfe has been discussed recently by Harris in the context of the fate of the literary remains of the former following his demise in April 1552.[18] Harris was concerned primarily with Leland's own autographic writings but points out that these formed only one portion of a large collection acquired by Royal authorisation. Subsequent to dissolution processes Leland had salvaged material from monastic and collegiate libraries, much of which went into royal libraries, some more into Leland's own library as recorded by Bale, with more unaccounted for.[19] Harris concluded that following Leland's mental collapse in 1547, Sir John Cheke was made responsible for his library, but with his fall from favour in 1533 the books began to be dispersed, although Leland's own writings stayed with Cheke until his death in 1557. By various routes they then passed to John Stow for transcription ,but the suggestion by Bagford in 1715 that 'most of [Leland's] Writings (after his Death) came into the hands of Reginald Wolfe the Printer, at whose House I believe Leland dyed.', in the absence of more substantive evidence, is deemed by Harris to be unlikely.[20] Nonetheless, Stow did know Wolfe and described him as ' a grave antiquary', who had 'collected the great Chronicles increased and published by his executors under the name of Ralph Holinshead.'[21] He also gives Wolfe's account of the clearing of the charnel house in St. Paul's cloisters preparatory to the expansion of Wolfe's publishing activities there.[22]

---

[17]Brewer DS (1952–1953) Sixteenth, seventeenth and eighteenth century references to to the voyage of Ohthere. Anglia 17:202–211, 202

[18]Harris O (2005) 'Motheaten, moldye and rotten': the early custodial history and dissemination of John Leland's manuscript remains. Bodleian Libr Rec XVIII(5):460–501

[19]loc. cit., 461–462

[20]Bagford J, A letter to the publisher. In: Hearne (ed) Collectanea, pp i.pplviii–lxxxvi, lxviii

[21]Stow J (1908) In: Kingsford CL (ed) A Survey of London, 2 vols. Clarendon press, Oxford, i, 293 note

[22]Ibid., 330,349

Whilst the fate of Leland's own writings seems clear, that of his collections of the manuscripts of other writers is unresolved. On this point Harris opines 'Bale clearly had reasonably free access to Leland's material during his lifetime, and so too, it may be assumed did Humanist friends like Robert Talbot, John Twyne, Sir Thomas Elyot and Sir John Pryse, all of whom gave him reciprocal help and advice.' That Recorde knew Talbot and Elyot seems very likely and he did at least have a non-antiquarian interest in common with Pryse. In 1553, at what must have been only shortly before his death, Pryse sent a letter to Queen Mary describing the perilous state of the coinage and urging her to restore the sterling standard. This could only have been months after Recorde had published his revised version of *The Grounde of Artes* in which he had expressed similar views and urged action on King Edward.

John Kyngston who printed *The Whetstone of Witte*, also printed the fourth edition of *Fabian's Chronicle* (1559). This edition of the latter work reinstated the contents of the second edition following their pruning in the third edition, but in addition it described itself as 'newley perused'. It seems probable that the peruser was Recorde and he was identified as responsible for a substantial annotation on p 25 listing Celtic kings presumably deriving from his earlier readings. An even more substantial contribution by the peruser on pps 33–34 references Geoffrey of Monmouth as his source.[23]

His interest in other civilizations and their writings may not have ended with those of Rome and Greece. In the *Grounde of Artes* , discussing the ordering of numerals from right to left in the place-value system, he says 'In that thinge all men do agree, that the Chaldays, which first invented thys arte did set these figures as the set all their letters. For they write backwarde as you tearme it, and so doo they reade. And that may appeare in all Hebrewe, Chaldaye and Arabike books. for they be not only wrytten from the ryghte hand to the lefte, and so must be readde, but also the ryghte ende of the booke is the begyinninge of it: where as the Greekes, Latines and all nations of Europe, do wryte and reade from the left hand towards the ryght. And al theyr bookes begyn at the lefte syde.' Which of these comment result from first-hand knowledge is not known, but his curiosity is very evident.

The Linguist

One consequence of Recorde's decision to write his books in the vernacular, perhaps unexpected more in its extent and long term effects than in its occurrence was the practice he eventually adopted with respect to the etymology of English scientific terms. He was uniquely qualified to take a view on such matters being a scholar competent in Greek, Latin and Anglo-Saxon as well as a scientific practitioner well in touch with those likely to benefit from his writings.

Any author writing in the vernacular had to deal with two main criticisms. Firstly how was the need for a vernacular version of the subject matter to be justified if one

---

[23] Bodleian Library, Douce F subl.16

already existed in another language; secondly, how if no vocabulary appropriate to the subject matter existed in the vernacular, how should it be generated? Recorde's approaches to these problems evolved over a period of some 13 years and are best dealt with chronologically.

Neither of these two issues seems to have occupied Recorde's attention in 1543 when his first book *the Grounde of Artes* was published. The Universities were not interested in teaching arithmetic, new or old, so no vested interests were being threatened. With one exception a specialist vocabulary was in place and the fact that it was of mixed etymological descent caused no comment. The number words were West Germanic in origin whilst the words for the four basic arithmetical operations had Latin sources. The exception was the word 'operation' which Recorde introduced and was also Latinate in origin.

By 4 years later when his second book the *Urinal of Physicke* was about to be published, Recorde felt it necessary in its Preface to launch a pre-emptive strike against perceived critics of his use of the vernacular. There was a sufficiency of vested interests across the medical profession to ensure that there was opposition to his move, one which indeed persisted generally for the rest of the century, although fighting a losing battle.[24] However he also gives in the Preface some indication of the problems he was beginning to face over the lack of an appropriate vocabulary in the vernacular '...... considering that it is much harder to translate into such a [native] tongue, wherein the art hath not been written before, than to write in those tongues which are accustomed, and (as I might say) acquainted with the terms of the science.' He presents no obvious examples of such difficulties in the text that follows, but raises the matter in a major way in his next work, *The Pathway to Knowledg* of 1552, which was the first book on geometry to be published in the vernacular.

The first part of this book which dealt with the definition of geometrical forms had, by Recorde's own account, been completed by 1547. By this time Roger Ascham had made his views known on how authors should aim to write in the vernacular (*Toxophilus*, 1545). An author should 'follow this council of Aristotle, to speak as the common people do, to think as wise men do, and so should every wise man understand him, and the judgement of wise men allow him. Many English writers have not done so, but using strange words, as Latin, French and Italian, do make all things dark and hard.' The vocabulary needed to define the geometric forms to be used provided Recorde with the opportunity to explore the precepts advocated by Ascham, but whether he did so deliberately can only be speculated upon. The task facing Recorde was one of finding a vocabulary for a text that was originally written in Greek but most widely available in Latin. He had demonstrated that he was methodical and logical in his approaches generally; he had knowledge of Anglo-Saxon as well as of Greek and Latin in addition to that of his subject matter, so was as well equipped to undertake the task. He would have been aware that the speech of the common people used a vocabulary that had its origins in Anglo-Saxon, Norman-French, Latin, to a minor extent in the Celtic languages and probably least

---

[24] Johnson FR (1944) Latin versus english: the sixteenth century debate over scientific terminology. Stud Philology XLI(2):109–135

of all in Greek. How was this to be reconciled with a requirement to avoid the use of 'strange words'? One solution could have been to rely on what he knew of the vocabulary of the craftsmen for whose benefit he was writing the book. Such knowledge would have been verbal rather than written, but from other written sources a basic list of technical words in common use may be winnowed out

Where terms of mixed parentage such as angle, point, prick, line, parallel, centre, circle, circumference, concentric, diameter, plumb-line, corner, and canticle were in common use he continued such practice. When qualifying the nature of a single line however he borrowed words from both Old French and Old English, yielding respectively spire line, twist line, straight line and touch line (O.F.) with grounde line and worm line (O.E.). For double lines, in addition to parallel lines he uses gemowe lines (O.E.) and for twin lines that converge he invents the term bought (O.E.) (bowed?) lines. He deals only briefly with the globe on which he says he has already a book written, introducing the English term 'axe-tree' and offering 'tourne point' in place of the Latin 'pole'. It is his offerings of English names, O.E. based, for three types of triangle and their failure to achieve permanence that have attracted most comment. He suggests threlike as the equivalent of isopleuron (Greek) and aequilaterum (Latin), similarly tweyleke for isosceles (G.) and aequicurio (L.) and noelikes for scalene (G.L.) These terms were offered but their use was not urged. Recorde continued in this vein for four–sided figures suggesting long square for oblong, like sides and like jammes for figures of equal sides. For figures with none of their four sides equal he offered borde forme (O.E.) for trapezium (G.), mensula (L.) and helmuariph (Arabic). For figures having five, six and seven sides he suggested the O.F. versions of cinkangle, siseangle and septangle instead of their Greek counterparts pentagon, hexagon and heptagon. Overall his intention seems to have been to introduce descriptive terms derived from O.E. or O.F. that would be meaningful to his craftsmen based audience and would also comply with Ascham's strictures. He found only stony ground for these constructive efforts, as made obvious in his last book, the *Whetstone of Witte* (1557), the Master answers the Student's query as to why he uses Latin terms for the types of proportions saying 'Bicause there are no soche names in the Englishe tongue. And if I should give them newe names, many would make a quarrelle against me, for obscuring the olde Arte with newe names: as some all redy have done'.

Who these critics were is not clear. The second, expanded edition of the *Grounde of Artes* that followed close on the heels of the *Pathway*, introduced only a few new terms such as 'alligation'(O.F.) and 'rebatement' (L.), both of which were in the English vocabulary already. The technical vocabulary used in the *Castle of Knowledge* is that found in Chaucer's *Treatise on the Astrolabe*. It is not known whether Recorde was aware of the Chaucerian text, hence whether he had to re-invent the English terms used therein, but it had been printed as part of Chaucer's collective works in 1532.[25] It must therefore have been the attempts he made to introduce neologisms based on O.E. that found no favour, which seems somewhat perverse.

---

[25] The works of Geffrey Chaucer newly printed with dyvers workes whiche were never in print before, (London. Thomas Godfray, 1532).

The debate on what constituted acceptable additions to the English vocabulary in a wider context, had occupied the attentions of some notable men of letters in the 1550s. Thomas Wilson and Sir John Cheke, both of whom would have been known to Recorde, are usually cited on the topic. Wilson had written his first English text, *Rule of Reason Containing the Arte of Logic*, in 1551. Dedicating it to the King, he gives as his motive 'I have assaid through my diligence to make Logique as familiar to Thenglishman, as by diverse menns industries the most part of the other liberal Sciences are'. He then proceeds to echo the opening words of the dedication of Recorde's *Grounde of Artes* very closely, so presumably he was familiar with it. Wilson's second English text, *The Arte of Rhetorique* (1553) dedicated to John Dudley, was more expansive on the topic of what additions to the English vocabulary were acceptable. It is difficult to establish whether Wilson intended his strictures to apply to the written as well as the spoken word, for in his introduction he makes it clear that oratory was not relevant to the teaching of arithmetic, astronomy and geometry. However, in deploring the use of 'inkhorn terms' he gives as examples of such practise as courtiers who would talk nothing but Chaucer, an auditor who would express his accounts in terms of French currency and ill-educated fanatics who would affect Latin in their speech to overawe the simple. Wilson also complains more generally 'Some seek so far for outlandish English that they forget their mother's language. And I do swear this that if their mothers were alive, they were not able to tell what they say, and yet these fine English clerks will say they speak in their mother tongue, if a man should charge them for counterfeiting the king's English'. He does not rule out the use of Greek or Latin terms 'Now whereas words be received, as well Greek as Latin, to set forth our meaning in the English tongue, either for lack of store or else because we would enrich the language, it is well done to use them, and no man therein can be charged for any affectation when all others are agreed to follow the same way.' He quotes the acceptance of 'Letters Patent' and the 'King's Prerogative' as examples. The neologisms that Recorde introduced in the *Pathway* do not seem to breach Wilson's ground rules, no more than Cheke, Wilson's friend and teacher at Cambridge had done in his English version of the Gospel according to St. Mathew, translated from the original Greek in about 1550.[26] In this version Cheke gave O.E.versions of certain Latinate words as alternatives to those used earlier by Tyndale and Wycliffe. Goodwin speculated that Cheke may have taken this step as a reaction to Bishop Gardiner's attempt in 1542 to weaken the existing English versions of the Bible by ring-fencing some 100 Latin words from any attempts at translation as part of his wider campaign to undermine the English Bible completely. Cheke is thus unlikely to have been one of Recorde's critics and this view is supported by the fact that Cheke had given the King a copy of the *Pathway* for his personal use. It seems that criticisms of Recorde's English neologisms had become trenchant during the reign of Mary, when Gardiner had become Lord Chancellor and Cardinal Pole was Archbishop of Canterbury.

[26]Cheke Sir J (1843) In: Goodwin J (ed) The Gospel according to St. Mathew..., William Pickering, London, pp 5–16

Unlike Cheke and Wilson, Recorde had not fled the country on Mary's accession. Whilst he had not as much to fear as they did, he still had testified against Gardiner at his trial, had written two religious texts that flew in the face of Catholic beliefs and retained his commitment to Protestantism. Within the mathematical community at the time, Thomas Digges had supported the Wyatt rebellion and had been imprisoned. He had also used a few of Recorde's O.E. neologisms in his as yet unpublished *Pantometria*. John Dee was imprisoned on dubious grounds but was eventually released into the custody of Bishop Bonner, to whom he became chaplain after having convinced him of his orthodoxy. Dee's much praised Mathematical Preface had no need of neologisms and most of his limited range of printed texts was in Latin. If Gardiner had need to consult specialist opinion on the topic of mathematical vocabulary for he had to hand the recently restored Bishop of Durham, author of the arithmetical text *de Arte Supputandi,* Cuthbert Tunstall. As this was the Latin text on Hindu-Arabic numbers for which Recorde's English text was intended as a replacement and which was the recommended book for the Universities, criticism from such quarters is likely to have been of the character that Recorde described. The nature of the times ensured that there would be no shortage of people seeking to curry favour with the new regime.

In addition to worrying about religious persecution Recorde had to cope with the consequences of the Clonmines affair, which had left him in severe financial straits. Nonetheless, without compromising his religious convictions he was beginning to find favour in Court circles. In 1556 the Queen accepted the dedication of Recorde's text on astronomy which also contained an Epistle to Cardinal Pole. That his complaint about the Earl of Pembroke was dealt with so expeditiously suggests that he had friends at Court. Under these circumstances, having lost the potential support of people like Cheke, Wilson and Digges, all of whom had been imprisoned, and assuming that Dee would at best have been neutral on the matter, he may have felt that championing of the cause of neologisms based on Anglo-Saxon roots was not worth the candle. Consequently, when the need for an original English nomenclature surfaced with the subject matter of the *The Whetstone of Witte*, particularly with respect to the classification of numbers he used Latin roots for his neologisms. Here are found for the first time the designations and definitions of numbers abstract, contract or denominate, relative, figural, figurate, rational, irrational,simple or uncompound, compound, perfect, superfluous or abundant, diminutive or defective, commensurable, incommensurable, linear, superficial, diametral, square, triple, quadruple, multiplex, manifold and absurd, of which only 'manifold' has an O.E. root. The cossike terms Recorde adopted from the German/Latin texts but he did introduce the terms equation, binomial and bimedial, which are again Latin based. He introduced the sign for equality but he also introduced the term 'equation' in a mathematical context. This list of new vocabulary items is not exhaustive. Whilst some of its members have fallen by the wayside many are still used. His introduction of the sign for equality has been well recognised. He did not invent the signs for plus and minus, but was the first to use them in any book published in England, whether in Latin or English. He also appears to have been first to use the term 'sign' in this context.

His successes in enriching the English vocabulary thus far outnumber his failures, and deserve to be recognised. But further, Johnson has drawn attention to the fact that 'After Recorde's change from the English to the Latin camp in the contest over scientific terminology, the forces which advocated using only English metal were quickly routed. Though there were a few isolated and ineffectual rear-guard actions by overzealous purists like Ralph Lever, the position taken by almost every scholar writing in English after 1557 was for resorting freely to Latin and Greek for technical terms whenever no suitable word existed in English.' Whether Recorde was the courageous experimenter who had found out that 'terms borrowed from the classical tongues could be more readily naturalised – in fact, would seem more English – than outlandish compounds created from native roots',[27] or whether his motive was of a baser variety, namely that of self-preservation for the greater struggles ahead also signalled in the *Whetstone*, he did not survive these struggles. The *Whetstone* was the last book he wrote so we will never know whether his next book published in Elizabeth's reign might have shown another reversal of attitude on neologisms. What can be said is that once again Recorde has made an appearance at a critical point on a critical issue in the history of English science

---

[27] Johnson FR, loc. cit, p 128,130

Chapter 12
His Readers and His Publisher

Abstract There were 16 printings of the *Grounde of Artes* in Tudor times so there
is no doubt that it was widely read. Recognition of sources was not a feature of
published works at that time, but debts to Record's arithmetic were acknowledged
by many authors of practical mathematical texts during that period. Probate inventories show that it had readers in the Universities, although never a recommended
book: Tunstall's Latin text was preferred. Recorde's version of Euclid was reprinted
only once. It was given to Edward VI by his tutor. Publication of the *Whetstone of
Witte* was similarly limited, but the sign for equality that it introduced was used
fairly promptly by Dee in his preface to Billingsley's Euclid: its origin was not
acknowledged. The *Castle of Knowledge* was republished only once, but had rendered obsolete the *de Sphera* of Sacrobosco and was supplanted eventually only by
a series of practical texts on navigation. Recorde's staunchest supporter and effectively his patron was Reyner Wolfe, his publisher. They both supported the
Reformation and had antiquarian interests in common. Wolfe was close to Cranmer
and probably was the means by which Recorde accessed many of the books by
Continental authors that he used.

His Readers

The success of Recorde's mathematical teaching methods and materials used can to
some extent be illustrated by the range of readers he attracted. By definition they
had to be able to read English. The opening paragraphs of *The Grounde of Artes*
which deal with numeration assume that the reader is familiar with Roman numerals
and their meaning, but possibly not with their manipulation. Further, they had to
have the wherewithal to buy the books. These constraints would have limited the
market somewhat.

The loosest of these constraints is likely to have been that of literacy. Reliable
estimates of the extent of literacy at the beginning of the sixteenth century have

J. Williams, *Robert Recorde: Tudor Polymath, Expositor and Practitioner
of Computation*, History of Computing, DOI 10.1007/978-0-85729-862-1_12,
© Springer-Verlag London Limited 2011

proved elusive. Thomas More in his *Apology* of 1523 dealing with the use for an English bible by the general population says '…of whiche people, far more than fowre partes of all the whole divided into tenne, could never reade englishe yet, and many now too olde to begynne to goe to schole.' Three years later, John Rastell was much more sanguine over the extent of the reading public, but then he was a book publisher. In 1547 Bishop Gardiner writes that 'not the hundredth part of the realme' could read, but he was writing in the context of a need for imagery in books for the illiterate. Bennett points out that the number of books recorded in the STC quadrupled from 1500 to 1550, as did the number of printers at work. The effect of the increase in the number of English bibles in numerous editions from 1535 can only have helped to spread the ability to read English, amongst the clergy as well as their parishioners.[1] Possibly the most that can be said about numeracy using Roman numerals is that it was far less widely spread than the ability to read English. University graduates would have been familiar with the numerals, but unlikely to have been as skilled in their manipulation as those professions who had practical need of the results of numeration or of numerical calculations.

The final and most rigorous constraint would have been the financial burden of cost. Printed books were much cheaper than the written word, but still not cheap and Recorde's books were lengthy. The account books of Frobisher for his voyage of 1576 show that the library possessed copies of *The Castle of Knowledge* and its slenderer imitation, the *Cosmographical Glasse* of William Cunningham. They were together valued at 10s., of which the lion's share would have gone to pay for Recorde's longer work. Probate inventories also provide some indication of the possible market value of secondhand books of the period. Inventories of Cambridge University staff and students analysed by Leedham-Green, show that the 'Castle' was valued at 20d in an inventory of 1559/60.[2] Values for what was presumably the second edition of the 'Grounde' vary from 4d. in 1565/6 up to 3s. in 1598/9; the 'Whetstone' ranged from 10d. in 1576/7 to 1s.4d. in 1605; the 'Pathway' varied little between values of 3d. and 4d. The weight to be placed on these values in absolute terms is debatable. They seem to be comparable to those assigned to other arithmetical works by Gemma Frisius and Tunstall found alongside them in these inventories.

The arguments that Recorde advanced in favour of the value of learning for its own sake as well as for its practical uses place no constraints on the types of potential user ranging from other contemporary English writers, academics, the monarch and courtiers, to those commoners who used practical mathematics in the pursuit of their living.

---

[1] Bennett HS (1970) English books and readers, 1475–1557, 2nd edn. Cambridge University Press, Cambridge, pp 27–29

[2] Leedham-Green ES (1986) Books in Cambridge inventories, 2 vols. Cambridge University Press, Cambridge

Contemporary Writers in English

Recorde's books, being written in English, passed unnoticed in mainland Europe. The comments of Bullein already mentioned related to his attributes in general and made no specific reference to the *Urinal of Physicke,* favourable or otherwise. It was unusual for authors of scientific texts to refer to living contemporaries by name in such texts. Recorde did not do so until his last book. One should therefore, with one possible exception, not be too surprised when he was not referred to in print by his English contemporaries.

William Cunningham was a fellow medical practitioner with other interests also in common with Recorde. In 1559 he published *The Cosmographical Glass,* which had much in common with the *Castle* although less extensive in coverage of most topics and less critical in choice of data. His debts to Recorde's writing were fully acknowledged by Cunningham.

Because he could find no works on astronomy written in English, William Salesbury, the Welsh lexicographer and translator of the Scriptures translated Linacre's Latin version of Proclus' *de Sphera* into English for the benefit of his cousin, John Edwards of Chirk. Salesbury claimed no competence in the subject himself but was sure that his cousin was versed well enough in the sciences to profit from the work. There are no examples of mathematical calculations in his version of the *de Sphera.* In the body of its text numbers are given in Roman form, but in the illustrative figures and tables Hindu-Arabic numbers are used which suggests that Edwards was familiar with both forms of numeration. There is no evidence that Salesbury had sought out copies of any of Recorde's works for his cousin.

William Thomas was a strong proponent of publishing books in English, concerning himself primarily, but not exclusively with the interface with Italian culture. He did however translate the *de Sphaera* of Sacrobosco into English, probably in 1551. It was never printed, possibly because Recorde's *Castle of Knowledge* rendered it redundant. Nonetheless, the sentiments expressed in its preface give a valuable insight into attitudes towards the use of the vernacular, particularly as the translation was intended for the use of the sons of the Earl of Suffolk who were young companions of the King. The main thrust of the argument propounded was that children should be taught to read English before they were introduced to Latin and other languages. He goes on to write 'But if the maister wolde firste teache his scholer tundrestande well his own tonge, and then divert him unto the maner of speache, and consequentlie unto the other liberall artes or sciences, that he may know how to reaken in nombres, what thearth is, what the water, ayre and fire, and what the other corporall operation of the creatures, then wolde I well allowe him to proceade to thinstruction not only of the latine tonge alone, but of all the tongues from one to another, unto the first tongue wherin paper and ynke were founde (if he think it goode). For with the sciences he shal always know what he readeth. And withoute the sciences, though well he had as many tongues as were founde at the confusion of Babell, yet shulde he be little or nothing the nearer any goode purpose.

And albeit I might be answered that because we lacke Aucthours in our owne tongue to direct the scholmaisters unto this trade of teaching, yet with more reason can I replie, that if the Scholmaister teach the latine tongue he ought to know the sciences, and knowing them in latine, it shulde seeme he shulde be sufficient to teach them in Englishe: and if he knew them not, than may he change his latine for pewter well enough. Wherefore if our nation desire to triumphe in Civile knowledge, as others nations do, the meane must be that eche man first covet to florishe in his own naturall tongue: withoute the whiche he shal have much ado to be excellent in any other tongue.'[3] The commonality of interests between Thomas and Recorde is obvious and it seems highly likely that Thomas was numerate. Thomas and Recorde had other interest in common, as has been discussed in Chap. 2. It is probable that they moved in similar circles and that Thomas was aware of Recorde's first edition of the *Grounde,* but there is no written evidence to this effect.

John Dee edited a few editions of the *Grounde of Artes* immediately after Recorde died. The augmentations that Dee made to the *Grounde* were slight, helpful in one place but confusing in two others. In the 1570 version he refers obliquely to Recorde in one of his pieces of doggerel, 'That whiche my freende hath well begonne For very love to commonweale...'. Dee became a pensioner of William Herbert in 1552 by his own telling.[4] At the end of that year Dee started an association with the household of the Duke of Northumberland. Pembroke was later instrumental in introducing Dee into Elizabethan court circles.[5] Whether Recorde would have regarded these affiliations of Dee as grounds for friendship at that particular stage of his own dealings with Herbert and Northumberland seems very doubtful. Dee's motives in waiting until 1570 before claiming friendship with Recorde could also be interpreted as being doubtful. Pembroke died in March of that year which also saw the acknowledgement of the debt of the Crown to Recorde's estate and its discharge in the following year. Perhaps Dee thought it had become safe to assume a little reflected glory from his averred association. The mathematical work for which Dee has been most widely praised is his *Mathematical Praeface* to Billingsley's translation of the full geometry of Euclid of Megara was published in February 1570. Much of the contents of this work are to be found distributed through Recorde's texts as views that he freely admitted were not his original contribution. There is however one unique contribution that Recorde made that was used by Dee several times and not acknowledged viz. the sign '='. Dee had demonstrably the largest collection of books of the time, but the only book of Recorde's that it contained was a 1567 edition of the *Urinal of Physicke*. Not even a copy of Dee's own edited version of the *Grounde* was catalogued. The friendship of Dee and Recorde, if it existed, seems to have been a somewhat asymmetric one.

---

[3] Adair ER (1924) William Thomas. In: Tudor studies. Longmans. Green & Co., London, pp 133–160, 159–60

[4] On Dee's copy of Cardano's *In Libelli Quinto*, held by the Royal College of Physicians, in Dee's hand is written 'Veni in Serviti comitis W. Pembrok, 1552 fine februarii die 28'

[5] French PJ (1972) John Dee, the world of an Elizabethan Magus. Routledge and Keegan Paul, London, Chapters 2, 7

Dee wrote an introduction to the *Ephemeris anni 1557* of Field, who was a competent mathematician. There is no evidence of any interaction between Field and Recorde although they obviously had an interest in common. The strong links between Dee and Thomas Digges must have originated with friendship of Dee with Thomas' father Leonard. It is conceivable that Recorde and Leonard Digges were acquainted. The latter had not published any work by the time Recorde had died but his treatise on surveying, *Pantometria* (1571) must have been complete and known to Recorde, probably pre-empting his own proposed work on the topic. Leonard had also written an arithmetical text according to his son who published it along with his own augmentations in the form of a text on military mathematics, *Stratioticos* (1579).[6] This latter is more a book of rules rather than a manual for self-instruction, but nonetheless showed competence in the subjects described, even if with only a limited depth of understanding.[7] Leonard and Recorde also had a dislike of the Government in common. If they met, they would not have been short of topics for discussion. Thomas Digges ends the treatment of arithmetic in *Stratioticos* without dealing with the rules of alligation and false position which were 'things frivolous, for the purpose of soldiering'.[8] It would have been interesting to know whether Digges jr. had anything to offer over and above the understanding shown by Recorde. In the treatment of algebra and cossike arithmetic, the former relied upon the treatment by Stifel, but indicated that he may have had access to Cardano's work and that he intended to treat the matter of *Cubical Equations* in due course. He did use Recorde's nomenclature for the cossike symbols, but used his own signs. By his own telling Thomas Digges had led an isolated childhood as far as his mathematical upbringing was concerned and seems unlikely to have met Recorde.[9] Unlike Dee he did not use the sign = when he could have done.

A favourable view on Recorde's publications, although not naming them is afforded by Thomas Wilson in the prefaces to his books *The Rule of Reason,*

---

[6]Leonard and Thomas Digges, 'An Arithmetical Militare Treatise, named *STRATIOTICOS Compendiously teaching the Science of Numbers,* as well in Fractions as in Integers, and so much of the Rules and Equations Algebraical and Arte of Numbers Cossicall, as are requisite for the Proffesion of a Soldiour.', Henrie Bynneman London 1579., sig. a.ij. [STC. No.6848]

[7]A comparison of the description of fractions by the two men is instructive. Leonard Digges writes, 'A fraction is a Distribution, appointed of a part or parts of an Integer. As the Integers take their beginning at 1, and continue in number without ende, even so the said Integers, by imagination from one second part, may be dissolved, or broken in portions or parts infinite. The partes of those simple or principall Fractions, have also to them parts following.' Recorde writes 'Now you must understand, that as no fraction properly can be greater than 1, so in smallness under one the nature of fractions doth extend infinitely as the nature of whole numbers is to increase above one infinitely so that not only one may be divided into infinite fractions or parts, but also every fraction may be divided into infinite fractions or parts, which commonly be called fractions of fractions....' Digges would have had more to learn from Recorde than vice versa.

[8]*Stratiaticos*, loc. cit., 31

[9]Johnston S (2006) Like father, like son? John Dee, Thomas Digges and the identity of the mathematician. In: Clucas S (ed.) John Dee: interdisciplinary studies in English Renaissance thought. International archive of the history of ideas/Archives internationals d'histoire des idées, vol 193. Springer, Dordrecht. See also *Stratioticos*, sig.A.iij.

conteyning the Arte of Logique, set forth in Englyshe by published in January 1552 and *The Arte of Rhetorique* of a year later. The existence of English texts on the liberal sciences of arithmetic, geometry and astronomy was deployed as an argument in support of his project. During the 1550s English texts on those five liberal arts that badly needed them were provided by Recorde and Wilson who had a mutual regard for the roles of the subjects on which they wrote and held a common view on the need for the use of the vernacular.

Academics

Academia, perhaps surprisingly, seems to have provided some productive terrain for Recorde's works. Following his graduation at Cambridge, Recorde is reported to have returned to Oxford where he gave public lectures on arithmetic. The sole source of this story seems to have been Bale, as quoted by Anthony à Woods No other evidence for continuance of such an activity has been found.

In 1549 there were visitations first to Cambridge University and later to Oxford by Crown commissioners bringing with them a new code of statutes intended to raise standards and revise the curriculum. Entrants were to be examined to ensure that they had reached a standard in Latin adequate to be able to learn dialectics and mathematics. The new arts course was to begin with a year's study of cosmography, arithmetic, geometry and astronomy, using texts by Pliny, Strabo, Ptolemy, Euclid, Cardano and Tunstall. This was to be followed by a year covering dialectic and rhetoric and a further 2 years spent reading philosophy, mainly Aristotle but also Plato. This provided qualification for a bachelor's degree; for a Masters degree a further year's study of philosophy, astronomy and Greek was required.[10]

It may have been this reform that Recorde was referring to in his Epistle to Edward in the 'Pathway', following his presentation of that book and the promise of an imminent publication of his second edition of *The Grounde of Artes*. 'And I for my poore ability considering your Majesties study for the increase of learnyng generally through all your highnes dominions, and namely in the universities of Oxforde and Cambridge, as I have an earnest good will as far as my simple service and small knowledg will suffice, to helpe towards the satisfiyng of your graces desire, so if I shall perceave that my service maybe to your majesties contentacion, I will not only put forth the other two books which should have been sette forth with these two, if misfortune had not hindered it, but also I will set forth other bookes of more exacter arte, both in the Latin tongue and also in the Englyshe, whereof parte be already written, and new instrumentes to theym devised, and the residue shall bee eanded with all possible speede.' He seems to have been offering a range of texts in both Latin and English that would have proved appropriate to the new courses. Clearly his offer was not taken up.

---

[10]Feingold M The statutes, Chapter I, 'The mathematicians apprentice', mathematicians' apprenticeship (C.U.P. 1984), pp 23–44

The extent to which the reforms proposed by the Commissioners were implemented with respect to mathematical teaching has been a matter of contention. The issue went into abeyance during Mary's reign but emerged again, in modified form, with the accession of Elizabeth. She granted sets of Statutes to Cambridge in 1558 and again, more formally, in 1570. They contain no reference to the teaching of mathematics. This has been interpreted by Hill as indicating a reversion to the conservative attitudes to science of pre-1549. Feingold has contested this view, pointing out that the 1570 statutes do specify explicitly that mathematics was one of the subjects to be disputed by undergraduates. Further, admission was not to be granted to those who had not acquired a proficiency in grammar sufficient to learn mathematics and logic.

The 1565 Statutes for Oxford were a mixture of injunctions and articles which were not formalised until Laud's code was imposed in 1634. The earlier version does however assign three terms of teaching for arithmetic and two for music. When Savile established the chairs of geometry and astronomy in 1619, he specified that the professor of geometry should as part of his broader duties, at least once a week teach simple arithmetic in his rooms, in English if necessary.

Feingold elaborates his contention that mathematics was part of the curriculum by examining the provision of mathematical instruction by lectureships and other forms of tuition. He has shown the existence of a more or less continuous sequence of mathematical lecturers at Cambridge from 1568 onwards, of whom Henry Briggs and Thomas Hood became distinguished mathematicians, albeit outside Cambridge. A similar list of lecturers for Oxford has not been found. However Feingold presents evidence that such lecturers did exist although not in the continuous sequence found for Cambridge.[11]

The most informative material on the state of mathematics at Oxford during the last quarter of the sixteenth century comes from the notebooks of Henry Savile detailing his lectures delivered in 1570 as part of his M.A. requirements. From his opening address Savile clearly expected his audience to have, or should have already completed three terms of arithmetic and two of geometry, by attending not only public lectures but also in private study. How he himself had reached his level of mathematical competence is not evident. He does append a list of *Auctores mathematici* from the ancients through to contemporary writers, numbering several hundred. Whether he was familiar with all of them is unclear as he had used previous compilers such as Diogenes Laertius, Gesner and Bale in assembling his list. John Chamber was university lecturer in astronomy in 1573. His lecture material relied very heavily on that of Savile so presumably he also expected the same level of mathematical competence of his audience. Feingold gives the names of a further eight lecturers in either astronomy or geometry dating to this period. Specific instruction in arithmetic is not mentioned.[12]

---

[11] Feingold M loc. cit, Chapter II, pp 45–85

[12] Goulding R (1999) Testimonia humanitas: the early lectures of Henry Savile. In: Ames-Lewis F (ed) Sir Thomas Gresham and Gresham College. Ashgate, Aldershot, pp 124–145

The use of Tunstall's *De supputandi* had been recommended in the statutes of 1549. Recorde's offer to produce his texts in Latin does not appear to have been taken up. So was *The Grounde of Artes* used at all in academe? Robert Recorde and Leonard Digges are listed together in Savile's 'Auctores', but only their names are given. At this stage of his professional life Savile was strongly opposed to the educational doctrine of a need for the practical application of mathematics, specifically as promulgated by Ramus. It is unlikely therefore that he would have acknowledged the same thesis propounded by Recorde who also had written it in the vernacular with the end of educating practitioners in mathematics. The lists of probate inventories for Cambridge college members, assembled by Leedham-Greene, throw some light on the question. Record's arithmetic was found in ten inventories between 1554/5 and 1598/9, possibly with another copy in 1545. None of their owners were involved with the teaching of arithmetic, so possibly they were acquired to help with private undergraduate studies. The 13 for Recorde (10 of the 'Grounde' and 3 of the 'Whetstone') can be compared with 16 copies of Tunstall and 20 for Gemma Frisius' 'Arithmetica' over the same time span. The numbers for Oxford are fewer as fewer inventories are available. Nonetheless, sprinkled through eight inventories in the decade from 1566 onward, many of the arithmetic texts of the century are to be found - Boethius, Sacrobosco, Apian, Tunstall, Gemma Frisius, Recorde, Cardano, Reisch, Ramus, and Baker. The best stocked library in this respect was that of Robert Hart of St.Johns College (d.1571) who had copies of Sacrobosco, Record, possibly Gemma Frisius, Reusch and Tunstall as well as a number of astronomical texts. This was matched in coverage of topic only by the major collection of Perne of Peterhouse, Cambridge.

In a relatively small cross-section of the community of the two universities therefore, there were a fair number of arithmetical texts to hand, of which, a relatively sizeable number were *The Grounde of Artes*. The subsidiary point needs to made, that apart from the latter, Gemma Frisius work was a practical arithmetic with examples of practical calculations, as were to a lesser extent those of Reusch, Apian, Cardano and Baker. If indeed a Platonic view of the eminence of pure geometric thought existed it was not adhered to so rigidly as to prevent advantage being taken of practical arithmetics for learning purposes. However copies of the 'Pathway' are to be found in the inventories along with those of Euclid. But only one copy of the 'Castle' is found, in the company of those of Orontius, Proclus, Apian and Mercator.

A fair assessment would seem to be that Recorde's works, although written in the vernacular, were not eschewed by university students and staff. The catalogue of the University of Cambridge for 1570, however lists no copies of Recorde's publications, nor does that of his own college All Souls College Oxford.

Mathematical Practitioners

The Grounde of Artes was published in at least 11 printings by the end of the century and soldiered on to a further 25 printings in the next century. The testimony of John Mellis in the Dedication to M.Robert Forth of his 1596 edition of this latter book is particularly explicit about the standing of this work. Mellis explains that his early natural talents lay in drawing, but that in the art of arithmetic 'also having great delight, I had no other instruction at my first beginning but onely this good Authors booke.' As this was prior to his service with Forth, which he left in 1568, Mellis must have been familiar with the earliest of the second edition versions of the book. He furthered his education in arithmetic at the common school in Cambridge during his service with Forth, following which he set up as a teacher of drawing, writing and of the 'infallible principles and brief practise of this worthie Science, having (I praise God for it) brought up a number to become faithfull and serviceable to their masters in great affaires, and many of them good numbers of a commonwealth, which is no small comfort to me in Christ.' He defends his succumbing to the blandishments of one of his past students, a publisher, to edit a further edition of *The Grounde of Artes* '...considering it a booke hath done many a thousand good.', the Author being '...the onely light & chief Lodestone unto the vulgar sort of Englishman in this worthy Science, that ever writ in our naturall tongue.' Similar sentiments were expressed by other contemporary writers of practical arithmetics such as Baker, Hylles, Billingsley and Cocker.

Recorde was also appreciated by practical navigators. William Bourne in his almanac for 1571 says 'And also of late wryters most famous men, in these our daies, as Johannes de Sacrobosco, and Horontius, Jewafritius and also in the English tong Doctor Record ....'. Norman in *The new attractive* of 1581 dealing with practical navigational matters using the magnetic compass writes of the contribution of 'Mechanicians' to scientific progress, '....notwithstanding the learned in those sciences, being in their studies amongst their bookes, can imagine great matters, and set downe their farre fetcht conceits, in faire shew, and with plausible wordes, wishing that all Mechanicians, were such as for want of utterance should be forced to deliver unto them their knowledge and conceits, that they might flourish uppon them, & applie them at their pleasures: yet there are in this land diverse Mechanicians, that in their severall faculties and professions, have the use of those arts effectuallie, and more readilie than those that would most condemne them. For albeit they have not the use of the Greeke and Latine tongues, to search the varietie of Aucthors in these artes, yet have they in English for Geometrie Euclids Elements, with absolute demonstrations: and for Arithmetike *Record's* workes, both his first and second part: and divers others, both in English and other vulgar languages, that have written of them, which books are sufficient to ye industrious Mechanician to make him perfect and readie in those sciences, but especiallie to apply the same to the art or facultie which he chieflie professeth. And therefore I would wish the learned to use modestie in publishing their conceits, & not disdainfullie to condemne men that will search out the secrets of their artes and professions and publish the same to the behoofe & use of others, no more than they would that others would judge of them for promising much and performing little or nothing at al.'[13]

---

[13] Johnson FR (1968) Astronomical thought in Renaissance England. Octagon Books, New York, pp 171–172

This situation had altered little by the time Richard Norwood compiled his Book on practical navigational matters, *The Sea-Mans Practice*, in 1636. His language was more moderate, but the sentiments were the same. 'It is true, that the Mathematicks afford large fields of delightful Speculations, wherein a man might walk far with much pleasure: But if from so many fair Flowers he brings home no honey, or from such large fields no sheaves: I meane, if he bring not those Speculations to some useful Practices; neither himself nor others are likely to receive much fruit by them...'.[14] Norwood's views are of interest because he was a man whose education in arithmetic he himself attributed to Recorde's book. Richard Norwood was the grandson of Roger Norwood, a fellow of Merton College in 1548. Richard, born in 1590 was the only son of Edward Norwood, who was the second son of Roger. Edward was not sufficiently financially sound for Richard's education at Berkhamsted School to be continued after the age of 14, when he was apprenticed to a London fishmonger. There he acquired an interest in matters maritime. The fishmonger's family and Richard were stricken with the plague when he was about 17 years old. He survived and took the opportunity to transfer his apprenticeship to a coastal sea-captain, a relative of the fishmonger based at Orford. Norwood writes, 'It was while I was in this employment that having by me Record's *Arithmetic* given me a little before by my father I went through it in numbers and fractions in some 3 weeks' time whilst we lay at Yarmouth, ..'. Norwood went on to be for a short time a teacher of mathematics himself and was obviously gifted in this direction. However the passage does illustrate that Recorde's book was still actively used, that it was passed from generation to generation and that it was so written as to prove capable of being mastered readily. Recorde's 'Arithmetic' seems to have been the only one of his books to have found its way into Norwood' hands. The latter wrote his own book on spherical geometry for which his sources seem to have been Euclid's *Elements* and Landsberger's *Doctrine of Triangles*. He references Clavius' *Algebra* and 'others', but did not mention *The Castle of Knowledge* in his navigational works. This is perhaps not surprising bearing in mind the haphazard way in which, by his own telling, Norwood obtained his books.[15] Johnson makes the point that the outstanding virtues of Recorde's book on astronomy were so strong as to ensure that Sacrobosco's *Sphaeri mundi* was never published in English during the second half of the sixteenth century. This was in stark contrast to the fact that it was translated and published in every other important Western European language during this time. Further, the most popular treatises on most other scientific subjects were translated into English during this time.

The only sour note on Recorde's arithmetic was struck by Worsop towards the end of the century, who said that it was amongst those works on the subject that were found difficult to understand by the common man.[16] A peculiar compliment was paid by the publisher and author John Tap in 1613, when he annexed the title of

---

[14]Norwood R (1655) The sea-mans practice. R & W, Leybourn, London, Bodleian Savile L10 (1), Preface

[15]Craven WF, Hayward WB (1945) The journal of Richard Norwood. Bermuda Historical Monuments Trust, New York, pp 12–17, 40–41, 50–51

[16]As quoted by Keith Thomas in his article 'Numeracy in early modern England', T.R.H.S., ser.5, XXVII(1987),118

Recorde's book on geometry for his own book on arithmetic, in which he followed Recorde's method of presentation by dialogue.[17]

There are few references to the use of *The Pathway of Knowledg*. As Recorde himself said in the 'Castle', a firm grounding in plane geometry was a prerequisite for an understanding of spherical geometry which, in turn, was needed by astronomers and navigators. The latter had however to wait until Edward Wright published his Tables in 1599 before mathematics really became finally established as of practical use. The 'Pathway' was issued in two further editions, the last in 1612, but it had to contend with Billingsley's English translation of the full Euclid of 1570. Further its direct relevance to land surveyors could have been diminished by the specific, virtually contemporaneous publications on that topic by the Digges, father and son.

It is apparent from the preface to 'The gentle Reader' in *The Whetstone of Witte*, his last book, that Recorde knew he was going to have trouble in interesting people in the topics with which it dealt. There is almost an element of desperate weariness in the way he writes 'Wherefore to conclude, I see moare menne to acknowledge the benefite of nomber, than I can espie wyllyng to studie, to attaine the benefites of it. Many praise it, but fewe dooe greately practise it: onlesse it be for the vulgare practise, concerning Merchauntes trade. Wherein the desire and hope of gain, maketh many willyng to sustaine some travell. For aide of whom, I did sette forth the firste part of *Arithmetike*. But if thei knewe how far this second part, doeth excell the firste parte they would not accoumpte any tyme loste, that were imploied in it. Yea thei would not thinke any tyme well bestowed, till thei had gotten soch habilitie by it, that it might be their aide in al other studies.' Whatever successes his writings on arithmetic may have achieved in terms of practical application, he still adhered to the Platonic view of the inherent value of 'nombre' and regretted the lack of appreciation of such value. However it also underlines that he had the practical utility of arithmetic clearly in mind, towards which application Plato was lukewarm, if not antipathetic.

Thomas Fale was a Cambridge – educated medical practitioner with mathematical inclinations. In the Preface to his book *Horologiographica; The Arte of Dialling* of 1593 he advises his readers to gain an understanding of mathematical principles from Recorde's three published books, not named. He also credits Recorde with the capability of writing on the subject of dialing had he chosen so to do.

The Monarchy, Courtiers and Surrounding Gentry

The notebooks of Edward VI show that the king was numerate and able to carry out calculations using Hindu-Arabic numbers. John Cheke intended that the monarch's mathematical abilities went further for he presented him with a copy of Recorde's

---

[17]Tapp J (1613) The Pathway to Knowledge; Containing the whole Art of Arithmeticke .... Digested into a plaine and easie methode by way of Dialogue, for the better understanding of the learners thereof. London

Pathway. In his Panizzi Lectures of 1956, Birrell notes, 'There are not many books on the sciences in Edward VI's library: there was certainly one of the rarest of early Euclids, folio Venice 1510. But what is more interesting is the inclusion of Robert Recorde's *Pathway of Knowledg* London 1551. The book is not at all rare. Edward's copy is in plain brown calf with the royal arms very simply stamped on the cover, in other words it was for use and not for ornament.'[18] It must have been a rare occurrence at this time for the dedicatee of a book to not only have been able to understand the contents of the Dedication, but also to have been able to profit from the contents of the book itself. Elizabeth was familiar with Hindu-Arabic numerals, using them for purposes of numeration in her translation into English in 1544 of Marguerite de Navarre's *The Mirror or Glass of the Sinful Soul.* There is no evidence that she could calculate using the numerals and it is on record that her tutor, Ascham, did not have a high regard for arithmeticians.

In his Epistle to the king, opening *The Pathway to Knowledg*, Recorde deplores the ignorance of the makers of Statutes and of their enforcers amongst the king's Ministers, as has been mentioned previously. It seems very likely, given the state of his relationships with Pembroke and Northumberland at this juncture, that he had them in mind particularly. It seems equally likely that they would have been neither mentally nor emotionally equipped to profit from Recorde's works. Of other members of the inner court circle, those recruited from Cambridge such as Cheke and Smith were highly likely to be numerate already. Mildmay probably was in course of becoming so, but he was using Tunstall's Latin text. Nevertheless he was clearly intending to become numerate and his espousal of the cause of mathematics became evident ultimately from his intents expressed in the establishment of Emmanuel College, Cambridge, at the turn of the century. In the meantime, from his private papers it is obvious that he worked with the new Hindu-Arabic numerals. The position of Cecil, as usual is not easily defined. He probably understood the new arithmetic, but used the old Roman numerals in his own correspondence.

In the same Epistle to the king in *The Pathway of knowledg*, Recorde mentions another class of person who might benefit from the study of mathematics, 'And I truste (as I desire) that a great numbre of gentlemen, especially about the courte, which understand not the the latin tong, or els for the hardnesse of the matter could not away with other mens writyng, will fall in trade with this easie form of teachyng in their vulgar tong, and so employe some of their tyme in honest studie, which were wont to bestowe most part of their time in triflyng pastime: For undoubtedly if the mean other your majesties service, other their own wisdome, they will be content to employ some time aboute this honest and wittie exercise. for whose encouragement to the intent they maie perceive what shall be the use of this science, I have written somewhat of Geometrie, but also have annexed to this boke the names and brefe argumentes of those other bokes whiche I will set forth hereafter, and that shall as shortly as it shall appeare unto your majestie by conjecture of their diligent usyng of this first boke, that they wyll use well the other bokes also.' Evidence that his

[18]Birrell TA (1956) English Monarchs and their books, The Panizzi lectures, pp 14–15

desires fell other than on stony ground will have to be accumulated from the study of the papers of individual nobility.

A significant contribution to an understanding of the status of Recorde's mathematical texts is provided by John Aubrey who included Recorde in his *Brief Lives*. By his time an English school of mathematics was well established and a century of historical perspective in place.

The fortunes of the Aubrey family had been laid by John's great-grandfather William Aubrey, Ll.D., a fellow of All Souls. The latter must have been aware of the difficulties between Recorde and Pembroke and, further was befriended by Pembroke. He also counted Dee as a friend as well as a distant relative. On grounds of family history, any bias that John Aubrey had towards Recorde might therefore be expected to be of an unfavourable kind. Aubrey was a competent mathematician, a Fellow of the Royal Society whose friends and acquaintances included Dury, Hartlib, Hobbes, Hooke, Pell, Petty, Ward and Wallis. It is unlikely that he would have written anything about Recorde that he would not have been prepared to defend to his peers and which therefore must also have reflected their opinions to some extent. The portion of his library that he devised to Gloucester Hall [Worcester College] contained three of Recorde's books, *The Grounde of Artes, (1615), The Whetstone of Witte (1577)* and *The Castle of Knowledge*. It is clear from his comments that Aubrey had read all of them.

Writing of what was the second edition of the *Grounde of Artes*, he says 'He was the first that wrote a good arithmetical treatise in English, which hath been printed a great many times'. He then goes on to quote from the Dedication to Edward VI the portion which deals with the corruption of the Statutes and the commentary by Recorde on the state of the coinage, coupled with a history of the realm which Recorde had felt inappropriate to publish without the king's approval. Aubrey comments 'Quare if ever published.' It was not. He also notes that Recorde promised further texts on geometry and cosmography, written in English and adds 'Quare of these'. Aubrey had the *Grounde* whilst at Trinity College and it is likely that this was his primary source of tuition on arithmetic. It may have been supplemented by Stifel's *Arithmetica Integra* which had been owned by Thomas Allen, but the date of acquisition of this book is not known. The first good book in English on arithmetic was thus appreciated as such some 90 years after it was first published, a view repeated by the same commentator 40 years later.

Aubrey's interest in *The Whestone of Witte* lies in the asides that Recorde makes about the 'trouble upon trouble' he is facing that force him to forego further teaching and writing.

The version of *The Castle of Knowledge* from which Aubrey quotes was the 1596 edition that he describes as '.... the first that was ever writ [on astronomy] in the English tongue'. This quotation concerns the sequence of five books that Recorde lists as making up the complete treatment of topics he had planned. Aubrey did not equate the *Treasure of Knowledge* with *The Whetstone of Witte* and indeed had no reason to do so. In common with us he did not have access to *The Gate of Knowledge*. Neither did he have a copy of *The Pathway to Knowledg* otherwise he would not have commented that 'All that I have seen of his are written in dialogues between master and scholar.' He concludes with a note to Wood 'Can you enform me where Dr. Record lies buried? Methinks Mr. Stow should mention him.' Mr. Stow mentioned Reyner Wolfe twice, but did not mention Recorde.

On the subject of Recorde, Anthony à Wood contented himself with paraphrasing John Bale ... 'he publicly taught arithmetic, and the grounds of mathematics, with the art of true accompting. All which he rendred so clear and obvious to capacities, that none ever did the like before him in the memory of man'.[19] Wood held no copies of any of Recorde's works in his extensive library.

A comment by Aubrey which reflects indirectly on Recorde's arithmetic is found in the former's correspondence with Pell.[20] Aubrey was dealing with the inheritance problem in which a dying man makes a will based on the premise that his pregnant wife gives birth to either a son or a daughter but not both, which latter is what happened. The problem lies in calculating the shares of wife, son and daughter according to the husband's dictat. A discourse given to the Royal Society in 1670, by a member John Collins, set out to prove that the equation can be solved by tables. On the topic, Aubrey writes 'Dr. Pell was wont to say that in the solution of equations the Maine Matter was the well stating of them, which requires mother witte, & Logicke as well as Algebra: for let the question be well stated and it will worke almost of itselfe. as for the example, the most Difficult Problem being thus.' Recorde had written in an identical vein a century earlier about a similar problem but provided a solution that did not need algebra, given that the problem was properly stated.

The most significant measure of the value of *The Grounde of Artes* is the number of times it was reissued, virtually every 3 years over a period of nearly a century and a half. The last edition was published in 1699. One is left with the feeling that for much of that century its presence and availability had been taken for granted. By this time the teaching of arithmetic in schools had probably made its method of presentation appear somewhat antiquated. Its job was done!

Reyner Wolfe: Robert Recorde's Publisher

Reyner Wolfe is mentioned as a bookseller at St. Paul's churchyard in April 1530. Whether Cranmer was responsible for bringing him to England before the former was made Archbishop is not known. However after taking up his appointment as Primate in 1533, he may have eased Wolfe's denization, when the latter was described as a native of Gelderland. Wolfe was admitted to membership of the Stationer's Company in 1536, possibly on the prompting of Anne Boleyn. The topic

---

[19]Bale J (1557) Scriptorum illustrium maioris Brytanniie, quam nunc Angliam & Scotiam vocant: Catalogus. Apud Ioannem Oporinum, Basle, p 695
   Anthony à Wood, Athenae Oxoniensis, ed. Bliss, I (1831), 255

[20]Powell A, loc. cit., 310

of the machinations involved in the ending of Henry VIII's first marriage and Cranmer's part in it is far too complex a matter of which to attempt a summary.[21] In his book on Cranmer, MacCulloch allocates a significant role to Wolfe as an interface for Cranmer with the continental reformers Bucer and Grynaeus. 'The Strassburg connection provided another enduring relationship for Cranmer; amongst Grynaeus other correspondents was a cultivated and enthusiastic evangelical Dutch printer based in Strassburg, Reyner Wolf. .... Wolfe played a major role both in acting as an English agent abroad during Henry's reign and in printing Cranmer's books under Edward VI; he also proved a true friend to Cranmer and his family in their time of disaster under Mary.'[22] MacCulloch suggests that he was an agent for Cranmer in the latter's exchanges with Bucer and Grynaeus. He certainly was an agent for Henry in Europe during the following decade.[23]

As a bookseller Wolfe would naturally have attended the book-fair at Frankfurt where would have met Froschauer the Swiss bookseller who acted as courier for the continental evangelicals. MacCulloch asks rhetorically 'What better cover for evangelical contacts?' In 1538 it was noted that 'our friend Rayner did not come to this fair by reason of the recent death of his wife.'[24] Feather makes the point that Wolfe made his money not from printing but from bookselling which in the case of Wolfe involved the importation of books.[25] In 1535/6 and 1538/9 a 'Raynarde bibliopolae London' provides book to the value of 29s. to Magdalen College, Oxford.[26] Wolfe had set up two shops and a printing house in St. Paul's Cross churchyard leasehold by the time he published Recorde's first book. In 1549 he bought the freehold of the site of the charnel house and chapel that had been levelled by Somerset acting for the Crown. Together with his abutting, existing leaseholds, it meant that he had a very substantial stretch of about 120 ft of premises fronting this prime site. It represented a block of eight book-sellers shops along Paternoster Row at the north entrance to St. Paul's.

His first known printing was of some of the writings of John Leland. In 1543 he printed Cheke's versions of the Homilies of St. John Chrysostom, both in Greek and Latin, the *Grounde of Artes* and Lily's grammar. Although perhaps not as welcome as the sales arising from his licence to print Lily's Latin grammar which was the

---

[21] The first half of Diarmaid MacCulloch's book, *Thomas Cranmer: A Life*, Yale University Press, 1996) deals with the matter.

[22] MacCulloch D, ibid., 66

[23] In the *Letters and Papers of King Henry VIII* series there are a number of references to Reyner Wolfe acting as a courier between Cranmer or members of Cranmer's entourage and their religious contacts, Bullinger, Theabold and Bucer, on the mainland during the years 1538–1540. [Vol, 12 Pt 2 no.969; Vol13, Pt I, no.754: Vol.13 Pt 2, no. 509: Vol 15, nos. 269, 458]. He also acted as the conduit between the King and Cromwell and Christopher Mounte their continental agent during 1539 [Vol. 14, Pt 2, nos. 580, 703, 781f.63]

[24] Duff EG A century of the english book trade, p 171

[25] Feather J (1988) A history of British printing. Routledge, London, p 35

[26] Ferdinand C (1997) Magdalen College and the book trade. In: Hunt A, Mandelbrote G, Shell A (eds) The book trade and its customers. Oak Knoll Press, New Castle, pp 175–187, 182

prescribed text for schools for the rest of the century, the success of Recorde's book must have been very acceptable. Cheke's book had been designed as a special offering to the King just prior to his appointment as tutor to Prince Edward. Wolfe was selected to print the description of the Scottish campaign of May 1544.[27] He published Recorde's *Urinal of Physicke* in 1547, in the April of which year he was granted a patent for life, by Edward as king's typographer, bookseller in Latin, Greek and Hebrew to print all such books in those languages. He was also to print grammars of Greek and Latin even if mixed with English and also maps and charts for the king and country. Two other editions of the *Grounde of Artes* were published by 1551. At the end of 1551 he printed Cranmer's tract *An Answer to a crafty and sophistical cavillation devised by Stephen Gardiner*. Wolfe printed the *Pathway to Knowledg* in January 1552, and at the end of that year he published the second, expanded version of the *Grounde of Artes*. In 1553 he lost a part of his printing privilege to John Day but produced the Latin edition of the 1553 catechism,[28] whilst Day printed the English version. In this year he published the first printed edition of *Pierces the Plowmans Creed*.[29] It was not until the end of 1556 that Wolfe published the *Castle of Knowledge*, which was to be the last book he printed for Recorde. The *Whetstone of Witte* was printed by Thomas Kyngston, who was never a member of the Company of Stationers but printed mainly for others and may have been delegated by Wolfe to do so in this instance.

Recorde' works sit anomalously amongst the other 40 or so works printed by Wolfe in the period 1542–1556. There was only one other technical work present in this corpus, an almanac in Latin by a foreign author. Leland's works were of a literary/antiquarian nature. The other educational works were Lily's grammar and a simple dictionary by Estienne. The remainder were concerned with Reformation religious matters expressing the views of the Cranmer faction. Recorde was of this religious persuasion.

Whilst Wolfe did not have as uncomfortable a time during Mary's reign as did Recorde, he nearly lost his privilege of 1547 to Cawood. The latter was granted its reversion but died before he could profit from it. Wolfe's promotion of the reformation was well known, as was his adherence to Cranmer. Following the latter's death in 1556, Wolfe was deeply involved in the preservation of the interests of Cranmer's family as well as of his papers.[30] It was therefore no trivial risk that Wolfe would have undertaken in 1557 in standing as surety for Recorde. Leland is reputed to have died in Wolfe's house, Wolfe was Lely's executor, he clearly supported Cranmer's family and his ties with Recorde must have been as strong as with these individuals. It seems reasonable to assume that their relationship was based on a common philosophy and not just on one of mutual service and that additionally a deep friendship existed between the two men.

---

[27] STC. 22270, *The late expedicion in Scotlande*,
[28] STC. 4807
[29] STC. 19904
[30] MacCulloch D loc. cit., pp 609, 635–636

In this situation what better person than Wolfe, whose main commercial activity was as an importer and seller of books, to ensure that Recorde was provided with the most up-to-date publications on the subjects in which he was interested, directly or indirectly. Cranmer would conceivably have been one of Wolfe's clients and the contents of Cranmer's library are of interest in this context.[31] Three versions of Euclid's *Elements* are listed, the Grynaeus Greek version of 1533 and the le Fevre Latin editions of Paris(1516) and Basle(1537). As already noted all of these versions are believed to have been sources for Recorde geometrical writings. There were also present the Grynaeus 1535 edition in Greek of the *Almagest* and a compendious volume of works, dated 1536, edited by Zeigler containing works on the sphere and planisphere by Proclus, Berosius, Aratus, Theon, Ptolemy and Jordanus, all of which were pertinent to the subjects of the *Castle*. Also relevant to this latter work was a set of commentaries on the *de Sphaera* of Sacrobosco printed in Venice in 1499. Geographical source material would have been found in Strabo's book and in Gemma Frisius' edition of Apian's text on the subject. The *de triangulis* of Regiomontanus would have reinforced Recorde's understanding of trigonometry, even if he deliberately never wrote about it overtly; he may have used the trigonometrical tables it contained for some of his calculations. There is no evidence that Cranmer ever had an interest in the subject matter of these books, but they could have provided Recorde with a sizeable fraction of the resources he used in the *Pathway* and the *Castle*. A number of the books that he used for his medical text are also to be found listed in Cranmer's library.

Stowe described Wolfe as a serious antiquarian and would have done so from first-hand experience. Wolfe had assiduously collected material for a Universal Cosmography since about 1548, but never completed the work. Along the way, he employed Holinshed and Harrison to help him. With the active support of Wolfe's widow and others such as Stow, they completed the more modestly sized history of Britain known as *Holinshed's Chronicles*.

---

[31] Selwyn DG (1996) The library of Thomas Cranmer. Oxford Bibliographic Society, Oxford

Part III
Finale

Chapter 13
Retrospect and Prospects

Abstract Recorde's objective in publishing his series of mathematical texts was to provide a means of self-education in such topics, little help being available from either Universities or schools. A century and a half after it was published John Aubrey, a member of the newly formed Royal Society, commended his as the first good arithmetic text to be printed in English. It had been reprinted at intervals of about 3–5 years during this time. His other books were not so durable, but between them they brought English mathematical texts at least up to standards comparable with those on the Continent. In his Crown appointments Recorde successfully applied the mathematics he had taught but his political acumen left much to be desired. His legacy to seventeenth century science is not clear cut. Commitment to the belief that knowledge was of intrinsic worth is repeated throughout his books. His views on what constituted knowledge and how it was acquired, are only to be found scattered throughout the bodies of his texts, largely as answers to posed questions. Collectively, they make it clear that Recorde was prepared to accept as true knowledge that which came from both 'ancient' and 'modern' sources, but only if it was in accord with experiential data or reason. His idea that knowledge is progressive in nature emerged with time. In these tenets he presaged aspects of the 'Scientific Revolution' to come.

Retrospect

Sore oftentimes have I lamented with myselfe the infortunate condition of England, seyng so many great clerkes to aryse in sundry other partes of the worlde, and so few to apere in this our nation: whereas for excellencye of witte (I thynke) fewe nations dooe matche englishemen. But I can not impute the cause to anye other thynge, than to the contempte or mysregarde of learnynge. These were the opening words of Robert Recorde's book, *the Grounde of Artes* (1543), the first of a series of texts in the vernacular designed for self education thereby removing at least one

J. Williams, *Robert Recorde: Tudor Polymath, Expositor and Practitioner of Computation*, History of Computing, DOI 10.1007/978-0-85729-862-1_13,
© Springer-Verlag London Limited 2011

barrier to the rectification of the situation he had described. He went further inasmuch as he practised what he preached and applied his learning in service to the Crown.

No doubt he would have been greatly pleased to have been remembered by John Aubrey, a member of the newly formed Royal Society, for having written the first good arithmetic book in English a century and a half after that book was first published. To have known that this book was to be reprinted at intervals of about 3.5 years over the same time-span would have pleased him even more, for it meant that his objective in writing it, to offer the prospect of self-taught numeracy to a wide cross-section of Englishmen, had been accomplished. The achievement was one of numeracy for a wide range of individuals based on the use of the hitherto virtually unknown Hindu-Arabic numbers, with little help coming from the Universities and still less from schools. The mountain to be climbed had been high. At the beginning of the sixteenth century, the standard arithmetic text based on Hindu-Arabic numbers and used in the Universities was still that of Sacrobosco, dating from the fourteenth century dealing only with integers.

The method of tuition by dialogue was skilfully deployed by Recorde, but of greater importance was the logical way in which he developed his subjects, insisting on moving on only when a firm base of understanding of the topic being discussed had been established. Illustrations of arithmetical procedures using practical examples to which the reader could relate, if not an innovation, was again made meaningful by his selection of subjects. In the choice of these topics Recorde was not unduly inhibited by considerations of political correctness. Recorde's examples were topical, posed in English contexts and used English units of measurement throughout. There is a distinction between the potential readers catered for in the examples chosen in the first and second editions of the *Grounde*. The first edition deals only with the arithmetic of integers, with examples chosen to be of interest to be of relevance to merchants and tradesmen. The augmentation presented in the second edition deals with the arithmetic of fractions and the examples chosen here treat of issues of significance to the State such as the Assize of Bread and Ale and the Statute for the Measuring of Land, both of which were in need of correction. A nod was made in the direction of reform of the currency, but was not enlarged upon due to its political sensitivity. No such timidity was present in an example dealing with the matter of enclosure and sheep-farming. If Bale is to be believed, Recorde was as gifted an expositor orally as he was in print, a born teacher.

Methods of teaching arithmetic move in and out of fashion, but the material taught has to be correct. The foundations of Recorde's arithmetic were as sound as he could make them comparing more than favourably with those of his Continental contemporaries, as evidenced by the meticulous corrections he made of some of their statements. Looked at from the perspective of today, he fell a little from grace only with respect to the complete mathematical treatment of alligation which was only reached nearly a century later, when Diophantine equations and their algebraic manipulation was understood. Even here he showed that he was aware of the limitations of his presentation of the subject.

Whilst it was *The Grounde of Artes* that played the main role in bringing about this educational achievement, it is the first two parts of *The Whetstone of Witte* which conclude his treatment of number theory that show not only his comprehensive understanding of his subject matter, but also that he had the whole presentational structure in place from the very beginning of his project. On the first page of his treatment of numeration in his first book, he has the Student deduce that he must know three things about a number, its value, its symbol and its place. The Master then says that a fourth property, that of order, has to be added. This precision of definition of an arithmetic as comprising both cardinal and ordinal numbers is unusual in contemporary mathematical text generally, but unique in texts by English authors, whether in Latin or English and shows his depth of understanding of the subject. This depth is overtly displayed again on returning to the theory of numbers at the start of *The Whetstone of Witte*, where an exposition of the subject as up-to-date and as clearly structured as possible was presented. He erred only in the value derived for the fifth perfect number as for once he did not follow his own teachings, query authority and check his result by back calculation. The latter would have involved a good deal of routine calculation and proofs of primacy of numbers so perhaps it is forgivable, but nonetheless an unusual exhibition of hubris on his part. He did not however fall into the error made by Stifel in assuming that the method of calculation of Pythagorean triplets attributed to Plato provided all such combinations. The unique contribution of Recorde in this field is his provision of a vocabulary of English terms, still in use today, for types of numbers. His attempt to provide a geometric visualisation for powers of numbers greater than three perhaps could be described as a brave try.

Recorde clearly regarded the 'Rule of Cose,' as the most important feature of this second book of arithmetic, as the terminal phase in his treatment of the subject: he singled it out as the subject of an introductory verse to the whole book. He had at hand when compiling his text the works of Stifel and Scheubel, which he used critically and selectively, taking those features that best fitted in with his overall presentational sequence, expressing and explaining them in his own way. The general methodical approach of Scheubel appears to held greater appeal for him than the rather less structured approach of Stifel. The latter was also more committed to a geometrical approach to the topic of surds for instance than Recorde and Scheubel, who seem to show signs of distancing themselves further from such an overtly geometrical mind set. There was therefore nothing novel in his material, but as with his previous publications great care is devoted to the logical, sequential, development of the topic and to the definition of terms. It is at this stage he introduces the word 'sign' into the English vocabulary, defining formally the uses of the + and − signs and making clear the distinction between a cossike number and a cossike sign. His designation of such signs is the same as those of his German predecessors up to and including that of a cubic number. After that he goes his own way, providing a system of symbols appropriate to his earlier attempt to provide a geometric visualisation for powers of numbers greater than three. Recorde deals comprehensively with the basic operations of numeration, addition, subtraction, multiplication, division and progression for cossike numbers uncompound and compound, as integers and as fractions. His last words on the 'works of cossike numbers' are 'And now will I applie them to practice in the rule of *equation* that is commonly called *Algebers* rule.' He then makes it clear which is the tail and which the dog. 'But now will I teache you that rule that is the principall in *Cossike* woorkes: and for which all the other dooe serve.' Whilst this rule is commonly called the rule of algeber, it should properly be called the rule of equation because by equating [cossike] numbers it solves problems. It does so by nominating an unknown [root] and working with it within the problem until an equation is arrived at. For ease of alteration of equations as they are in turn worked upon, Recorde introduces the equals sign =. He then demonstrates the solution of properly constructed first- and second- order equations in a manner which agrees with that of Scheubel.

On the basis of his invention and use of this sign to construct such equations, Hughes has described Recorde as the first of all writers on mathematical topics to have published a symbolic statement that can properly be named an equation.[1] Heeffer has emphasised further the importance of the step that Recorde took, describing the equality symbol as the crown jewel of symbolic algebra.[2] He says further 'As the minus sign facilitated the acceptance of negative numbers, so did the equation sign contribute towards the study of the structure of equations.' Heeffer defines symbolic algebra as algebra using a symbolic model, which allows for manipulations on the level of symbols only. All the worked examples given in this section demonstrate manipulation at the level of symbols [signs]. Whilst Recorde therefore was practicing algebra, it is not clear whether or not he was still thinking of algebra as the ultimate form of arithmetic, or of arithmetic as a sub-set of algebra. The final problem posed in this section results in a cubic equation to which Recorde provides one answer, without indicating how he deduced it. He then closes the discourse saying 'But of these and many other verie excellente and wonderfulle woorkes of equation, at an other tyme I will instructe you farther…'. It is possible that he meant to attempt some clarification of Cardano's *Ars Magna* which, in passing, might have given some further insight into his thinking about algebra.

As anticipated somewhat ruefully by Recorde in his dedication to this book, interest in number theory was going to be very limited and such proved to be the case. In the seventeenth century, interest in this topic was overtaken by interests in algebra, analytical geometry and the approaches to the infinitesimal calculus. Fermat provided the exception to this generalisation. The unique contribution of Recorde in this field is his introduction of the equality symbol and provision of a vocabulary of English terms, still in use today, for types of numbers.

Knowledge of elementary geometry was an essential precursor to the development in arithmetic that was to take place between the first and second editions of *The Grounde of Artes. The Pathway to Knowledg, containing the first principles of Geometrie, as they moste aptly be applied unto practice, both for use of instrumentes Geometricall, and astronomicall and also for the projection of plattes in*

---

[1] Hughes B (1993) Robert Recorde and the first published equation. In: Folkerts M, Hogendijk JP (eds) Vestigia Mathematica. Rodopi, Amsterdam, pp 163–173.

[2] Heeffer A (to appear) On the nature and origin of algebraic symbolism. In:Van Kerhove B (ed) New perspectives on mathematical practices. Essays in philosophy and history of mathematics. World Scientific Publishing (Singapore 2009), pp 1,21,23.

everye kinde, and therefore much necessary for all sortes of men was published at
the beginning of 1552 and the aforesaid, augmented second edition of Recorde's
arithmetic followed at the end of that year. The geometry presented was that of the first
four books of Euclid's *Elements*. The text used was almost certainly of the Greek
version prepared by Grynaeus which also gave Proclus' commentary on the first of
Euclid's books. Recorde had also consulted other contemporary geometrical texts,
but showed his general preference, to be displayed in other contexts for the use of
original Greek texts, where possible. The presentational sequence he adopted was
not that of the *Elements*. A series of Constructions were given in the first half of the
book and the associated Theorems in the second half. The reasons for these rear-
rangements were to make the subject easier for the reader to grasp. Recorde gave
precedents for this approach, claiming no originality for his presentation. He also
eschewed the dialogue form used in his other books. It would be wrong to think that
Recorde's own geometrical knowledge and interests stopped with this selection of
material. From asides he made in his later books, it is clear that he had read and used
sections of the *Elements* up to and including Book 13. It is also clear that the rigour
of proof to which he aspired generally, if not always achieved, was that of geometric
theorems. The sub-title of The Pathway to Knowledg: 'Containing the first principles of Geometrie, as they may moste aptly be applied unto practise, both for use of
instrumentes Geometricall, and astronomicall and also for the projection of plattes
in everye kinde, and therefore much necessary for all sortes of men,' makes his
intentions very clear. In the first part of his book, items grouped under the title
Constructions, in essence constitute practical applications of Geometry. They were
made using instruments with which practitioners would have been familiar. The
relevant proofs are then given in the following section dealing with 'Demonstrations'.
This latter may be construed as providing the practitioners described in the sub-title
with formal justification for their well established practices and possibly indicating
how they might legitimately extend such processes. It was not Euclid, it was applied
Euclid and followed in the footsteps of Continental authors of practical geometries
such as Durer and Reisch. The limited success of the book compared with that of the
'Grounde' may be attributable to practitioners feeling secure in the correctness of
their existing practices and that extension of such practices had to move to matters
involving quantitative planar and spherical trigonometry which was the domain of
astronomers and had been so without break since at least the time of Ptolemy.
Nonetheless Cheke thought sufficiently of the *Pathway* to present Edward VI with a
copy for his personal use

In England, the study of astronomy had fallen into desuetude in the fifteenth
century, the last practitioner of any distinction being Lewis of Caerleon. The reference astronomical text in use was that of Sacrobosco dating back to the thirteenth
century. The *Castle of Knowledge* was the first published treatise on astronomy in
the English language. It was comprehensive, up-to-date and critical in its selection
of material. Where appropriate Recorde consulted the original Greek and Latin
texts, but also used the works of contemporary Continental writers that were no
more than a few years old. Where they were in error he corrected them firmly but
charitably. Recorde spent much time and effort in exposing and correcting

Sacrobosco's text in the light of the discoveries of fifteenth and sixteenth century explorers, which shortcomings were therefore excusable. He was also forgiving of errors resulting from the mistranslation of Greek by Latin writers. A good deal of space was devoted to the rebuttal of the 'flat earth' hypothesis, far more than was the wont of his contemporaries. The reason for his undue sensitivity to this topic is not obvious, but did provide him with the opportunity to demonstrate the need to argue logically and to establish the need for hypothesis to accord with experiential data. Recorde generated some experiential data himself and also provided tables of astronomical calculations that he had made, although he seems not to have relished time spent on such number crunching activities. He expected his readers to have mastered his books on arithmetic and geometry before attempting to learn astronomy. He specifically excluded any treatment involving trigonometric functions and also postponed any discussion of the 'Theorike of Planetes'. The first, third and fourth books present material in ascending degrees of difficulty of understanding. When Recorde came to write the *The Castle of Knowledge,* he was able to list some thirteen instruments available to astronomers, two of which were English in origin. His instrument of choice was the sphere and the sub-title of *The Castle of Knowledge* reads '*containing the explication of the sphere both celestial and material and divers other thinges incident thereto*'. The problem that Recorde faced was how to make his presentations sufficiently quantitative to be of practical value without taking his readers into mathematical realms well beyond their reach. He did this by having his readers construct both solid and armillary spheres, with the appropriate numeration appended, giving detailed instructions on how to do so. These instruments, together with astronomical observations and tables, he used to sidestep the difficult issue of the dealing with the mathematics involved. He did not attempt to show how the tables of astronomical data were derived; the reader had to accept them. Some of them he had computed himself, others he had derived from printed sources. Readers were expected to be able to add and subtract measurement given in degrees and minutes, although such operations had not specifically been dealt with in his arithmetical texts. He did use such computational material in a very practical way when describing how the construction of an armillary sphere could be eased by the use of templates for the manufacture of the six hoops that make up sphere. The dimensions of these templates he had calculated using '... the tables of Cordes and Arkes, called commonly in latine Tabulae Sinuum.' Recorde was clearly not only a teacher of astronomy but also an instrument maker. The second and last edition of the *Castle* was published in 1596 but as evidenced by Aubrey, who had a copy, was being used well into the following century. Although many classic scientific texts were translated from their original languages into English during the sixteenth century, this did not happen with astronomical texts. Seemingly the *Castle* had satisfied all needs for such information and Sacrobosco's *De Sphera*, the text that was subjected to most comment and correction by Recorde, was never printed in English.

Within his astronomy, the equivalent to the debate on Recorde's role in introducing algebra into England is that of his role in introducing and supporting Copernican theory. The discussion he provided, limited though it was, still was more extensive than the notice of Copernican theory provided by Dee in his preface to Field's revision of the Prutenic Tables published at almost the same time. Recorde was demonstrably familiar with the *De Revolutionibus* up to and including Part IV Chap. 21 that dealt with the distance of the sun from the earth and thus with Copernicus' espousing of the heliocentricity hypothesis. Recorde did not commit himself either way at this juncture, refusing to dismiss the hypothesis, promising to return to it later. It would presumably have formed part of his book on the 'Theoricke of the Planetes', promised but not written. This cautious approach was typical of his general attitude that the grounds on which a teaching text was based had to be demonstrably firm. Positive support for the hypothesis from astronomical observation was to prove a long time coming and in the meantime Ptolemaic theory was adequate for instructional purposes. Recorde held this stance in common with most other writers on astronomical matters during the sixteenth century. He was aware of Copernican theory, drew attention to it, but neither rejected or supported it.

His involvement with navigational matters preceded publication of the *Castle* by at least 3 years and he was clearly known to such notable navigators as Sebastian Cabot, Harry Estrege, Pintagio and Chancellor.[3] Contact with the Muscovy Company was maintained and he dedicated his last book the *Whetstone* to them. In this dedication he promised them a book on Navigation that would include a translation of the account, written originally in Saxon by King Alfred, of the voyages of Ohthere towards the North-East passage. He also gracefully declined any offers of help in the difficulties with Pembroke that he was facing at the time.

One can only bewail the losses of the *Gateway* and the promised book on navigation.

He would have been pleased to have had the North-East Passage demonstrated finally, even if it was four and a half centuries after his involvement.

There had always been a demand for the services of medical practitioners from the lowest to the highest persons on the social scale. The activity had always been a profitable one and attracted a wide range of practitioners ranging from the university educated to the totally unqualified. The sixteenth century saw an attempt to regulate these activities, which by Elizabethan times had developed into territorial struggles between various factions over who had authority to practice. In a number of ways Recorde bestrode the factions who were involved in these struggles. He was a physician initially licensed to practice by Oxford University, but he gained his Doctorate in Physicke from Cambridge University, primarily on the grounds of having practiced, presumably successfully, for 12 years. His book on Urology, which subject was the preserve of Physicians, he dedicated to The Company of Barber-Surgeons, but he was a member of neither body. He seems to have practiced the profession of physician successfully and without harassment for the rest of his life, but his concept of cooperation between the various factions certainly was not followed. He had this occupation in common with many of his mathematician

---

[3] On page 188 of the *Castle*, Recorde fully acknowledges his debt to the Muscovy Company for the astronomical information they had provided up to latitude 75° North. He encourages them to explore still further North and promises to try to shed further light on the subject.

contemporaries on mainland Europe. The proportion of his living that he derived from his medical activities is not clear.

Urology was one of the few non-invasive diagnostic medical techniques available at the time that was also 'scientific' in its approach. Superficially it appears that the approach adopted by Recorde in the *Urinal of Physicke* was conventional. However he did comment on possible errors and abuses in its application. More importantly presented an ordered and logical approach to the subject that was appreciated as such at the time. As a result, whilst the use of Urology for diagnostic purposes fell into disrepute during the seventeenth century, Recorde's book proved to be robust in its following, the last edition being published in 1679. The book also was in the van with respect to the increased use of the printed word, in English, by the medical profession in Elizabethan times. This was a process that had started to accelerate in the 1540's although not universally accepted at the time, as evidenced by the preface to Recorde's book on Urology.

He had wide and varied antiquarian interests as his collection of manuscripts shows. He was also an Anglo-Saxon scholar as his annotations to extant manuscripts illustrates. Bale places him in distinguished company as a preserver of old manuscripts during the dark period of the destruction of the libraries. His preservation of mathematical and astronomical writings by English authors from the fourteenth and fifteenth centuries was not only particularly valuable in both establishing the status of such contributors to the subjects but also in registering the English tradition in such subjects. His contribution to establishing the scientific vocabulary of English mathematics was also important.

Recorde's earnings as an author are hardly likely to have supported him but, his Crown appointments were well remunerated on the occasions that they were paid. As comptroller of the Mints at Durham House and Bristol he received £133 6s 8 day per annum, the same as an Under-treasurer, which was quite a handsome reward. He repaid this with a show of technical competence in achieving both outstanding quality of coinage produced and also strict financial control and probity. His accounts for their operations showed them to be in surplusage, a unique happening for such activities at the time when losses due to dubious accounting or blatant fraud were the norm. Whilst his mastery of the arithmetic of alloying would have played a large part in control of the composition of the coinage, a good understanding of the full process, which was one of the most sophisticated manufacturing activities of the time, was needed to control costs. Recorde's portion of the surplusage, a matter of more than £100, was diverted by Pembroke as payment for a spurious debt resulting from the iron-making at Pentyrch. No evidence on how this latter operation was funded or authorised has emerged.

His remuneration for the operations on silver mining and smelting at Clonmines was of a similar order amounting to over £400 for a little more than 21 month's work. At closure of this abortive exercise he was owed close to £1,000 by the Crown according to their accounts. His accounts had been kept meticulously. How Recorde had accumulated this amount of his own money, which effectively subsidised the activities in Ireland, adds to the mystery of his wealth or lack of same. The damages for slander, awarded to Pembroke against Recorde, were £1,000.

He had no offspring and was close to both his mother and his brother's family as evidenced by the odd remark he made in his books about the need to provide for them. His will showed that he had little in the way of wealth to distribute, which is not surprising as he died in a debtor's prison. When the Crown eventually paid its debt to Recorde over a decade later his nephew, also called Robert, benefited greatly and was enabled to establish himself as one of the gentry in Pembrokeshire.

His circle of friends and acquaintances were influential in shaping his career, but the extent of their individual influences has been difficult to quantify. They were all of Protestant inclinations, as was Recorde. Whether he had these views before or after he met them has not been determined. He could have been exposed to such teachings at either or both Oxford and Cambridge. His first book was dedicated to Richard Whalley who was to become Chamberlain to his kin by marriage, Protector Somerset. Rumour has it that Recorde was tutor to Whalley's children. It is conceivable therefore that Recorde's first Crown appointment arose from this connection and his attachment to the Somerset faction followed. There would have been an intellectual attraction to this faction in their espousal of the ideals of the 'Commonweal' to the furtherance of whose ends Recorde's books were partially directed in that they were deliberately written in the vernacular. The learned Doctor in Thomas Smith's radical book *A Discourse concerning the Commonweal* had many characteristics in common with Robert Recorde. His books contained social comment either overtly, or covertly in the form of the practical examples of arithmetical calculations chosen that leave no doubt over Recorde's predilections in this arena. He was avowedly a god-fearing man but it is clear from his discussion of the cause of darkening of the sky at the time of the crucifixion that it was not an unquestioning faith. He was not prepared to accept the common explanation that it was associated with an eclipse, but also did not doubt that an unexplained astronomical event of particular significance had happened. Similarly, whilst prepared to accept much Greek scholarship he preferred technological explanations for myths surrounding their Gods. He could properly be described as a man in whom 'faith and reason stood in delicate balance, both strong, neither denying the other, integrated in a personality at peace with itself'.

His intellectual appetites would however have been more completely fed from his acquaintances that moved in the circles surrounding Thomas Cranmer. There is no evidence that Recorde ever received any favours directly from Cranmer, but Reyner Wolfe was part of Cranmer's inner circle and his friendship with Recorde proved critical to the latter's career. Wolfe was one of the few English printers capable of printing books that demanded the use of Hindu-Arabic numbers, mathematical symbols, Greek type-face and a variety of tables, diagrams and other illustrations. In terms of sophistication of presentation, Recorde could not have been better served.[4] There were errors, but not in excessive amounts when compared with those found in similar contemporary Continental publications.

---

[4] Wolfe was not above experimenting a little with presentation. *The Castle of Knowledge* is one of the earliest examples of the use of pagination in an English printed book. It was inaccurate in places but nonetheless had been attempted. The experiment was not extended to *The Whetstone of Witte*, but this was printed by Kyngston not Wolfe, although commissioned by the latter.

Recorde was perhaps a little more open in apologising in advance for such errors than many other writers. Wolfe attended the Frankfurt book fairs and so was in a position to not only act as a courier for Cranmer, but also as a source of information and channel whereby Recorde was kept up to date on the progress of his topics of interest and of relevant publications. He did not travel to the Continent himself. Wolfe stood as surety for Recorde prior to his trial. Wolfe had antiquarian interests and was in close contact with Leland some of whose works he published, and likely with others such as Talbot and Bale whose religious stances were akin to his own, as well as those of Recorde. The manuscripts listed by Bale as held by Recorde may have been gathered from such contacts. Directly or indirectly, Wolfe could have been instrumental in providing Recorde with information, ancient and modern on the subjects of his books. Other prominent figures who were both antiquarians and Cambridge men of Recorde's vintage were John Cheke and Thomas Smith.

John Dee claimed Recorde as a friend. There is no evidence that Recorde gained anything from this friendship, if it existed. In what little mathematical work he published Dee makes no reference to Recorde's publications. None of Recorde's mathematical books were catalogued in Dee's library. Dee, must have been aware of them for he used the sign for '=' uniquely attributable to Recorde. and had edited his first arithmetical text. Much of the respectable material presented by Dee in his *Mathematical Preface* for which he has been praised and cited as a forerunner to Francis Bacon had already been given by Recorde, albeit in dispersed form across several sites in his books. Many of the books by Continental authors used by Recorde in compiling his books were also found in the earliest listing for Dee's library which collection was compiled after Recorde's books were published. There are no indications that there was any interchange of material between them and no indications that Dee used the books referenced by Recorde. Dee was also on the payroll of Pembroke and Northumberland at various times, which was unlikely to have endeared him to Recorde.

Pembroke and Recorde clashed following the latter's first public assignment at Pentyrch. They clashed again subsequent to Pembroke's opportunistic activities respecting the Bristol Mint following his role in the downfall of Protector Somerset. His behaviour towards Recorde over payments for his work at the Bristol Mint was vindictive. As Pembroke became one of the Mint Commissioners it was inevitable that they clashed once more. It was foolhardy for Recorde to have laid himself open to the charge of libel against Pembroke and why he followed this course of action is unclear. He had been accused of intellectual arrogance by the German mining engineer, Gundelfinger, but that was over technical matters and no other evidence of him being prone to precipitate action has been found. His letter to William Rice that triggered the whole debacle suggests that his action was defensive subsequent to actions to discredit him having already been set in train by Pembroke. Consequent upon Mary's accession very many people were uncertain of their future, Pembroke and Recorde amongst them. Pembroke had the most to lose and probably the most to hide. If as suggested by Recorde, Pembroke became worried by Recorde coming back into favour, it was possible that he may have feared that the latter knew where some of the bodies were buried. Recorde duly reacted to Pembroke's provocation. There were probably a number of other eminent men who had their fingers in the same pies as Pembroke. The intervention by Cecil in support of Pembroke is noteworthy. Robert Recorde's friends were not sufficiently influential to save him.

Despite his early death, Robert Recorde achieved much of what he set out to do. He provided mathematical texts in English that enabled those who wished to do so to bring themselves to the same level of competence as their counterparts elsewhere in Europe. Two centuries of stagnation were provided with remedies in less than two decades. Foundations to structures rarely if ever win applause, attracting publicity only when they fail in their function. Robert Recorde provided foundations for English mathematics that were sound and enduring as well as accessible for others to build on safely. His works provided essential transitional stages between the distinguished contributions of medieval English mathematicians and the accomplishments of their counterparts, the practical and theoretical mathematicians of the seventeenth century. He was as original in the ideas he expressed as the ends of his books required and the limitations of his intended readers imposed. It would be wrong to assess his capability for original mathematical thought solely on the basis of these books. The absence of an archive of personal documentation makes it impossible to characterise adequately either him or his potential. Of his own telling books were written or planned that were never published. Such works were also listed by Bale in his compilation of manuscripts by English authors. Recorde must have had working documents but no trace of them has been found. One has to be satisfied with what he did accomplish, which was considerable. Polymath and communicator he was, politician he was not.

Prospects

And I have goode hope that Englande wyll (after she hath taken some sure taste of learnynge) not only brynge foorthe more favourers of it, but such learned men that she shall be able to compare with any Regime in the world. Recorde's hope, expressed on the second page of the Preface to his first book of 1543, was not fulfilled until the following century. So what effect, if any, did his achievements, as described in the preceding section have upon the events of the seventeenth century? Given the extent of his contributions as scholar and as practitioner of the 'Sciences' and the 'Artes' during his lifetime, some evidence of his influence on events that were to take place in the following century might be expected to remain. Distinctions between scholar and practitioner became more exaggerated in the seventeenth century compared with those found in the previous century. It seems appropriate therefore to seek evidence of Recorde's influence, if any, separately within his own categorizations of *Intrinsic Worth* and of *Profite and Commoditie.*

Intrinsic Worth

As a repeatedly affirmed champion of the intrinsic worth of knowledge, Recorde surely qualifies as a 'philosopher' in an etymological sense at least. He did not write of *a philosophy*; he did however express a succession of opinions that together constituted an attitude to knowledge and how it was to be obtained and tested, which can be construed as a philosophy which he both practiced and preached.

Commitment to the belief that knowledge was of intrinsic worth is reiterated throughout his books primarily, but not exclusively, in their Dedications and Prefaces. He relied on the authority of Greek philosophers, primarily Plato, to support his case. His views on what constituted knowledge and how it was acquired, are only to be found scattered throughout the bodies of his texts, largely as answers to purported questions from his pupil.

Recorde was prepared to accept as true knowledge that which came from both 'ancient' and from 'modern' sources, provided it was in accord with experiential data or reason: if such criteria were not met then it was to be discarded. Edition A of his first book on arithmetic opens with the Scholar saying 'Syr such is your auctority in mine estimation, that I am content to your saying and to receive it as truth, though I see none other reason thereunto.' The Scholar is not upbraided for this attitude immediately but following a lengthy discussion on the commodities of arithmetic the Scholar repeats himself saying 'And I to your authoritie my wyt do subdue, what so ever you say, I take it for trewe.' This time the Master does correct him 'That is to muche, and meete for no man, to be believed in all things without shewyng of reason. Thoughe I myghte of my Scholar some credence require, yet excepte I shewe reason I do it not desyre.' The Master was the only authority to appear in this text and expected to justify his teachings.

As Kaplan has pointed out, this contrasts sharply with the attitude found in his next book *The Urinal of Physicke* which is rife with uncritical presentations of the views of ancient authorities on uroscopy.[5] Kaplan concludes, 'In his Uroscopy Robert Recorde was working in one of the most backward areas of medicine, itself one of the most backward areas of science. It is this that is the most likely explanation for the marked contrast between his easy acceptance of the conventional wisdom of the authorities in uroscopy, as against his intelligently critical analysis of the authorities in astronomy and mathematics.' Where Recorde did use his critical abilities in this text was in ordering his subject and in setting out the parameters that had to be taken into account in collecting data which only then could be used for diagnosis. Given the material available to him he probably would have felt unable to present soundly based, justifiable criticisms of his sources. As evidenced by the foreword to the seventeenth century editions of this work, it was the ordering of information that was gratefully acknowledged by his successors.

---

[5] Kaplan E (1963) Robert Recorde and the authorities of uroscopy. Bull Hist Med 37:65–71.

The *Urinal* was not presented in dialogue format and neither was his next work *The Pathway to Knowledge* which was the work of a single authority, Euclid. Needless to say, Recorde did not criticise Euclid and though he used the latter's first four books, the work therein was presented in a different sequence from the original and some of the definitions used were post-Euclid in their derivation. Both modifications were made in the belief that they would aid the teaching process and there were precedents for the changes. There was no criticism of the given knowledge as such.

The next book in the sequence was Edition B of the *Grounde of Artes*. This was essentially the first edition augmented with a second part that introduced the arithmetic of fractions based on Hindu-Arabic numerals and their practical applications. No ancient authorities are therefore invoked, but an unnamed recent writer who acted as a source for Recorde was cited and criticised mildly. The issue involved four applications of the rule of fellowship, posed in fact by Gemma Frisius. The Master shows that the first question, in the form presented, was unanswerable and dealt with it by reinterpretation. 'But these learned men did not meane anye other thing by this question, then to find such numbres as should beare the same proportion togyther, as those numbres in the question proponed dyd beare one to an other.' He then illustrates his point using a further three examples from the same source. The Master prefaced his apologia by the general observation, 'It is commonly sene, that when men wyll receave thinges from elder writers, and wyll not examine the thynge, they seme rather willing to erre with these ancients for company, then to be bolde to examine their workes or wrytinges. Which scrupulositie hath ingendered infinite errors in all kyndes of knowledge … .'

This observation presages the critical attitude to authority that suffuses Recorde's next book, the *Castle of Knowledge*, concentrated in the Fourth Treatise of this work. It is re-stated, bluntly, early on in this Treatise. Whilst praising Ptolemy, the Master warns '…. yet you and all men take heed, that both in him and in al such mennes workes, you be not abused by their authority, but evermore attend to their reasons, and examine them well, evermore regarding what is saide, and how it is proved, then who saieth it: for authoritie often times deceiveth many menne …. .' We assume the message has struck home for later it is the Scholar says 'Here by I perceive, that who so ever will travail in these sciences with profit, must lean to reason, then to authority, els he may be deceived.' Recorde practiced what he preached and there are numerous examples of this approach. A general feature of his presentation is that where he quotes an 'authority' in English, he also provides the Latin text and if that was derived from a Greek source he provides that as well. This is particularly obvious in the First Treatise where he defines terms, relying heavily on Euclid and Proclus, but not uncritically so in the latter case, reserving criticism of his attitude to the properties of the Arctic and Antarctic circles. The Master takes a somewhat jaundiced view of many of the 'ancient' writers on Astronomy, '… for as the number of writers are infinite, so have I found great tedious payne in readinge a greate multitude of them.' In addition to Proclus, Sacrobosco and Orontius, he recommends Euclid's *Phaenomena*, Stoffler's commentaries on Proclus and Cleomedes in Greek. Ptolemy is deemed too difficult for a beginner, Pliny, Hyginus and Aratus are set aside as being of variable reliability. He deals with the 'flat earth hypothesis of Lactantius at length, destroying it both by showing the illogicality of the arguments used to support it and by demonstrating that it was not supported by the evidence of ones eyes. He supports the arguments advanced by Cleomedes to refute all shapes for the earth other than that of a sphere, whilst also correcting Cleomodes' text where it had been corrupted in transcription. In doing so the writings of Aristotle, Plutarch, Galen and Eusebius on the same matter are criticised. His criticism is not confined to only that of ancient authors. He points out gently a minor error made by Erasmus Reinhold, only lately deceased, in using a Great Circle to trace part of the course of the Rhine where a lesser circle was more appropriate. His treatment of Sacrobosco's argument in support of a model of distribution of land and sea was less generous. 'That reason savoureth more of the determinations theological, then of the demonstration mathematical, whereof I will add thereto a proof by good demonstration.' This latter was hardly a proof by present standards, but it did use measurements and actual observations as its basis. A number of his types of criticisms are conflated in the issue of the true nature of the Arctic and Antarctic circles as defined by some ancient Greek writers. It follows from this latter definition that the extent of these two Circles varied with latitude and hence the Circles could not be used to delineate the extent of specific latitudes such as the Temperate Zones. The false assertion was started by Parmenides and then followed unquestioningly by Aristotle, Cleomedes and Proclus. The error was noted by Posidonius according to Strabo, but the Latin version of the latter's work was so 'evil expressed' as to obscure the matter. Proclus is further criticised for his discussion of the ways in which the hours of the day changed with longitude and latitude. Proclus said that the meridians changed every 300 leagues in going from east to west but every 400 leagues in going from south to north which was clearly wrong. His second error was that, traversing the Climates from south to north, he said that the length of the longest day increased proportionately with distance.

Recorde's treatment of the Ascension of the Planets is essentially a critical appraisal of the so-called rules of Sacrobosco and, inasmuch as it includes all latitudes up to the polar circle, both more comprehensive and more correct. The crucial evidence he uses was not available to Sacrobosco as none of the latter's possible sources provided information for the oblique sphere below 12 and a half degrees of latitude, territory which had become of significance to sixteenth century explorers. Finally, Recorde's criticism embraces his near-contemporaries, Stoffler, Schoner, Copernicus and Reinhold for having followed a mistranslation of shield for spear in relation to certain stars in the Centaur's spear (Centaurus Constellation). His standards applied to the new as well as to the old authorities.

Whilst Recorde's views on how knowledge was to be acquired were with him from the earliest of his writings, his idea that knowledge is progressive in nature emerged only with time. Lilly assigns this emergence to the Address to the Reader which opens the ...."Pathway to Knowledg [where] we do find him expressing the 'contribution' idea as the justification of his work. He writes in a lyrical and poetic style which seems to indicate that this new conception was something that he now felt deeply; he puts small value on his own work, but swells with pride at the thought of all that may be built upon it by his successors."[6] By the time he came to write the *Castle of Knowledge* any doubts he may have had about the concept of progression in knowledge had disappeared. In his opening paragraph, responding to the pleasure expressed by the Student at obtaining knowledge the Master says 'And sometimes we see, that when the desire [for knowledge] is partly perfourmed, and the pleasant-nes of the same ones tasted of, the desire nothinge asswageth , but contrarye ways greatly increaseth: and the more it getteth the more it desireth. So that in this point may knowledge well be compared to covetousnes; for as the covetous mynd with getting is never satisfied, so knowledge by knowing doth covet styll more: And as it increaseth, so doth it still lerne the vilenes of Ignorance, and profite of Sciences, and therfore cannot rest from searching more knowledge, as long as it spyeth any spot of ignorance.' Lilly tentatively puts forward the thought that the 'contribution' idea of Recorde was, "the central idea which explains the relation of his work to the world around him" and as such an achievement far greater than that embodied in his contributions to arithmetic.

Where does Recorde's 'philosophy' as described in the preceding paragraphs stand in relation to those events of the following century commonly referred to as 'The Scientific Revolution'? In his book of that name, Steven Shapin describes the intellectual struggle between the 'moderns' and the 'ancients'. The latter generally represented a 'scholastic' approach to what constituted true knowledge derived pri-marily from ancient texts; the attitude of the former Shapin encapsulates as follows. 'No seventeenth-century maxims seem more self-evidently sound than these: rely not on the testimony of humans but on the testimony of nature; favor things over words as sources of knowledge; prefer the evidence of your own eyes and your own reason to what others tell you.'[7] The concept of scientific knowledge being progres-sive in nature was a sine qua non to the 'moderns'. Humanist scholars were much concerned with the re-examination of ancient texts for evidence of corruption in translation and transcriptions and its correction. Observational science, so favoured by the 'moderns', provided additional evidence to help in such studies and vice versa. From the evidence scattered throughout Recorde's writings it seems clear that he was a moderate Humanist by the standards of the seventeenth century and also a moderate 'Modern'. On both counts he presaged attitudes that were fully developed in the following century but his contribution passed without comment. This is not surprising as there is little indication that Elizabethan 'Science' had any concern at all for an underlying philosophy and thus was not a likely channel for the propaga-tion of Recorde's ideas. The fact that his advanced views on such matters were sparsely distributed through four books, more likely to be read by practitioners of science than by academics, would not have aided their recognition and appreciation some seven decades later.

If Recorde had any influence on the fulfilment of his hope that Englishmen would take their places alongside the best 'learned clerkes' in the Western world it could

---

[6] Lilley S (1957) Robert Recorde and the idea of progress. Renaiss Mod Stud 2:3–37, 28.

[7] Shapin S (1998) The scientific revolution. University of Chicago Press, Chicago, p 69.

therefore only have been exercised through his teaching and practicing of practical mathematics and their applications as transmitted through his successors in Elizabethan times.

'Profite and Commoditie'

In his book 'Intellectual Origins of the English Revolution', Hill wrote that 'the science of Elizabeth's reign was the work of merchants and craftsmen, not of dons; carried on in London not in Oxford and Cambridge.' In her most comprehensive study of such activity in Elizabethan London, 'The Jewel House', Harkness provides strong evidence to support Hill's contention. Covering the same period, dealing in depth with the activities of a more limited category of a small number of related Mathematical Practitioners, Johnston illustrates the same thesis.[8] He found that amongst some dozen or so practitioners, excluding the wilder extravagances of John Dee, only Thomas Bedwell took up philosophical issues and then only in the limited context of the subject of ballistics. The science of Elizabeth's reign was what Recorde had previously both categorised and practiced as being for 'profite and commoditie'.

Harkness' thesis is that 'The foundations of the Scientific Revolution in Elizabethan London depended on three interrelated social activities: forging communities, establishing literacies, and engaging in hands on practices.'[9] To establish this thesis, amongst other topics she examines the 'Contest over Medical Authority', 'Mathematics and Instrumentation in Elizabethan London' and 'Big Science in Elizabethan London'. Recorde had been active in all these areas during the reigns of Elizabeth's predecessors. The influence that he might have exerted on subsequent developments, post-mortem, varies from area to area, but was never negligible.

The unique position that Recorde held both as medical practitioner and as a writer has already been discussed and will not be elaborated further other than to reiterate that his book on Urology was sufficiently well-regarded for it to be republished several times over a century after it had been written.

The fourth chapter of Harkness's book is entitled 'Big Science in Elizabethan London'. Elizabeth's inheritance from her father were the same economic crises and Crown bankruptcy that her siblings had inherited. Harkness explores the ways in which the Queen, or more particularly her chief minister William Cecil, tried to alleviate the situation with the aid of large technological projects aimed at exploiting natural resources and the part that inhabitants of London played in these activities. Perversely, it was the economic abuse of what can be viewed as one of the largest and most technologically advanced processes of the time that gave rise to the economic problems. The coinage had been the most closely specified product of any

---

[8] Johnston S (1994) Making mathematical practice. Doctoral thesis, University of Cambridge, Cambridge, Chap. 4, part 2.

[9] Harkness DE (2007) The jewel house. Yale University Press, New Haven, p 11.

large scale process for centuries, was under direct control of the Crown and remained so. Weights and compositions of the coinage had tolerances placed upon them and they were measured for samples of coins taken using defined sampling procedures. That this process was capable of being operated efficiently was demonstrated by Recorde during his time at the Bristol Mint. Pirry also seems to have been successful in his time as farmer of the Dublin Mint.

Abuse of the monetary system arose at two levels. The first followed from the setting of the difference between the intrinsic and face value of the coins, which was a prerogative of the Crown made manifest as debasement; the second resulted from either bad or deliberately false accounting at various levels of management. Cecil would have been fully aware of both Recorde's success at the Bristol Mint and of his concerns with abuses of the system. The exceptional abuse which had been practiced for centuries and awaited a technical solution was that of clipping, but its resolution by the introduction of milled edges lay in the distant future. King Edward had been made aware of the need to resume sterling standard by a number of people including Robert Recorde. The lure of a quick short term profit, made by following his father's lead and continuing with the debasement of the Irish currency was too attractive for Edward to eschew. Cecil was well aware of this venture. When Mary came to the throne she was made aware of issues surrounding abuse of the coinage. She set actions in place to evaluate the best course for reinstitution of the sterling standard and sent a proposal to this end to certain of her counsellors in 1557. Its recommendations were never implemented. The stage had thus been largely set for Elizabeth to undertake a recoinage.

The only financially sound option available was to melt down the existing coinage and refine it. The necessary technology had to be imported from Germany and working through Gresham and Thomas Parry, Daniel Ulstadt and Company were contracted to undertake this refinement. Recoinage was undertaken at the Royal mints and as the starting material for the recoinage would have been refined silver, the process to be followed would have been that which had been practised there for centuries. Success would have been predicated by the proficiency and integrity of the people who managed it and Cecil may have had a part in choosing these individuals, people who would have known Recorde and his achievements.

Recorde's influence on Mint operations through his mathematical treatment of alligation may well have extended into the following century. Arguably the most influential individual involved in minting matters in the first half of the seventeenth century was John Reynolds.[10] He was either assistant to the assay master or deputy assay master nearly continuously from 1607 to 1665 and in 1651 published a treatise *A Brief and Easie way by Tables To cast up Silver to Standard of XI.Ounces ij Peny-weight. And Gold To the Standard of XL.Carracts, with Questions wrought by the Golden-Rule: Also by Decimal Tables.* Notwithstanding this title, numeration in the text was in Hindu-Arabic numbers and there was a short section that relied implicitly on the arithmetic of alternate alligation. By defining composition in terms

---

[10] Challis CE (1992) Presidential address. Br Numismatic J 62:237–246.

of betterness or worseness than standard and dealing only with the alloying of only two components one better and the other worse than standard, Reynolds simplified the calculations needed in practice to only a few lines.[11] It seems unlikely that Reynolds would have spent so much time and effort on his treatise if it was not to have been of practical use and the examples he gives not of real significance. Practical Arithmetics by authors other than Recorde usually included a section on alternate alligation and the topic did not depart the scene until the end of the seventeenth century.

Exploitation of mineral resources in the gift of the Crown had been practiced by both Henry and Edward. With the exception of the iron industry the scale of operations was not large. Towards the end of Henry's reign the iron-masters of the Weald were about to introduce the blast furnace for iron smelting and during the reigns of Edward and Mary the industry became well established at a number of locations in England and Wales. The end use of the iron was of a military nature and reached its zenith with the successful and critical introduction of cast iron cannon towards the end of Elizabeth's reign. This technology was a result of private initiatives from its inception and would have been very well known to the Crown. The latter's only direct foray into the field was possibly that described and operated by Recorde at Pentyrch, which was terminated abruptly by Pembroke. Other mineral resources such as tin, lead and copper had been exploited for many years. For obvious reasons the Crown zealously guarded its sole rights to the exploitation of silver and gold bearing minerals. The attraction of the mining and refining of silver ores at Clonmines in Ireland was not lost on Henry who instituted the opening gambits in this affair. It was carried through to its conclusion by Recorde in Edward's reign and finally terminated in that of Mary. Recorde's accounts demonstrated clearly that no matter how good the management or the technical ability of the work-force, if the success of the venture was predicated on false information regarding the true silver content of the ore, failure was inevitable. The Crown lost a lot of money and Recorde was bankrupted. The memory of this event was still with William Humfrey, one of Cecil's chief technical advisers, when he cautioned against belief in the richness of mineral deposits in Ireland. The memory should also have been with Cecil, who was close to the king for the duration of the venture.

Were lessons to be learned from Recorde's experiences that were relevant to Cecil's approach to promotion of big technological projects? Good technical advice on the feasibility of a project was obviously highly desirable. Such native competences were in short supply and had therefore to be imported, as they had been under previous administrations. Good management was based on good and honest accountancy. Whilst this might be enforceable in projects that were wholly Crown funded, the introduction of private venture capital would necessarily complicate control. The inevitable clash between the greater good and the private profit motive led to the abandonment of the patent system introduced by Cecil as a means of encouraging

---

[11] Williams J (1995) Mathematics and the alloying of coinage1202–1700; Part II. Ann Sci 52: 235–263, 254–256.

'Big Science'. The technical competences of the country would however have been increased. It was an expensive learning process.

The influences that Recorde's mathematical publications had on development in the seventeenth century as transmitted by Elizabethan science differ from topic to topic.

Recorde's second book on arithmetic *The Whetstone of Witte* was too advanced in its contents to be of interest to other than a few readers then, and now is most remembered for the introduction of the = sign. Number theory, which it introduced, held little of practical interest and, Fermat apart, was not a fashionable subject for theorists. However the English vocabulary that Recorde devised for the subject persists largely to the present day. Cossike arithmetic proved to be of transitory interest, mutating into algebra. But the work did find at least one important reader, the shipwright Mathew Baker who worked with the cossike symbols and used square and cubic roots in his design of ships.[12] His disciples Borough and Bourne may have followed his lead in use of the text.

The title of Recorde's book of Euclidian geometry, 'The Pathway to Knowledg: Containing the first principles of Geometrie, as they may moste aptly be applied unto practise, both for use of instrumentes Geometricall, and astronomicall and also for the projection of plattes in everye kinde, and therfore much necessary for all sortes of men', makes his intentions very clear. In the first part of his book, items grouped under the title Constructions, in essence constitute practical applications of Geometry. They were made using instruments with which practitioners would have been familiar. The relevant proofs are then given in the following section, 'Demonstrations'. This latter may be construed as providing the practitioners described in the sub-title with formal justification for their well established practices and possibly indicating how they might legitimately extend such processes. It was not Euclid, it was applied Euclid and followed in the footsteps of Continental authors of practical geometries such as Durer and Reisch. It was printed three times. The limited success of the book compared with that of the 'Grounde' may be attributable to practitioners feeling sufficiently secure in the correctness of their existing practices and that extension of such practices had to move to matters involving quantitative planar and spherical trigonometry. This latter was the domain of astronomers and had been so without break, since at least the time of Ptolemy.

The second and last edition of the *Castle* was published in 1596 but as evidenced by Aubrey, who had a copy, was being used well into the following century. Although many classic scientific texts were translated from their original languages into English during the sixteenth century, this did not happen with astronomical texts. Seemingly the *Castle* had satisfied all needs for such information and Sacrobosco's *De Sphera*, the text that was subjected to most comment and correction by Recorde, was never printed in English. In the 40 years that passed between the first and last editions of the *Castle* there was no publication in English that rivalled its status as

---

[12] Johnston S (1994) Making mathematical practice. Doctoral thesis, University of Cambridge, Cambridge, Chap.11. Mathew Baker and the art of the shipwright.

an introductory text to astronomy and cosmology. Its only real competitor was William Cunningham's *The Cosmographical Glass*, published in 1559. Where the two works overlap, the *Glasse* is a much reduced' uncritical version of the *Castle*, even including errors of other authors that had been corrected by Recorde. Debts to Recorde's works were however fully acknowledged by Cunningham. It did not achieve a second edition, but it was written in English and did accompany a copy of the *Castle,* with Frobisher to the New World in 1576.

A number of practical aids to mariners in the form of almanacs were published in successive, updated editions during the rest of the century. Arguably the most important of these were those of William Bourne starting in 1567 with *An Almanacke and Prognostication for three yeares*. Bourne declined to expand on matters astronomical as being unworthy to do so in the face of a number of authors on the subject, both ancient and modern including, 'in the English tong Doctor Recorde'. This deference was continued in his next publication *A Regiment for the Sea*, published in 1574.

The last decade of the century saw the publication of a flush of books whose contents overlapped with those of the 'Castle' to differing extents and were essentially concerned with the application of astronomy and its associated techniques to Navigation. In 1590 *The Use of the celestial Globe in Plano (1590)* by Thomas Hood was printed, followed 2 years later by *The Use of both the Globes* and *The Mariners Guide*. The second publication was effectively a manual for the use of the celestial and terrestrial globes constructed by Emery Molyneux. Written in the form of a dialogue, it was a pale imitation of the *Castle*, being an improvement only inasmuch as it incorporated new experiential data.

Robert Hues' *Tractatus de Globis et eorum usu*, written in Latin, was published in 1594, as was Thomas Blundeville's *His Exercises, containing six Treatises*.The next year *The Seamans Secrets* by John Davis was printed. *The Arte of Navigation* by Martin Cortes had first been translated into English from Portugese by Richard Eden in 1561. In 1596, it was reprinted by John Tap, who brought it up-to-date and augmented it. Tap's important work, *The Seamans Kalender, or an Ephemerides of the Sun, Moone and of the most notable fixed Starres* was not published until 1602. Edward Wright's important work *Certaine Errors in Navigation* was published in 1599, but his subsequent treatise, *The Description and Use of the Sphere*, although written in 1600 was not published until 1637.

Wright's Tables as well as those of contemporary writers of his ilk were calculated using trigonometric functions. Thomas Blundeville was the first Englishman to publish tables of such functions in his 'Exercises', printed 4 years before 'Certain Errors'. He gave twenty examples of astronomical calculations using a variety of such functions Most of these calculations relate to individual cases of relationships that are to be found in more general form in Tables of related values to be found in the *Castle*. Blundeville gives the calculations in extenso, exposing their laborious nature, possibly explaining why Recorde limited the amount data he displayed in this way.

The accuracy provided by navigational charts was generally acknowledged and appreciated, but was not easy to use on the high seas. Interest in practical navigation therefore moved increasingly in the direction of providing instruments that incorporated both the data found in such charts and its manipulation to practical ends.

Recorde would have been comfortable with this change of emphasis and possibly even anticipated it, for he was a competent instrument maker as displayed in the *Castle*. He had signalled his awareness of shortcomings in astronomical instrumentation in the outline he had provided for *The Pathway to Knowledge* that preceded the *Castle*. Sufficient was said to show that he had the means of correcting at least some of these deficiencies. There were also indications that he realised that instruments could be devised to circumvent some calculations at least. He certainly had the necessary mathematical and practical skills to do so, skills which would have been appropriate also for devising shipboard instruments.

The rise in mathematical literacy through the medium of the printed book was well under way by the time that Elizabeth ascended the throne. The first successful printed English arithmetic text was *An Introduction for to Lerne to Recken with the Pen or with the Counters*, a compilation taken from Dutch and French texts, five editions of which were published before Elizabeth's reign. Its first and second editions were published in 1537 and 1539, preceding Recorde's first edition of the *Grounde of Artes* by 5 years. There were two printing of the first edition of this latter and two of its augmented second edition, by 1558. It was printed repeatedly until near the end of the next century. Both books provided examples of calculations couched in terms that were intended to appeal to readers seeking practical applications of arithmetic. In Recorde's case, as quoted earlier these examples embraced a wide range of people likely to have interests in mercantile types of calculations at personal and communal levels. Thus during the last few years of the reign of Henry VIII and those of Edward and Mary, nine texts teaching elementary arithmetic using Hindu-Arabic numbers were printed. The average size of a print run in Elizabethan times was over a 1,000 copies, so even if the run length for such specialised mathematical texts was only a quarter of that average, 2,000 copies had been purchased. A thirst for numeracy was therefore well established before Elizabeth's time. Whether demand preceded supply or vice-versa cannot be established but the speed at which the market developed suggests that the former was the more likely cause.

The last edition of the *Introduction* was printed in 1629, by which time arithmetical texts, in the vernacular, had been published by a number of authors including Humfrey Baker (1562), Dyonisius Gray (1577) and Thomas Hylles (1600) who acknowledged their debt to Recorde. None of them proved to have the enduring popularity of the *Grounde*, which was last reprinted in 1696. It had not remained unchanged over the near century and a half of its life. The changes made by successive editors took the form of supplements rather than modifications to the original text. John Dee did attempt some modifications which were arguably less than helpful but his additional material was perhaps indicative of a perceived need by prospective purchasers. In the 1570 edition he included sections on exchange rates for major Continental currencies and values of weights and measures. This trend towards the emphasis on mercantile relevance was continued in a more substantial way by John Mellis (1582) who included sections on loss and gain, rules of payment, barter, exchange and, significantly, interest. John Wade (1610) did little, but the next editor, Robert Norton (1615) added '...*the Arte and Application of Decimal Arithmeticke'*, together with other material he had translated from the Dutch author Stevin. Robert Hartwell (1618) reinforced the drift towards meeting commercial needs with additions dealing with interest and annuities. Edward Hatton (1699) was the editor of the final edition and followed it with his own book on commercial arithmetic, which made no pretensions to serving a scholarly audience and was but a pale successor to Recorde.

It is clear therefore that Recorde's texts had, as he had wished, brought forth 'favourers of learning' within the mercantile community at least and possibly within the larger portion of English society that was concerned with the management and generation of wealth - including the State. It is unclear what effect Recorde's works had, if any, on the generation of 'such learned men that she shall be able to compare with any Regime in the world'.

Starting with Savile, the academic community had adopted a very offhand attitude towards mathematical practitioners, an attitude that was reciprocated. The schism seemed to have deepened as the century progressed, academe having Latin as its language of choice. Nonetheless, 'Arithmetic made Easie' by Wingate(1630), written in the vernacular seems to have been found intellectually acceptable, possibly helped by the author having the social cachet of having written texts in French and English on Gunter's logarithmic slide rule and having been tutor in English to the Princess Henrietta Maria. This Arithmetic, as modified by Kersey was to hold sway well into the next century. Scholarly interest in arithmetic was to move towards being subsumed in Algebra. In the introduction to his book A New System of Arithmetick Theoretical and Practical ...(1730) Alexander Malcolm wrote 'An Algebra is nothing but an universal method of representing Numbers and reasoning about them, so it very naturally belongs to Arithmetik. And in the opinion of the Great Sir Isaac Newton, who calls it the Universal Arithmetic makes with what in distinction from it he calls the Vulgar Arithmetick, but one complete Arte of Computation.' There is no evidence that Recorde's texts had any direct influence on this community and hence on such 'learned men' that it might potentially provide. There is however evidence that he had an indirect influence in the form of the formal and symbolic language that certain influential members of the community used. The most obvious example is the equality sign. Harriot used a modified version of this but Oughtred and his followers used Recorde's invention unmodified; whether this adoption can be assumed to have resulted from direct exposure to Recorde's text is a moot point. Support for this view based on an examination of vocabulary common to the *Whetstone of Witte* and the publications of Oughtred and Harriot is not possible, as the original versions of the 'Praxis' and the 'Clavis' were written in Latin. However the English version of the latter does contain the words binomial, figurate, bimedial, medial and rational, which originated with Recorde, which usage presumably Oughtred had approved. Jonas Moore's *Arithmeticae* of 1650, written in the vernacular, adopted similar terms, but preferred Billingsley's term 'prime number' to Recorde's 'uncompound number', and 'species' to 'symbol'. Moore's book comes the closest in terms of content to Recorde's two books on arithmetic. As only one edition of the *Whetstone* was published, Moore and Robert Wood,

Oughtred's translator and an Oxford scholar, must have had access to it directly or indirectly. As noted elsewhere Aubrey had a copy of the book which he donated along with two other of Recorde's books to Worcester College, Oxford. No search for other seventeenth-century owners of the *Whetstone* has been undertaken.

The blossoming of natural philosophy in seventeenth century England was concerned with more than that of mathematics alone, embracing what would now be considered as physics, chemistry and biology, with an emphasis on the generation of knowledge by experimentation. Prior to this time most knowledge had derived from observation yielding experiential data. Recorde had nothing to offer on chemical and biological matters. Neither had he anything to say on experimentation on a laboratory scale. He would have had to carry out on a large scale the type of experimentation that would be necessary for any successful manager of an industrial process, but no accounts of such activities have survived. Nor did he possess either of the two interlinked necessities for the pursuit of experimental activities, namely time and money. Patronage in the form of monetary support was never available to him. Such financial support as may have come to him was in return for undertaking specific tasks. This was typical of the approach of the later Tudor monarchs.

Only towards the end of Elizabeth's reign or shortly afterwards are to be found examples of patronage that fostered in England, free-thinking research activities such as were commonly found on the Continent during the previous century and a half. In seventeenth century England, new wealth came mainly from increased mercantile activity. Robert Recorde's texts had played a major part in introducing a wide cross-section of the community to the financial benefits potentially arising from competence in practical mathematics. Were there amongst such beneficiaries individuals who appreciated that the knowledge underlying this competence had come from the pursuit of knowledge for it's 'intrinsic worth'; and might they in consequence be prepared to patronise such activities? If so, even if his basic thinking exercised no discernable direct influence on seventeenth century science, perhaps Robert Recorde contributed indirectly to the developments in this era by influencing the availability of patronage for studies aimed at generating knowledge of intrinsic worth.

Chapter 14
His Testament and His Religion

Abstract Robert Recorde was raised in the Catholic faith, but died a confirmed Protestant, as demonstrated by the form of his will. It seems likely that he received his early education at a chantry school attached to the Church of St. Mary the Virgin in Tenby. This church had putative links with the Tudors and was the only religious house in Southern Wales that housed a chantry to the Holy Name of Jesus. The latter was a cult much favoured by the ruling classes, Margaret de Beaufort in particular, one which much advocated preaching and teaching functions. Recorde's known associates were also largely of the Protestant persuasion and he joined a number of them in testifying against Stephen Gardiner, Bishop of Winchester. The two theological texts that Recorde wrote are not extant, but their titles are clearly anti-Catholic in tone. Unlike many of his intellectual associates, on Mary's accession he neither fled to the Continent nor recanted his beliefs. He did not seem to have suffered overly from persecution, as did other Protestants

His Last Will and Testament

Probate for this was first granted in June 1558, was not executed and was granted again in November 1570. It reads:-

In the name of god amen. fforasmuche as nothing is more certaine to man then deathe and nothing more uncertain, the houre and tyme thereof, knowe you m^r Robert Recorde doctor of Phisicke though sicke in bodye yet whole of mynde thankes be to god make my last Will and testament in manner and forme following. ffirst I committ my soule unto thandes of the same almighti god my only maker and Redeemer trusting by the merrites of his passion to be one of his electe in glorye forever. my bodye as receyved from the erthe I bequeathe thither again to be buryed among other christians according to the solempne usage of the churche, my temporall goodes and chattalles I order will and dispoas in manner and forme following. Secondly I give to Arthure hilton undermarshall of the Kings benche where I now

J. Williams, *Robert Recorde: Tudor Polymath, Expositor and Practitioner of Computation*, History of Computing, DOI 10.1007/978-0-85729-862-1_14,
© Springer-Verlag London Limited 2011

remayn prisonner xxs. Item to his wife other xxs. Item to the gents now prisonners with me xxxs. Item I give other xxs. to the Arthure hilton to be by him distributed amonge thofficers according to his discreation. Item to his wif to be distributed amonge her women vjs. viijd. Item I give generally to the comon gaole Item I give other xxs. to the said of the saide prisonne xls. to be distributed among the prisonners there. Item I give and bequeathe to my widowe mother and to my father in lawe her husband xx li. Item I giveto my servante John vi li. Item I give and bequeathe children of John Battyn xls. to be distributed at the discreation of their saide father. The Residue of all my goods and chattalls moveable and unmoveable reall and personall. I give and beqeath unto my brother Richard Recorde and Robert Recorde his sonne my nephewe whom I make and ordayn my full and hole executors to thende that they beside my ------- of the same shall truly and faithfully pay my dettes. Whyche are to Nicolas ffulysham, Citizen and merchanntaillor of London fiftie poundes. to Mr. Battyn xl s.

   Memorandum. that the saide testator on the morrowe next after the making of his testament aforesaide being then of his parfite mynde and memorye adding to the said testament gave and bequeathed to Alice and Rose Recorde daughters of the saide Richard Recorde, and also unto Julian Baye all his utensils and household stuff to be equally divided betwene them Item he willed and divised that Nicolas Adames then being prisoner in the Kings bench shuld have all his bookes concerning the lawes of this Realm at the price of iiij li. Witnesses hereunto Richard Corbett George Marten and Richard Thymylby.' [Transcribed from PRO. PROB 11/40. Courtesy of National Records U.K.]

His Religion

The opening words of Robert Recorde's last testament show that he was at this time Protestant in his religious persuasions, seemingly embracing the concept of predestination. The circles in which he moved during late Henrician times and throughout the reign of Edward VI were those of a reformist nature, revolving around Cranmer.

   Following the advent of Queen Mary, unlike many of his acquaintances he chose not to leave the country and dedicated his penultimate book *The Castle of Knowledge* to her. The effusiveness of his dedication was relatively restrained and concerned only with the legitimacy of her claim to the throne. He owed no allegiance to the Northumberland faction that had supported Lady Jane Grey so his declaration in favour of the legitimacy of Mary's succession would not have troubled his conscience. The Epistle to Cardinal Pole is written in Latin and peppered with words and phrases in Greek. It savours of a scholarly communication between one scholar and another senior, admired scholar, debating matters relating to the antiquity of Astronomy, somewhat flattering and deferential but raising no controversial religious issues. The address to the Reader is in English, discusses issues surrounding judicial astronomy and the practical values of Astronomy in relation to agriculture, navigation and the church calendar. Pervading the whole dissertation is the theme 'The Heavens declare the Creator's glory ... ' couched in terms that would have been acceptable to Catholic and Protestant alike. Recorde was careful in his presentation, but did not compromise his religious principles.

Probably he had little choice in choosing to stay in England. The affair at Clonmines had left him in much reduced financial circumstances and, as a scholar, he was little known on the mainland of Europe. During his dispute with Pembroke, Recorde implied that one of the reasons for the latter's antipathy was that Recorde was beginning to find acceptance at Court again. Recorde probably had returned to his medical practice as a source of income, evidenced by an excerpt from Underhill's anecdotes. Shortly after the accession of Mary, Underhill found himself imprisoned accused of issuing a ballad against the papists. He writes 'My very frende mr. Recorde, doctor of phisicke, singularly sene in all the seven syences, and a great devyne, visited me in the preson, and after I was delyvered, to his great parralle yff it had been knowne, who longe tyme was att charges and payne withme gratis. By meanes whereoff and the provydence off god I receaved my health.'[1] Underhill had known Recorde as a friend for at least 5 years and holding extreme Protestant views, was in a position to comment on Recorde's abilities as a theologian , unlikely to categorise him lightly as such. Perhaps he was aware of the religious texts written by Recorde, listed by Bale but no longer extant, viz. *De auricula confessione, De negotio Eucharistie, atque alia.,* and may have been amongst those Protestant tracts that he was accused of having hidden.[2] Whether it was his theological standing that was being called upon in the evidence he tabled at the trial of Stephen Gardiner in January 1550, or whether it was his wider status as an intellectual that was being recognised is difficult to judge. His evidence concerned only the contents of the sermon delivered by Gardiner to the king and his court on 29 June 1549. He stated that he was well placed to hear it, so probably situated not too far from the monarch. He was listed ninth in the order of depositions and was preceded by Wingfield, Cecil, Sadler, Challenor, Throgmorton, Wroth, Cheke, Coxe, Watson, Honing and Ayre. With the exception of Watson, Gardiner's chaplain and to be the last Catholic bishop of Ely and Ayre, dean of Chichester, the remainder were all members of the close circle of attendants on the king. His deposition asserted that he had seen the writings that had been given to Gardiner instructing him to deliver the sermon and outlining what its contents were expected to be viz. to assert the monarch's authority in religious matters, to support the measures that the king had introduced, but not to make reference to the mass and to trans-substantiation. Recorde testified that Gardiner's sermon did not conform to such requirements, to the offence of himself and other listeners.[3] Gardiner reacted by saying of Coxe, Ayre, Honing, Cheke, Challenor, Recorde and Smith, all either 'depose[d] generally or by heresay, not concluding any proofe or els so utter their own affection, as they be worthie no

---

[1] Nichols G, loc. cit., 150–151.
Nichols JG (ed) (1848) The diary of Henry Machyn ....'. Camden Society, London, xlii, pp 150–151.
[2] Index Scriptorum no. 387.
[3] Foxe J The Unabridged Acts & Monuments Online at TAMO (1563 edition). HRI Online Publications. Sheffield 2011. http//www.johnfoxe.org. pps.867, 920.

credite, or els show themselves so loth to remember any thing that might relieve the bishop, as they might be reputed not indifferent.' Specifically, 'Maister Recorde saying that the Bishop is yet disobedient, & so wrongfully judging of the bishop in his private prejudice is unworthy of all fayth in the matter.'[4] Possibly the most that can be abstracted from this passage of words is that at the time of the sermon Recorde was a committed Protestant. His antipathy to the Papacy was evident in the preface to his *Urinal of Physicke*, of 1547 .... 'Though the Pope, Cardinals and Monkes have practised to poison men even with the very sacrament of the Supper of the Lord, yet no man will be so mad as to eschew the use of that blessed Sacrament.' ..... . A slight oddity is that his doctorate in medicine at Cambridge of 1545 was granted under the general category 'De Doctoribus In theologia'. His patronage by Whalley, acknowledged by his dedication of the first edition of the *Grounde of Artes* in 1543, implies that he was by that time at the latest, of the Protestant persuasion, and of course his publisher, Reyner Wolfe was a dedicated Reformer. More was perhaps not to be expected in terms of expression of his commitment to reform at this time, for 1543 was a year when a number of anti-reformist measures were introduced.

The possibility that he was converted to Protestantism prior to entry into University c. 1525 seems remote, but perhaps some of the groundwork for such a transition had been laid during his early schooling in Tenby. Chantry priests are the most likely source of early education for the young Robert Recorde and there were three of such associated with the Church of St. Mary in that town. Such information as still exists on these institutions comes from the documentation arising from their dissolution. Chantry certificates exist for both the Henrician (1546) and Edwardian (1548) exercises. A singularity with respect to the nature of the chantries in Southern Wales is found at Tenby which deserves consideration because of its possible effect on Recorde's religious views.

Of the 100 or more chantries listed for Southern Wales in Chantry certificate No. 74 there is only one Chapel and Altar of Jesus listed, viz. that at Tenby. The cult of the Holy name of Jesus had become one of the most popular amongst the highest ranks of society in the second half of the fifteenth century and early sixteenth centuries. It was favoured by Henry VII and particularly by his mother, Margaret de Beaufort who became patron of the feast of the cult. The cult has been described as '....a sign of the revitalised Christocentrism that was the hallmark of evangelical Catholic piety in the decades that preceded the reforms.' and as 'an uneasy nexus between the dynamism of the Catholic Church and emerging Protestanism. The use of the Holy name is a paradigm of the bifurcation of Western Europe. On the continent, it achieved its apogee in the creation of the Society of Jesus, while in England some of its features survived, in adapted form, to suit a Protestant role.'[5] The chantry priest who served the altar of Jesus in Tenby clearly found no difficulty in making the

[4]Ibid., p 864.

[5]Wabuda S (2002) Preaching during the English reformation. Cambridge University Press, Cambridge, p 148.

transition to Protestantism, for he was appointed by the Commissioners who carried out the survey to the cure of the church at Tenby at the same stipend that he was already receiving. Richard Thomas the priest concerned was 36 years old and therefore a contemporary of Recorde and not his mentor. In certain court cases over the source of income that provided the funding for the stipend, the original endowment was for a priest to sing and serve and to say masses in the chapel for its founders. Commonly in England, a duty laid on the priest serving such an altar would have been to preach and to teach, but no such stipulations have been found in this case. The date of the endowment is not known, other than it was before 1531. The source of the income is known, but the names of the donors are not, although there is a hint that can be interpreted as inferring that they moved in royal circles. There is room for a great deal of speculation about royal links with Tenby through associations with the West Porch, the College and the Altar of Jesus, but not enough hard evidence is available to more than speculate about potential effects that such a singularity may have had upon the enquiring mind that Robert Recorde clearly possessed.

Index

J. Williams, *Robert Recorde: Tudor Polymath, Expositor and Practitioner
of Computation*, History of Computing, DOI 10.1007/978-0-85729-862-1,
© Springer-Verlag London Limited 2011